America's National Park System

America's National Park System

The Critical Documents

Edited by Lary M. Dilsaver

ROWMAN & LITTLEFIELD PUBLISHERS, INC.

ROWMAN & LITTLEFIELD PUBLISHERS, INC.

Published in the United States of America
by Rowman & Littlefield Publishers, Inc.
4720 Boston Way, Lanham, Maryland 20706

3 Henrietta Street
London WC2E 8LU, England

Copyright © 1994 by Rowman & Littlefield Publishers, Inc.

The documents in this book are reproduced in a manner
that maintains the integrity of the original spelling
and punctuation except for obvious errors corrected
by the editor.

British Cataloging in Publication Information Available

Library of Congress Cataloging-in-Publication Data

America's national park system : the critical documents / edited by
Lary M. Dilsaver.
p.   cm.
Includes bibliographical references and index.
1. National parks and reserves—United States—History—Sources.
2. National parks and reserves—Law and legislation—United States.
I. Dilsaver, Lary M.
E160.A579 1994   973—dc20   94-7666 CIP

ISBN 0-8476-7922-5 (alk. paper)

Printed in the United States of America

 ™ The paper used in this publication meets the minimum requirements of
American National Standard for Information Sciences—Permanence of
Paper for Printed Library Materials, ANSI Z39.48–1984.

Contents

The National Park System

The national park system of the United States is a complex aggregate of some 370 units falling into 20 separate categories. They represent, in principle, the finest America has to offer in scenery, historical and archaeological relics, and cultural definition. The system attempts to explain America's history, interpret its culture, represent and preserve its varied ecosystems, and provide, incidentally, for the recreation of its 250 million people. The parks have been justifiably called "the crown jewels" of America and "the best idea America ever had."

Officially, the purposes of the park system, and by inference its management policies, are best summed up in the 1916 act that established the National Park Service. In now familiar words to all who have studied the parks and their management, the new agency was charged "to conserve the scenery and the natural and historic objects and the wild life therein and to provide for the enjoyment of the same in such manner and by such means as will leave them unimpaired for the enjoyment of future generations."

The origin of the movement to create this system is shrouded in the early nineteenth century while the confusion of managing such a system for both preservation and recreation has been a frustrating puzzle since the beginning. An increasing scholarly attention to the history of the agency, its parks, and the culture of conservation has arisen in the past two decades among environmental historians, historical geographers, park scientists, and others interested in these most beloved and enigmatic of American treasures.

Two processes must be concurrently addressed to understand the park system. First, there is the growth of the system itself. Officially it began with the founding of Yellowstone National Park on March

1, 1872. Another eighteen years passed before a second park that would survive the nineteenth century was established. As the twentieth century approached, Congress begrudgingly took steps to conserve other resources by other means. However, after the creation of the Park Service, the haphazard, individual preservation efforts were paralleled by direct efforts to shape a meaningful system, much of this impetus provided by the initial director, Stephen Mather. The Park Service and the public have continued to shape the contents of the system through varied initiatives and sets of interacting or sometimes opposing beliefs.

A second process is the maturing framework of management policy for the parks. The 1916 Organic Act tersely outlined a seemingly contradictory set of preservation and recreation mandates. In subsequent decades new interpretations, changes in public and government attitudes, and the maturation of ecological science have added to the confusion.

Although both these processes have been linear and evolutionary, distinct periods can be identified. Prior to 1918, the major component categories of the system were created. National parks, monuments, and battlefields were established and historic and prehistoric relics protected. However, there was no "system" per se. Each component was administered separately by the secretaries of War, Agriculture, and the Interior, depending on their location and purpose. The Organic Act brought order to at least the parks and monuments under the control of the Department of the Interior. In 1918, the secretary outlined a code of rules and policies that further defined management of the new park system.

From 1918 to 1932 the system and its parent agency matured, rose in public esteem, and fended off challenges to their existence and policies. A collection of motivated and able men, "Mather men," laid the foundations of National Park Service (NPS) image and operations. Studies of potential new areas coincided with establishment of system-wide policies. Yet these policies were based on relatively simple concepts of object preservation, the protection of individual trees, animals, or at best, scenes, with little knowledge of or attention to ecological complexity. Only at the beginning of the 1930s did concerns for scientific management become manifest in a series of policy statements about the biological resources of the system.

The 1930s brought a radical change and expansion of the national park system and its duties. Sweeping in with Franklin Roosevelt came a variety of government streamlining and public relief initiatives that would benefit and forever alter the parks. Two of the most profound occurred in 1933 with the reorganization of the federal executive branch which lumped all the parks, monuments, battlefields, and memorials under the Park Service's jurisdiction and the act that led to the Civilian Conservation Corps, a public works army of unemployed men, many of whom would transform the built environments of the parks. Indeed, while the Park Service wrestled with how and in which directions it should expand the system, some, both inside and outside the government, challenged the developmental mania that flush work budgets had brought.

All this came to a crashing end with the arrival of World War II. It heralded the onset of a grim and financially strapped postwar period. While park popularity and attendance continued to rise rapidly and damage from wartime neglect festered, funding hovered at prewar levels. At the same time, the growth of the system itself slowed and early political checks on hitherto unrestrained expansion appeared. Finally, the crisis of management forced such an outcry that a funding bonanza, Mission 66, ended the drought.

The ensuing decade of development radically transformed, again, the built environment of the park system. Before Mission 66 ended, however, questions of purpose and management priorities would also arise again. Spurred by the focus on construction and development that Mission 66 provided, and by advances in ecological science, many questioned how well the Park Service was doing at "preserving" the park system resources. Prodded by university scientists and conservation groups, the agency undertook a series of studies and policy formulations culminating in the Leopold and Robbins reports which redirected NPS goals to ecosystem preservation and scientific inquiry. Coincident with this redefinition of the traditional parks, attention to expanding the nation's recreation opportunities also rose, culminating in new studies of seashores, recreation areas, and parkways.

As Mission 66 wound down and long-term NPS employees tried to accommodate the concepts of the Leopold report, both the national park system and its managing agency continued to expand and diversify. Ecological preservation became the entrenched policy focus while new initiatives in both recreation and preservation led to exciting new categories in the park system. Among them were wild and scenic rivers and national trails. New recreation areas and seashore parks also appeared. The public itself developed an environmental consciousness and an intense interest in both using and saving the nation's parklands. By 1968, management had become so complex that three separate systems of administration seemed to be needed—one set each for historical, natural, and recreational areas. A web of environmental legislation culminating in the National Environmental Policy Act (NEPA) further protected the parks and constrained their planning and development.

The 1970s, if anything, seemed to intensify and accelerate both the processes of system expansion and management redefinition. Both the public and Congress seemed to take a heightened interest in shaping the system. The decade saw new charges to preserve archaeological and historic sites, novel urban recreation areas, expansion of a park purely for ecological integrity, and the doubling of the system size by the addition of huge units in Alaska. Despite these successes, the decade ended on an ominous note as a survey disclosed that the system, while ever larger and more popular, faced serious threats in nearly every unit.

Unfortunately, the decade of the 1980s heralded the onset of recession and government spending cuts even as threats to the parks multiplied. External development, issues of serious overcrowding, deterioration of infrastructure, and decay of Park Service morale rose nearly as quickly as did the popularity of and visitation to the parks. The 1980s saw less expansion and fewer changes in the system, but a flurry of reports at the end of the decade and the start of the 1990s demonstrated heightened public interest ignited by the Yellowstone fire of 1988 and the swollen ranks of conservation groups frustrated by administration challenges to existing environmental laws.

Today the national park system continues to face new threats and old problems. It has a hierarchy of policy controls that begin with the government-wide controls such as NEPA, then descends to the Organic Act, individual park founding legislation, NPS management policies which reflect and elaborate the above legislation, and finally individual park policies. The latter often reflect unwritten agency tradition of what a national park should be and how it should be run.

The purpose of this volume is to help explain this complex system and its ill-defined management culture by reproducing, in their original texts, the key documents that have shaped them. Laws that established and molded the system, policy statements, explanations of policy in action, self-evaluations, scientific studies, and the comments of outside observers from the scientific and conservation communities have all contributed to the composition of the park system and its current administration. Sixty-seven documents are presented in their entirety if possible or abbreviated to include only pertinent abstracts and recommendations if necessary. Six more exceptionally long but fundamental laws are summarized in the final section of the book. Most of the documents led directly to policy formulation or additions to the park system. A few are contemporary efforts to justify what policies were already in place or at least demonstrate the minds of park administrators of the time. Taken in total, these documents show most directly, in the original language, the evolution of the national park system, the National Park Service, and conservation in America through the early 1990s. It is hoped that this volume will assist park officers, scholars, and the interested general reader to understand how the parks and their problems of today have arisen from the policies and acts of the past.

Most of my thanks for the completion of this volume go to the National Park Service for giving me the opportunity to search its files and for its support during the research process. Many knowledgeable and enthusiastic NPS people helped in the identification and acquisition of these documents. Among the most important were Carol Aten, Larry Bancroft, Ed Bearss, Harry Butowsky, Gordon Chappell, John Dennis, Dennis Fenn, Dave Graber, William Gregg, David Jervis, Roger Kelly, J. Mike Lambe, Thomas Mulhern, Dave Nathanson, Tom Nichols, David Parsons, Tom Ritter, Jerry Rogers, and James Stewart. For his exceptional assistance in the complex procedure of identifying which documents to include, my thanks go to Richard Sellars. And for his consistent encouragement, ready assistance, and leadership by example I particularly wish to note and thank NPS Bureau Historian Barry Mackintosh.

For their monumental help in retyping all the documents I thank Debra Moberg, Jerry Dixon, and Deanna Bowers. I also wish especially to thank Patricia Janssen for organizing the manuscript and preparing it for submission to press, and for the careful editing she so expertly provided. Finally I thank Dean Lawrence Allen and the University of South Alabama for their financial and research time support, Jonathan Sisk and Rowman & Littlefield Publishers for their faith in the project, and my wife, Robin, for her love and confidence.

1

The Early Years

1864–1918

The origins of the national park idea are the subject of considerable academic speculation. Suffice it to say, however, the concept did not originate over a Wyoming campfire. The processes that led up to national parks are more readily identified but also far more complex. From the arts and literature came the Romantic movement which encouraged the experience of mountains and wilderness. Authors like Henry David Thoreau and Washington Irving exhorted Americans to pursue nature even as the frontier rolled away from them. Landscape artists culminating with Thomas Moran and Albert Bierstadt presented awesome spectacles that received huge public interest.

The rise in attention to nature coincided with the search for identity and pride among American literati. When compared to Europe's thousands of years of history, its fabric of ancient structures and sites, its rich cultural legacy built on many centuries of interchange, the United States appeared a rude, uncultured backwater. Stung by caustic criticism and snobbery from Europe, Americans looked for elements in their own land to flaunt. In Yellowstone and Yosemite, and indeed the whole western wilderness, Americans had what they needed. America was new, rugged, spectacular, and could be proud of its splendor and its clean slate upon which to develop the human experience.

Yet another motive for national parks came from the American experience at Niagara Falls. The famous falls were America's paramount scenic wonder during the first half of the nineteenth century. However, local landowners had, in their frenzy to maximize profits, gone so far as to erect fences and charge viewers to look through holes at the spectacle. Tawdry concessions and souvenirs, filth, and squalor attended a visit to this most sublime of eastern American features. Clearly government control of such a feature to assure its availability to the public was in order.

The first movement to create a park came amidst the Civil War. Yosemite Valley had been first entered by Americans chasing a band of Indians in 1851. Within five years the situation at Niagara Falls began to repeat itself. Claims on the valley lands were filed and tolls charged. Haphazard tourism began even as the fame of the valley spread to a wondering and suspicious East. Concern for this amazing spectacle and its availability to all comers led Congress to withdraw the lands from alienation in 1864 and turn over the valley and a nearby grove of giant sequoias to the state of California as a public park. The state would continue to manage this first federal withdrawal for a park until 1906 when it was merged with Yosemite National Park.

Eight years later Congress established the world's first true national park. Instrumental in its creation was the Northern Pacific Railroad, beginning a fifty-year period during which railroads became the most profound influence on the establishment of these reserves and on the development of tourism in them. Where the Yosemite withdrawal consisted of a pair of relatively small areas, Yellowstone was an enormous tract of more than 3,400 square miles. The creation of Yellowstone National Park marked the first serious challenge to the culture of land alienation and consumptive use in American history.

The Yellowstone withdrawal was so massive and to many land users apparently purposeless that it was another eighteen years before a second successful park was created. By the end of the century still only five national parks existed, three of them in California. However, Congress was not idle in its preservation efforts and in forming reserves that would later become part of the national park system. During the latter portion of the century antiquities of the Southwestern Indians became a source of interest and gain for many Americans. Vandals and pot-hunters looted Anasazi and other sites, often destroying structures, not to mention the archaeological record, in their greedy haste. The earliest steps to protect an archaeological site came at Casa Grande Ruin in Arizona. Reacting to vandalism of this ancient adobe structure, Congress attached a rider to a civil appropriations bill in 1889 calling for the repair and protection of the first of what would become the national monuments.

A year later, a separate movement led to the establishment of the nation's first national battlefield park. Although some national cemeteries had been established even during the Civil War, no substantial protection had been provided to an entire battlefield. Initially, of course, the South saw no reason to preserve and celebrate its traumatic defeat. However, by the late 1880s, efforts to heal the nation's wounds and commemorate all who fought in the war led to various associations to protect major battlefields. At the behest of one such group Congress set aside in 1890 the Chickamauga and Chattanooga battlefields.

Finally, in 1906 the early spate of preservation efforts culminated with the Antiquities Act. The outgrowth of continued clamor for protection of archaeological sites, its chief impact came in section two where the president was given the power to unilaterally declare national monuments on federal lands in order to protect items of historic or scientific interest. The national park system today includes more than fifty national monuments deriving from this legislation (out of a total of more than seventy).

The establishment of the roots of the national park system was matched by a slower but no less important definition of management policies and priorities. In a consumptive society, these parks were novel and, to many, uncomfortable interruptions of business as usual. Much of the ground-laying of national park policy came from recommendations on the management of Yosemite Valley by eminent landscape architect Frederick Law Olmsted, creator of New York's Central Park. In his 1865 report to the governor of California, he laid the philosophical foundations of preservation for inspirational purposes and made explicit recommendations on such matters as concession operations, development, scientific protection, and interpretation.

Despite Olmsted's ideas, the pressure to use park resources in traditional consumptive ways was substantial, especially in the huge Yellowstone reserve. Acting to forestall hunting and trapping and further define the degree of protection afforded in a national park, Congress in 1894 passed the Yellowstone Game Protection Act.

By 1912 the parks were well established and reasonably safe from hunting, logging, and mining. Still, in a rapidly changing nation, uses of and threats to the parks evolved, and answers to new questions had to be found. By the turn of the century, automobiles had appeared in several national parks. However, no definitive policy had been established. Instead, reflecting perhaps the piecemeal management that preceded the 1916 Organic Act, auto use in each park was a separate issue. During 1912, Department of the Interior officials, conservationists, and others met in Yosemite to discuss auto use in the valley. Their comments indicate the prevalent attitudes of the time—that all forms of access to parks should be encouraged; that the primary concern is for the safety of drivers on the rough and twisting roads; and that no damage either to the park or to the park experience is expected from the admission of automobiles.

In 1916 came the most important document in this entire collection. For some years the parks were run as independent units lost in the bureaucratic maze of the Interior Department. In a concerted campaign by Robert Sterling Yard, future directors Stephen Mather and Horace Albright, the National Geographic Society, and many others, Congress was encouraged to establish a National Park Service, place all the existing parks under its management, and spell out the purposes for their preservation. The ensuing act, often known as the Organic Act, and its difficult charge to both preserve park resources and make them available to tourists, form the legal foundation stone of the system.

As a final management statement culminating this long and complex formative period, Secretary of the Interior Franklin Lane issued a detailed policy statement on management of the parks in 1918. Albright actually wrote the letter reflecting the ideas of Mather and others. The Lane letter addresses concession policy, priorities of protection, and many other issues not elaborated in the Organic Act. With these details of policy established and aired, Mather was set to consolidate and refine the national park system.

AN ACT AUTHORIZING A GRANT TO THE STATE OF CALIFORNIA OF THE "YO-SEMITE VALLEY," AND OF THE LAND EMBRACING THE "MARIPOSA BIG TREE GROVE,"

Approved June 30, 1864 (13 Stat. 325)

Be it enacted by the Senate and House of Representatives of the United States of America in Congress assembled, That there shall be, and is hereby, granted to the State of California the "cleft" or "gorge" in the granite peak of the Sierra Nevada Mountains, situated in the county of Mariposa, in the State aforesaid, and the headwaters of the Merced River, and known as the Yo-Semite Valley, with its branches or spurs, in estimated length fifteen miles, and in average width one mile back from the main edge of the precipice, on each side of the valley, with the stipulation, nevertheless, that the said State shall accept this grant upon the express conditions that the premises shall be held for public use, resort, and recreation; shall be inalienable for all time; but leases not exceeding ten years may be granted for portions of said premises. All incomes derived from leases of privileges to be expended in the preservation and improvement of the property, or the roads leading thereto; the boundaries to be established at the cost of said State by the United States surveyor-general of California, whose official plat, when affirmed by the Commissioner of the General Land Office, shall constitute the evidence of the locus, extent, and limits of the said cleft or gorge; the premises to be managed by the governor of the State with eight other commissioners, to be appointed by the executive of California, and who shall receive no compensation for their services.

SEC. 2. *And be it further enacted,* That there shall likewise be, and there is hereby, granted to the said State of California the tracts embracing what is known as the "Mariposa Big Tree Grove," not to exceed the area of four sections, and to be taken in legal subdivisions of one quarter section each, with the like stipulation as expressed in the first section of this act as to the State's acceptance, with like conditions as in the first section of this act as to inalienability, yet with same lease privilege; the income to be expended in preservation, improvement, and protection of the property; the premises to be managed by commissioners as stipulated in the first section of this act, and to be taken in legal subdivisions as aforesaid; and the official plat of the United States surveyor-general, when affirmed by the Commissioner of the General Land Office, to be the evidence of the locus of the said Mariposa Big Tree Grove. (U.S.C., title 16, sec. 48.)

THE YOSEMITE VALLEY AND THE
MARIPOSA BIG TREE GROVE

By Frederick Law Olmsted
1865

It is a fact of much significance with reference to the temper and spirit which ruled the loyal people of the United States during the war of the great rebellion, that a livelier susceptibility to the influence of art was apparent, and greater progress in the manifestations of artistic talent was made, than in any similar period before in the history of the country.

The great dome of the Capitol was wholly constructed during the war, and the forces of the insurgents watched it rounding upward to completion for nearly a year before they were forced from their entrenchments on the opposite bank of the Potomac; Crawford's great statue of Liberty was poised upon its summit in the year that President Lincoln proclaimed the emancipation of the slaves. Leutze's fresco of the peopling of the Pacific States, the finest work of the painter's art in the Capitol; the noble front of the Treasury building with its long colonnades of massive monoliths; the exquisite hall of the Academy of Arts; the great park of New York, and many other works of which the nation may be proud, were brought to completion during the same period.

Others were carried steadily on, among them our own Capitol; many more were begun; and it will be hereafter remembered that the first organization formed solely for the cultivation of the fine arts on the Pacific side of the nation was established in California while the people of the state were not only meeting the demands of the government for sustaining its armies in the field but were voluntarily making liberal contributions for binding up the wounds and cheering the spirits of those who were stricken in the battles of liberty.

It was during one of the darkest hours, before Sherman had begun the march upon Atlanta or Grant his terrible movement through the Wilderness, when the paintings of Bierstadt and the photographs of Watkins, both productions of the war time, had given to the people on the Atlantic some idea of the sublimity of the Yosemite, and of the stateliness of the neighboring Sequoia grove, that consideration was first given to the danger that such scenes might become private property and through the false taste, the caprice or the requirements of some industrial speculation of their holders, their value to posterity be injured. To secure them against this danger Congress passed an act providing that the premises should be segregated from the general domain of the public lands, and devoted forever to popular resort and recreation, under the administration of a Board of Commissioners, to serve without pecuniary compensation, to be appointed by the Executive of the State of California.

His Excellency the Governor in behalf of the state accepted the trust proposed and appointed the required Commissioners; the territory has been surveyed and the Commissioners have in several visits to it, and with much deliberation, endeavored to qualify themselves to present to the legislature a sufficient description of the property, and well considered advice as to its future management.

The Commissioners have deemed it best to confine their attention during the year which has elapsed since their appointment to this simple duty of preparing themselves to suggest the legislative action proper to be taken, and having completed it, propose to present their resignation, in order to render as easy as possible the pursuance of any policy of management the adoption of which may be determined by the wisdom of the legislature. The present report, therefore, is intended to embody as much as is practicable, the results of the labors of the Commission, which it also terminates.

As few of the legislature can have yet visited the ground, a short [description is given.]

"The main feature of the Yo Semite is best indicated in one word as a chasm. It is a chasm nearly a mile in average width, however, and more than ten miles in length. The central and broader part of this chasm is occupied at the bottom by a series of groves of magnificent trees, and meadows of the most varied, luxuriant and exquisite herbage, through which meanders a broad stream of the clearest water, rippling over a pebbly bottom and eddying among banks of ferns and rushes; sometimes narrowed into sparkling rapids and sometimes expanding into placid pools which reflect the wondrous heights on either side. The walls of the chasm are generally half a mile, sometimes nearly a mile in height above these meadows, and where most lofty are nearly perpendicular, sometimes overjutting. At frequent intervals, however, they are cleft, broken, terraced and sloped, and in these places, as well as everywhere upon the summit, they are overgrown by thick clusters of trees.

"There is nothing strange or exotic in the character of the vegetation; most of the trees and plants, especially those of the meadows and waterside, are closely allied to and are not readily distinguished from those most common in the landscapes of the Eastern States or the midland counties of England. The stream is such a one as Shakespeare delighted in, and brings pleasing reminiscences to the traveller of the Avon or the upper Thames.

"Banks of heartsease and beds of cowslips and daisies are frequent, and thickets of dogwood, alder and willow often fringe the shores. At several points streams of water flow into the chasm, descending at one leap from five hundred to fourteen hundred feet. One small stream falls, in three closely consecutive pitches, a distance of two thousand six hundred feet, which is more than fifteen times the height of the falls of Niagara. In the spray of these falls superb rainbows are seen.

"At certain points the walls of rock are ploughed in polished horizontal furrows, at others moraines of boulders and pebbles are found; both evincing the terrific force with which in past ages of the earth's history a glacier has moved down the chasm from among the adjoining peaks of the Sierras. Beyond the lofty walls still loftier mountains rise, some crowned by forests, others in simple rounded cones of light, gray granite. The climate of the region is never dry like that of the lower parts of the state of California; even when, for several months, not a drop of rain has fallen twenty miles to the westward, and the country there is parched, and all vegetation withered, the Yo Semite continues to receive frequent soft showers, and to be dressed throughout in living green.

"After midsummer a light, transparent haze generally pervades the atmosphere, giving an indescribable softness and exquisite dreamy charm to the scenery, like that produced by the Indian summer of the East. Clouds gathering at this season upon the snowy peaks which rise within forty miles of each side of the chasm to a height of over twelve thousand feet, sometimes roll down over the cliffs in the afternoon, and, under the influence of the rays of the setting sun, form the most gorgeous and magnificent thunder heads. The average elevation of the ground is higher than that of the highest peak of the White Mountains, or the Alleghenies, and the air is rare and bracing; yet its temperature is never uncomfortably cool in summer, nor severe in winter.

"Flowering shrubs of sweet fragrance and balmy herbs abound in the meadows, and there is everywhere a delicate odor of the prevailing foliage in the pines and cedars. The water of the streams is soft and limpid, as clear as crystal, abounds with trout and, except near its sources, is, during the heat of the summer, of an agreeable temperature for bathing. In the lower part of the valley there are copious mineral springs, the water of one of which is regarded by the aboriginal inhabitants as having remarkable curative properties. A basin still exists to which weak and sickly persons were brought for bathing. The water has not been analyzed, but that it possesses highly tonic as well as other medical qualities can be readily seen. In the neighboring mountains there are also springs strongly charged with carbonic acid gas, and said to resemble in taste the Empire Springs of Saratoga.

"The other district, associated with this by the act of Congress, consists of four sections of land, about thirty miles distant from it, on which stand in the midst of a forest composed of the usual trees and shrubs of the western slopes of the Sierra Nevada, about six hundred mature trees of the giant Sequoia. Among them is one known through numerous paintings and photographs as the Grizzly Giant, which probably is the noblest tree in the world. Besides this, there are hundreds of such beauty and stateliness that, to one who moves among them in the reverent mood to which they so strongly incite the mind, it will not seem strange that intelligent travellers have declared that they would rather have passed by Niagara itself than have missed visiting this grove.

"In the region intermediate between the two districts the scenery generally is of a grand character, consisting of granite mountains and a forest composed mainly of coniferous trees of great size, yet often more perfect, vigorous and luxuriant than trees of half the size ever found on the Atlantic side of the continent. It is not, however, in its grandeur or in its forest beauty that the attraction of this intermediate region consists, so much as in the more secluded charms of some of its glens, formed by mountain torrents fed from the snow banks of the higher Sierras.

"These have worn deep and picturesque channels in the granite rocks, and in the moist shadows of their recesses grow tender plants of rare and peculiar loveliness. The broad parachute-like leaves of the peltate saxifrage, delicate ferns, soft mosses, and the most brilliant lichens abound, and in following up the ravines, cabinet pictures open at every turn, which, while composed of materials mainly new to the artist, constantly recall the most valued sketches of Calame in the Alps and Apennines.

"The difference in the elevation of different parts of the district amounts to considerably more than a mile. Owing to this difference and the great variety of exposure and other circumstances there is a larger number of species of plants within the district than probably can be found within a similar space anywhere else on the continent. Professor Torrey, who has given the received botanical names to several hundred plants of California, states that on the space of a few acres of meadow land he found about three hundred species, and that within sight of the trail usually followed by visitors, at least six hundred may be observed, most of them being small and delicate flowering plants.

"By no statement of the elements of the scenery can any idea of that scenery be given, any more than a true impression can be conveyed of a human face by a measured account of its features. It is conceivable that any one or all of the cliffs of the Yo Semite might be changed in form and color, without lessening the enjoyment which is now obtained from the scenery. Nor is this enjoyment any more essentially derived from its meadows, its trees, its streams; least of all can it be attributed to the cascades. These, indeed, are scarcely to be named among the elements of the scenery. They are mere incidents, of far less consequence any day of the summer than the imperceptible humidity of the atmosphere and the soil. The chasm remains when they are dry, and the scenery may be, and often is, more effective, by reason of some temporary condition of the air, of clouds, of moonlight, or of sunlight through mist or smoke, in the season when the cascades attract the least attention, than when their volume of water is largest and their roar like constant thunder.

"There are falls of water elsewhere finer, there are more stupendous rocks, more beetling cliffs, there are deeper and more awful chasms, there may be as beautiful streams, as lovely meadows, there are larger trees. It is in no scene or scenes the charm consists, but in the miles of scenery where cliffs of awful height and rocks of vast magnitude and of varied and exquisite coloring, are banked and fringed and draped and shadowed by the tender foliage of noble and lovely trees and bushes, reflected from the most placid pools, and associated with the most tranquil meadows, the most playful streams, and every variety of soft and peaceful pastoral beauty.

"The union of the deepest sublimity with the deepest beauty of nature, not in one feature or another, not in one part or one scene or another, not in any landscape that can be framed by itself, but all around and wherever the visitor goes, constitutes the Yo Semite the greatest glory of nature. No photograph or series of photographs, no paintings ever prepare a visitor so that he is not taken by surprise, for could the scenes be faithfully represented the visitor is affected not only by that upon which his eye is at any moment fixed, but by all that with which on every side it is associated, and of which it is seen only as an inherent part. For the same reason no description, no measurements, no comparisons are of much value. Indeed the attention called by these to points in some definite way remarkable, by fixing the mind on mere matters of wonder or curiosity presents the true and far more extraordinary character of the scenery from being appreciated."

It is the will of the nation as embodied in the act of Congress that this scenery shall never be private property, but that like certain defensive points upon our coast it shall be held solely for public purposes.

Two classes of considerations may be assumed to have influenced the action of Congress. The first and less important is the direct and obvious pecuniary advantage which comes to a commonwealth from the fact that it possesses objects which cannot be taken out of its domain, that are attractive to travellers and the enjoyment of which is open to all.

To illustrate this it is simply necessary to refer to certain cantons of the Republic of Switzerland, a commonwealth of the most industrious and frugal people in Europe. The results of all the ingenuity and labor of this people applied to the resources of wealth which they hold in common with the people of other lands have become of insignificant value compared with that which they derive from the price which travellers gladly pay for being allowed to share with them the enjoyment of the natural scenery of their mountains. These travellers alone have caused hundreds of the best inns in the world to be established and maintained among them, have given the farmers their best and almost the only market they have for their surplus products, have spread a network of rail roads and superb carriage roads, steamboat routes and telegraphic lines over the country, have contributed directly and indirectly for many years the larger part of the state revenue and all this without the exportation or abstraction from the country of any thing of the slightest value to the people.

The Government of the adjoining kingdom of Bavaria undertook years ago to secure some measure of a similar source of wealth by procuring with large expenditure, artificial objects of attraction to travellers. The most beautiful garden in the natural style on the Continent of Europe was first formed for this purpose, magnificent buildings were erected, renowned artists were drawn by liberal reward from other countries, and millions of dollars were spent in the purchase of ancient and modern works of art. The attempt thus made to secure by a vast investment of capital the advantages which Switzerland possessed by nature in its natural scenery has been so far successful that a large part, if not the greater part of the profits of the rail roads, of the agriculture and of the commerce of the kingdom is now derived from the foreigners who have been thus attracted to Munich, its capital.

That when it shall become more accessible the Yosemite will prove an attraction of a similar character and a similar source of wealth to the whole community, not only of California but of the United States, there can be no doubt. It is a significant fact that visitors have already come from Europe expressly to see it, and that a member of the Alpine Club of London having seen it in summer was not content with a single visit but returned again and spent several months in it during the inclement season of the year for the express purpose of enjoying its winter aspect. Other foreigners and visitors from the Atlantic states have done the same, while as yet no Californian has shown a similar interest in it.

The first class of considerations referred to then as likely to have influenced the action of Congress, is that of the direct pecuniary advantage to the commonwealth which under proper administration will grow out of the possession of the Yosemite, advantages which, as will hereafter be shown, might easily be lost or greatly restricted without such action.

A more important class of considerations, however, remains to be stated. These are considerations of a political duty of grave importance to which seldom if ever before has proper respect been paid by any government in the world but the grounds of which rest on the same eternal base of equity and benevolence with all other duties of republican government. It is the main duty of government, if it is not the sole duty of government, to provide means of protection for all its citizens in the pursuit of happiness against the obstacles, otherwise insurmountable, which the selfishness of individuals or combinations of individuals is liable to interpose to that pursuit.

It is a scientific fact that the occasional contemplation of natural scenes of an impressive character, particularly if this contemplation occurs in connection with relief from ordinary cares, change of air and change of habits, is favorable to the health and vigor of men and especially to the health and vigor of their intellect beyond any other conditions which can be offered them, that it not only gives pleasure for the time being but increases the subsequent capacity for happiness and the means of securing happiness. The want of such occasional recreation where men and women are habitually pressed by their business or household cares often results in a class of disorders the characteristic quality of which is mental disability, sometimes taking the severe forms of softening of the brain, paralysis, palsy, monomania, or insanity, but more frequently of mental and nervous excitability, moroseness, melancholy or irascibility, incapacitating the subject for the proper exercise of the intellectual and moral forces.

It is well established that where circumstances favor the use of such means of recreation as have been indicated, the reverse of this is true. For instance, it is a universal custom with the heads of the important departments of the British government to spend a certain period of every year on their parks and shooting grounds or in travelling among the Alps or other mountain regions. This custom is followed by the leading lawyers, bankers, merchants and the wealthy classes generally of the empire, among whom the average period of active business life is much greater than with the most nearly corresponding classes in our own or any other country where the same practice is not equally well established.

For instance, Lord Brougham, still an active legislator, is eighty-eight years old. Lord Palmerston, the Prime Minister, is eighty-two, Earl Russell, Secretary of Foreign Affairs, is 74, and there is a corresponding prolongation of vigor among the men of business of the largest and most trying responsibilities in England, as compared with those of our own country which physicians unite in asserting is due in a very essential part to the advantage they have possessed for obtaining occasional relief from their habitual cares, and for enjoying reinvigorating recreation.

But in this country at least it is not those who have the most important responsibilities in state affairs or in commerce, who suffer most from the lack of recreation; women suffer more than men, and the agricultural class is more largely represented in our insane asylums than the professional, and for this, and other reasons, it is these classes to which the opportunity for such recreation is the greatest blessing.

If we analyze the operation of scenes of beauty upon the mind, and consider the intimate relation of the mind upon the nervous system and the whole physical economy, the action and reaction which constantly occur between bodily and mental conditions, the reinvigoration which results from such scenes is readily comprehended. Few persons can see such scenery as that of the Yosemite and not be impressed by it in some slight degree. All not alike, all not perhaps consciously, and amongst all who are consciously impressed by it, few can give the least expression to that of which they are conscious. But there can be no doubt that all have this susceptibility, though with some it is much more dull and confused than with others.

The power of scenery to affect men is, in a large way, proportionate to the degree of their civilization and the degree in which their taste has been cultivated. Among a thousand savages there will be a much smaller number who will show the least sign of being so affected than among a thousand persons taken from a civilized community. This is only one of the many channels in which a similar distinction between civilized and savage men is to be generally observed. The whole body of the susceptibilities of civilized men and with their susceptibilities their powers, are on the whole enlarged.

But as with the bodily powers, if one group of muscles is developed by exercise exclusively, and all others neglected, the result is general feebleness, so it is with the mental faculties. And men who exercise those faculties or susceptibilities of the mind which are called in play by beautiful scenery so little that they seem to be inert with them, are either in a diseased condition from excessive devotion of the mind to a limited range of interests, or their whole minds are in a savage state; that is, a state of low development. The latter class need to be drawn out generally; the former need relief from their habitual matters of interest and to be drawn out in those parts of their mental nature which have been habitually left idle and inert.

But there is a special reason why the reinvigoration of those parts which are stirred into conscious activity by natural scenery is more effective upon the general development and health than that of any other, which is this: The severe and excessive exercise of the mind which leads to the greatest fatigue and is the most wearing upon the whole constitution is almost entirely caused by application to the removal of something to be apprehended in the future, or to interests beyond those of the moment, or of the individual; to the laying up of wealth, to the preparation of something, to accomplishing something in the mind of another, and especially to small and petty details which are uninteresting in themselves and which engage the attention at all only because of the bearing they have on some general end of more importance which is seen ahead.

In the interest which natural scenery inspires there is the strongest contrast to this. It is for itself and at the moment it is enjoyed. The attention is aroused and the mind occupied without purpose, without a continuation of the common process of relating the present action, thought or perception to some future end. There is little else that has this quality so purely. There are few enjoyments with which regard for something outside and beyond the enjoyment of the moment can ordinarily be so little mixed. The pleasures of the table are irresistibly associated with the care of hunger and the repair of the bodily waste. In all social pleasures and all pleasures which are usually enjoyed in association with the social pleasures, the care for the opinion of others, or the good of others largely mingles. In the pleasures of literature, the laying up of ideas and self-improvement are purposes which cannot be kept out of view.

This, however, is in very slight degree, if at all, the case with the enjoyment of the emotions caused by natural scenery. It therefore results that the enjoyment of scenery employs the mind without fatigue and yet exercises it; tranquilizes it and yet enlivens it; and thus, through the influence of the mind over the body, gives the effect of refreshing rest and reinvigoration to the whole system.

Men who are rich enough and who are sufficiently free from anxiety with regard to their wealth can and do provide places of this needed recreation for themselves. They have done so from the earliest periods known in the history of the world, for the great men of the Babylonians, the Persians and the Hebrews, had their rural retreats, as large and as luxurious as those of the aristocracy of Europe at present. There are in the islands of Great Britain and Ireland more than one thousand private parks and notable grounds devoted to luxury and recreation. The value of these grounds amounts to many millions of dollars and the cost of their annual maintenance is greater than that of the national schools; their only advantage to the commonwealth is obtained through the recreation they afford their owners (except as these extend hospitality to others) and these owners with their families number less than one in six thousand of the whole population. The enjoyment of the choicest natural scenes in the country and the means of recreation connected with them is thus a monopoly, in a very peculiar manner, of a very few, very rich people. The great mass of society, including those to whom it would be of the greatest benefit, is excluded from it. In the nature of the case private parks can never be used by the mass of the people in any country nor by any considerable number even of the rich, except by the favor of a few, and in dependence on them.

Thus without means are taken by government to withhold them from the grasp of individuals, all places favorable in scenery to the recreation of the mind and body will be closed against the great body of the people. For the same reason that the water of rivers should be guarded against private appropriation and the use of it for the purpose of navigation and otherwise protected against obstruction, portions of natural scenery may therefore properly be guarded and cared for by government. To simply reserve them from monopoly by individuals, however, it will be obvious, is not all that is necessary. It is necessary that they should be laid open to the use of the body of the people.

The establishment by government of great public grounds for the free enjoyment of the people under certain circumstances, is thus justified and enforced as a political duty.

Such a provision, however, having regard to the whole people of a state, has never before been made and the reason it has not is evident.

It has always been the conviction of the governing classes of the old world that it is necessary that the large mass of all human communities should spend their lives in almost constant labor and that the power of enjoying beauty either of nature or of art in any high degree, requires a cultivation of certain faculties, which is impossible to these humble toilers. Hence it is thought better, so far as the recreations of the masses of a nation receive attention from their rulers, to provide artificial pleasure for them, such as theatres, parades, and promenades where they will be amused by the equipages of the rich and the animation of crowds.

It is unquestionably true that excessive and persistent devotion to sordid interests cramps and distorts the power of appreciating natural beauty and destroys the love of it which the Almighty has implanted in every human being, and which is so intimately and mysteriously associated with the moral perceptions and intuition, but it is not true that exemption from toil, much leisure, much study, much wealth, are necessary to the exercise of the esthetic and contemplative faculties. It is the folly of laws which have permitted and favored the monopoly by privileged classes of many of the means supplied in nature for the gratification, exercise and education of the esthetic faculties that has caused the appearance of dullness and weakness and disease of these faculties in the mass of the subjects of kings. And it is against the limitation of the means of such education to the rich that the wise legislation of free governments must be directed. By such legislation the anticipation of the revered [Andrew Jackson] Downing may be realized.

The dread of the ignorant exclusive, who has no faith in the refinement of a republic, will stand abashed in the next century, before a whole people whose system of voluntary education embraces (combined with perfect individual freedom) not only schools of rudimentary knowledge, but common enjoyments for all classes in the higher realms of art, letters, science, social recreations and enjoyments. Were our legislators but wise enough to understand, today, the destinies of the New World, the gentility of Sir Philip Sidney, made universal, would be not half so much a miracle fifty years hence in America, as the idea of a whole nation of laboring men reading and writing was, in his day, in England.

It was in accordance with these views of the destiny of the New World and the duty of the republican government that Congress enacted that the Yosemite should be held, guarded and managed for the free use of the whole body of the people forever, and that the care of it, and the hospitality of admitting strangers from all parts of the world to visit it and enjoy it freely, should be a duty of dignity and be committed only to a sovereign state.

The trust having been accepted, it will be the duty of the legislature to define the responsibilities, the rights and the powers of the Commissioners, whom by the Act of Congress, it will be the duty of the Executive of the state to appoint. These must be determined by a consideration of the purposes to which the ground is to be devoted and must be simply commensurate with those purposes.

The main duty with which the Commissioners should be charged should be to give every advantage practicable to the mass of the people to benefit by that which is peculiar to this ground and which has caused Congress to treat it differently from other parts of the public domain. This peculiarity consists wholly in its natural scenery.

The first point to be kept in mind then is the preservation and maintenance as exactly as is possible of the natural scenery; the restriction, that is to say, within the narrowest limits consistent with the necessary accommodations of visitors, of all artificial constructions and the prevention of all constructions markedly inharmonious with the scenery or which would unnecessarily obscure, distort or detract from the dignity of the scenery.

In addition to the more immediate and obvious arrangements by which this duty is enforced there are two considerations which should not escape attention.

First: the value of the district is in its present condition as a museum of natural science and the danger, indeed the certainty, that without care many of the species of plants now flourishing upon it will be lost and many interesting objects be defaced or obscured if not destroyed.

To illustrate these dangers, it may be stated that numbers of the native plants of large districts of the Atlantic states have almost wholly disappeared and that most of the common weeds of the farms are of foreign origin, having choked out the native vegetation. Many of the finer specimens of the most important trees in the scenery of the Yosemite have been already destroyed and the proclamation of the Governor, issued after the passage of the Act of Congress, forbidding the destruction of trees in the district, alone prevented the establishment of a saw mill within it. Notwithstanding the proclamation many fine trees have been felled and others girdled within a year. Indians and others have set fire to the forests and herbage and numbers of trees have been killed by these fires; the giant tree before referred to as probably the noblest tree now standing on the earth has been burned completely through the bark near the ground for a distance of more than one hundred feet of its circumference; not only have trees been cut, hacked, barked and fired in prominent positions, but rocks in the midst of the most picturesque natural scenery have been broken, painted and discolored by fires built against them. In travelling to the Yosemite and within a few miles of the nearest point at which it can be approached by a wheeled vehicle, the Commissioners saw other picturesque rocks stencilled over with advertisements of patent medicines and found the walls of the Bower Cave, one of the most beautiful natural objects in the state, already so much broken and scratched by thoughtless visitors that it is evident that unless the practice should be prevented not many years will pass before its natural charms will be quite destroyed.

Second: it is important that it should be remembered that in permitting the sacrifice of anything that would be of the slightest value to future visitors to the convenience, bad taste, playfulness, carelessness, or wanton destructiveness of present visitors, we probably yield in each case the interest of uncounted millions to the selfishness of a few individuals.

It is an important fact that as civilization advances the interest of men in natural scenes of sublimity and beauty increases. Where a century ago one traveller came to enjoy the scenery of the Alps, thousands come now, and even where forty years ago one small inn accommodated the visitors to the White Hills of New Hampshire, half a dozen grand hotels, each accommodating hundreds, are now overcrowded every summer. In the early part of the present century the summer visitors to the Highlands of Scotland did not give business enough to support a single inn, a single stage coach or a single guide. They now give business to several rail road trains, scores of steamboats and thousands of men and horses every day.

It is but sixteen years since the Yosemite was first seen by a white man, several visitors have since made a journey of several thousand miles at large cost to see it, and notwithstanding the difficulties which now interpose, hundreds resort to it annually. Before many years if proper facilities are offered, these hundreds will become thousands and in a century the whole number of visitors will be counted by the millions. An injury to the scenery so slight that it may be unheeded by any visitor now, will be one of deplorable magnitude when its effect upon each visitor's enjoyment is multiplied by these millions. But again, the slight harm which the few hundred visitors of this year might do, if no care were taken to prevent it, would not be slight if it should be repeated by millions.

At some time, therefore, laws to prevent an unjust use by individuals, of that which is not individual but public property must be made and rigidly enforced. The principle of justice involved is the same now that it will be then; such laws as this principle demands will be more easily enforced, and there will be less hardship in their action, if the abuses they are designed to prevent are never allowed to become customary but are checked while they are yet of unimportant consequence.

It should, then, be made the duty of the Commission to prevent a wanton or careless disregard on the part of anyone entering the Yosemite or the Grove, of the rights of posterity as well as of contemporary visitors, and the Commission should be clothed with proper authority and given the necessary means for this purpose.

This duty of preservation is the first which falls upon the state under the Act of Congress, because the millions who are hereafter to benefit by the Act have the largest interest in it, and the largest interest should be first and most strenuously guarded.

Next to this, and for a similar reason preceding all other duties of the state in regard to this trust, is that of aiding to make this appropriation of Congress available as soon and as generally as may be economically practicable to those whom it is designed to benefit. Had Congress not thought best to depart from the usual method of dealing with the public lands in this case, it would have been prac-

ticable for one man to have bought the whole, to have appropriated
it wholly to his individual pleasure or to have refused admittance to
any who were unable to pay a certain price as admission fee, or as
a charge for the entertainment which he would have had a monopoly
of supplying. The result would have been a rich man's park, and for
the present, so far as the great body of the people are concerned, it
is and as long as the present arrangements continue, it will remain,
practically, the property only of the rich.

A man travelling from Stockton to the Yosemite or the Mariposa
Grove is commonly three or four days on the road at an expense of
from thirty to forty dollars, and arrives in the majority of cases quite
overcome with the fatigue and unaccustomed hardships of the journey. Few persons, especially few women, are able to enjoy or profit
by the scenery and air for days afterwards. Meantime they remain
at an expense of from $3 to $12 per day for themselves, their guide
and horses, and many leave before they have recovered from their
first exhaustion and return home jaded and ill. The distance is not
over one hundred miles, and with such roads and public conveyanc-
es as are found elsewhere in the state the trip might be made easily
and comfortably in one day and at a cost of ten or twelve dollars.
With similar facilities of transportation, the provisions and all the
necessities of camping could also be supplied at moderate rates. To
realize the advantages which are offered the people of the state in
this gift of the nation, therefore, the first necessity is a road from the
termination of the present roads leading towards the district. At present
there is no communication with it except by means of a very poor
trail for a distance of nearly forty miles from the Yosemite and twenty
from the Mariposa Grove.

Besides the advantages which such a road would have in reducing
the expense, time and fatigue of a visit to the tract, to the whole
public at once, it would also serve the important purpose of making
it practicable to convey timber and other articles necessary for the
accommodation of visitors into the Yosemite from without, and thus
the necessity, or the temptation, to cut down its groves and to prepare its surface for tillage would be avoided. Until a road is made
it must be very difficult to prevent this.

The Commissioners propose also in laying out a road to the Mariposa
Grove that it shall be carried completely around it, so as to offer a
barrier of bare ground to the approach of fires, which nearly every
year sweep upon it from the adjoining country, and which during the
last year alone have caused injuries, exemption from which it will
be thought before many years would have been cheaply obtained at
ten times the cost of the road.

Within the Yosemite the Commissioners propose to cause to be constructed a double trail, which, on the completion of our approach road, may be easily made suitable for the passage of a single vehicle, and which shall enable visitors to make a complete circuit of all the broader parts of the valley and to cross the meadows at certain points, reaching all the finer points of view to which it can be carried without great expense. When carriages are introduced it is proposed that they shall be driven for the most part up one side and down the other of the valley, suitable resting places and turnouts for passing being provided at frequent intervals. The object of this arrangement is to reduce the necessity for artificial construction within the narrowest practicable limits, destroying as it must the natural conditions of the ground and presenting an unpleasant object to the eye in the midst of the scenery. The trail or narrow road could also be kept more in the shade, could take a more picturesque course, would be less dusty, and could be much more cheaply kept in repair. From this trail a few paths would also need to be formed, leading to points of view which would only be accessible to persons on foot. Several small bridges would also be required.

The Commission also propose the construction of five cabins at points in the valley conveniently near to those most frequented by visitors, especially near the foot of the cascades, but at the same time near to convenient camping places. These cabins would be let to tenants with the conditions that they should have constantly open one comfortable room as a free resting place for visitors, with the proper private accommodations for women, and that they should keep constantly on hand in another room a supply of certain simple necessities for camping parties, including tents, cooking utensils and provisions; the tents and utensils to be let, and the provisions to be sold at rates to be limited by the terms of the contract.

The Commissioners ask and recommend that sums be appropriated for these and other purposes named below as follows:

For the expense already incurred in the survey and
    transfer of the Yosemite and Mariposa Big Tree
    Grove from the United States to the State of
    California . . . . $ 2,000
For the construction of 30 miles, more or less, of
    double trail and footpaths 3,000
For the construction of bridges 1,600
For the construction and finishing five cabins,
    closets, stairways, railings, etc. 2,000
Salary for Superintendent (2 years) 2,400
For surveys, advertising and incidentals 1,000
For aid in the construction of a road 25,000

                                                     $37,000

The Commissioners trust that after this amount shall have been expended the further necessary expenses for the management of the domain will be defrayed by the proceeds of rents and licenses which will be collected upon it.

The Yosemite yet remains to be considered as a field of study for science and art. Already students of science and artists have been attracted to it from the Atlantic states and a number of artists have at heavy expense spent the summer in sketching the scenery. That legislation should, when practicable within certain limits, give encouragement to the pursuit of science and art has been fully recognized as a duty by this state. The pursuit of science and of art, while it tends more than any other human pursuit to the benefit of the commonwealth and the advancement of civilization, does not correspondingly put money into the hands of the pursuers. Their means are generally extremely limited. They are likely by the nature of their studies to be the best counsellors which can be had in respect to certain of the duties which will fall upon the proposed Commission, and it is right that they should if possible be honorably represented in the constitution of the Commission.

Congress has provided that the Executive shall appoint eight Commissioners, and that they shall give their service gratuitously. It is but just that the state should defray the travelling expenses necessarily incurred in the discharge of their duty. It is proposed that the allowance for this purpose shall be limited in amount to four hundred dollars per annum, for each Commissioner, or so much thereof as shall have been actually expended in travelling to and from the ground and while upon it. It is also proposed that of the eight Commissioners to be appointed by the Executive, four shall be appointed annually and that these four shall be students of natural science or landscape artists. It is advised also, in order that it may be in the power of the Governor when he sees fit to offer the slight consideration, represented in the sum of $400 proposed to be allowed each Commissioner for travelling expenses, as an inducement to men of scientific note and zealous artists to visit the state, that he should not necessarily be restricted in these appointments to citizens of the state. The Yosemite being a trust from the whole nation, it seems eminently proper that so much liberality in its management should be authorized.

As reprinted in *Landscape Architecture*, 43, 1952, 12–25.
Note: The description of Yosemite was taken from Olmsted's narrative in *The Saturday Evening Post* (June 18, 1868) and added to the handwritten transcript of his report from which the original description was missing.

AN ACT TO SET APART A CERTAIN TRACT OF LAND LYING NEAR THE HEADWATERS OF THE YELLOWSTONE RIVER AS A PUBLIC PARK,

Approved March 1, 1872 (17 Stat. 32)

Be it enacted by the Senate and House of Representatives of the United States of America in Congress assembled, That the tract of land in the Territories of Montana and Wyoming, lying near the headwaters of the Yellowstone River, and described as follows, to wit, commencing at the junction of Gardiner's river with the Yellowstone river, and running east to the meridian passing ten miles to the eastward of the most eastern point of Yellowstone lake; thence south along said meridian to the parallel of latitude passing ten miles south of the most southern point of Yellowstone lake; thence west along said parallel to the meridian passing fifteen miles west of the most western point of Madison lake; thence north along said meridian to the latitude of the junction of Yellowstone and Gardiner's rivers; thence east to the place of beginning, is hereby reserved and withdrawn from settlement, occupancy, or sale under the laws of the United States, and dedicated and set apart as a public park or pleasuring-ground for the benefit and enjoyment of the people; and all persons who shall locate or settle upon or occupy the same, or any part thereof, except as hereinafter provided, shall be considered trespassers and removed therefrom. (U.S.C., title 16, sec. 21.)

SEC 2. That said public park shall be under the exclusive control of the Secretary of the Interior, whose duty it shall be, as soon as practicable, to make and publish such rules and regulations as he may deem necessary or proper for the care and management of the same. Such regulations shall provide for the preservation, from injury or spoilation, of all timber, mineral deposits, natural curiosities, or wonders within said park, and their retention in their natural condition. The Secretary may in his discretion, grant leases for building purposes for terms not exceeding ten years, of small parcels of ground, at such places in said park as shall require the erection of buildings for the accommodation of visitors; all of the proceeds of said leases, and all other revenues that may be derived from any source connected with said park, to be expended under his direction in the management of the same, and the construction of roads and bridle-paths therein. He shall provide against the wanton destruction of the fish and game found within said park, and against their capture or destruction for the purposes of merchandise or profit. He shall also cause all persons trespassing upon the same after the passage of this act to be removed therefrom, and generally shall be authorized to take all such measures as shall be necessary or proper to fully carry out the objects and purposes of this act. (U.S.C., title 16, sec. 22.)

PROTECTION OF CASA GRANDE RUIN

1889

CHAP. 411.—An act making appropriations for sundry civil expenses of the Government for the fiscal year ending June thirtieth, eighteen hundred and ninety, and for other purposes.

Be it enacted by the Senate and House of Representatives of the United States of America in Congress assembled, That the following sums be, and the same are hereby, appropriated for the objects hereinafter expressed for the fiscal year ending June thirtieth, eighteen hundred and ninety, namely:

UNDER THE INTERIOR DEPARTMENT.

REPAIR OF THE RUIN OF CASA GRANDE, ARIZONA: To enable the Secretary of the Interior to repair and protect the ruin of Casa Grande, situate in Pinal County, near Florence, Arizona, two thousand dollars; and the President is authorized to reserve from settlement and sale the land on which said ruin is situated and so much of the public land adjacent thereto as in his judgement may be necessary for the protection of said ruin and of the ancient city of which it is a part.

50th Congress, Session II, Chapter 411, 1889.

AN ACT TO ESTABLISH A NATIONAL MILITARY PARK AT THE BATTLE-FIELD OF CHICKAMAUGA,

Approved August 19, 1890 (26 Stat. 333)

Be it enacted by the Senate and House of Representatives of the United States of America in Congress assembled, That for the purpose of preserving and suitably marking for historical and professional military study the fields of some of the most remarkable maneuvers and most brilliant fighting in the war of the rebellion, and upon the ceding of jurisdiction to the United States by the States of Tennessee and Georgia, respectively, and the report of the Attorney General of the United States that the title to the lands thus ceded is perfect, the following described highways in those States are hereby declared to be approaches to and parts of the Chickamauga and Chattanooga National Military Park as established by the second section of this act, to wit: First. The Missionary Ridge Crest road from Sherman Heights at the north end of Missionary Ridge, in Tennessee, where the said road enters upon the ground occupied by the Army of the Tennessee under Major-General William T. Sherman, in the military operations of November twenty-fourth and twenty-fifth, eighteen hundred and sixty-three; thence along said road through the positions occupied by the army of General Braxton Bragg on November twenty-fifth, eighteen hundred and sixty-three, and which were assaulted by the Army of the Cumberland under Major-General George H. Thomas on that date, to where the said road crosses the southern boundary of the State of Tennessee, near Rossville Gap, Georgia, upon the ground occupied by the troops of Major-General Joseph Hooker, from the Army of the Potomac, and thence in the State of Georgia to the junction of said road with the Chattanooga and Lafayette or State road at Rossville Gap; second, the Lafayette or State road from Rossville, Georgia, to Lee and Gordon's Mills, Georgia; third, the road from Lee and Gordon's Mills, Georgia, to Crawfish Springs, Georgia; fourth, the road from Crawfish Springs, Georgia, to the crossing of the Chickamauga at Glass' Mills, Georgia; fifth, the Dry Valley road from Rossville, Georgia, to the southern limits of McFarland's Gap in Missionary Ridge; sixth, the Dry Valley and Crawfish Springs road from McFarland's Gap to the intersection of the road from Crawfish Springs to Lee and Gordon's Mills; seventh, the road from Ringold, Georgia, to Reed's Bridge on the Chickamauga River; eighth, the roads from the crossing of Lookout Creek across the northern slope of Lookout Mountain and thence to the old Summertown Road and to the valley on the east slope of said mountain, and thence by the route of General Joseph Hooker's troops to Rossville, Georgia, and each and all of these herein described roads shall, after the passage of this act, remain open as free public highways, and all rights of way now existing through the grounds of the said park and its approaches shall be continued.

SEC. 2. That upon the ceding of jurisdiction by the legislature of the State of Georgia, and the report of the Attorney-General of the United States that a perfect title has been secured under the provisions of the act approved August first, eighteen hundred and eighty-eight, entitled "An act to authorize condemnation of land for sites of public buildings, and for other purposes," the lands and roads embraced in the area bounded as herein described, together with the roads described in section one of this act, are hereby declared to be a national park, to be known as the Chickamauga and Chattanooga National Park; that is to say, the area inclosed by a line beginning on the Lafayette or State road, in Georgia, at a point where the bottom of the ravine next north of the house known on the field of Chickamauga as the Cloud House, and being about six hundred yards north of said house, due east to the Chickamauga River and due west to the intersection of the Dry Valley road at McFarland's Gap; thence along the west side of the Dry Valley and Crawfish Springs roads to the south side of the road from Crawfish Springs to Lee and Gordon's Mills; thence along the south side of the last named road to Lee and Gordon's Mills; thence along the channel of the Chickamauga River to the line forming the northern boundary of the park, as hereinbefore described, containing seven thousand six hundred acres, more or less.

SEC. 3. That the said Chickamauga and Chattanooga National Park, and the approaches thereto, shall be under the control of the Secretary of War, and it shall be his duty, immediately after the passage of this act to notify the Attorney General of the purpose of the United States to acquire title to the roads and lands described in the previous sections of this act under the provisions of the act of August first, eighteen hundred and eighty-eight; and the said Secretary, upon receiving notice from the Attorney-General of the United States that perfect titles have been secured to the said lands and roads, shall at once proceed to establish and substantially mark the boundaries of the said park.

SEC. 4. That the Secretary of War in hereby authorized to enter into agreements, upon such nominal terms as he may prescribe, with such present owners of the land as may desire to remain upon it, to occupy and cultivate their present holdings, upon condition that they will preserve the present buildings and roads, and the present out-

lines of field and forest, and that they will only cut trees or under-
brush under such regulations as the Secretary may prescribe, and
that they will assist in caring for and protecting all tablets, monuments, or such other artificial works as may from time to time be
erected by proper authority.

SEC. 5. That the affairs of the Chickamauga and Chattanooga
National Park shall, subject to the supervision and direction of the
Secretary of War, be in charge of three commissioners, each of whom
shall have actively participated in the battle of Chickamauga or one
of the battles about Chattanooga, two to be appointed from civil life
by the Secretary of War, and a third, who shall be detailed by the
Secretary of War from among those officers of the Army best acquainted with the details of the battles of Chickamauga and Chattanooga, who shall act as Secretary of the Commission. The said
commissioners and Secretary shall have an office in the War Department building, and while on actual duty shall be paid such compensation, out of the appropriation provided in this act, as the Secretary
of War shall deem reasonable and just.

SEC. 6. That it shall be the duty of the commissioners named in
the preceding section, under the direction of the Secretary of War,
to superintend the opening of such roads as may be necessary to the
purposes of the park, and the repair of the roads of the same, and to
ascertain and definitely mark the lines of battle of all troops engaged in the battles of Chickamauga and Chattanooga, so far as the
same shall fall within the lines of the park as defined in the previous
sections of this act, and, for the purpose of assisting them in their
duties and in ascertaining these lines, the Secretary of War shall have
authority to employ, at such compensation as he may deem reasonable and just, to be paid out of the appropriation made by this act,
some person recognized as well informed in regard to the details of
the battles of Chickamauga and Chattanooga, and who shall have
actively participated in one of those battles, and it shall be the duty
of the Secretary of War from and after the passage of this act, through
the commissioners, and their assistant in historical work, and under
the act approved August first, eighteen hundred and eighty-eight,
regulating the condemnation of land for public uses, to proceed with
the preliminary work of establishing the park and its approaches as
the same are defined in this act, and the expenses thus incurred shall
be paid out of the appropriation provided by this act.

SEC. 7. That it shall be the duty of the commissioners, acting
under the direction of the Secretary of War, to ascertain and substantially mark the locations of the regular troops, both infantry and
artillery, within the boundaries of the park, and to erect monuments
upon those positions as Congress may provide the necessary appro-

priations; and the Secretary of War in the same way may ascertain and mark all lines of battle within the boundaries of the park and erect plain and substantial historical tablets at such points in the vicinity of the Park and its approaches as he may deem fitting and necessary to clearly designate positions and movements, which, although without the limits of the Park, were directly connected with the battles of Chickamauga and Chattanooga.

SEC. 8. That it shall be lawful for the authorities of any State having troops engaged either at Chattanooga or Chickamauga, and for the officers and directors of the Chickamauga Memorial Association, a corporation chartered under the laws of Georgia, to enter upon the lands and approaches of the Chickamauga and Chattanooga National Park for the purpose of ascertaining and marking the lines of battle of troops engaged therein: Provided, That before any such lines are permanently designated the position of the lines and the proposed methods of marking them by monuments, tablets, or otherwise shall be submitted to the Secretary of War, and shall first receive the written approval of the Secretary, which approval shall be based upon formal written reports, which must be made to him in each case by the commissioners of the park.

SEC. 9. That the Secretary of War, subject to the approval of the President of the United States, shall have the power to make, and shall make, all needed regulations for the care of the park and for the establishment and marking of the lines of battle and other historical features of the park.

SEC. 10. That if any person shall willfully destroy, mutilate, deface, injure, or remove any monument, column, statue, memorial structure, or work of art that shall be erected or placed upon the grounds of the park by lawful authority, or shall willfully destroy or remove any fence, railing, inclosure, or other work for the protection or ornament of said park, or any portion thereof, or shall willfully destroy, cut, hack, bark, break down, or otherwise injure any tree, bush, or shrubbery that may be growing upon said park, or shall cut down or fell or remove any timber, battle relic, tree or trees growing or being upon such park, except by permission of the Secretary of War, or shall willfully remove or destroy any breast-works, earth-works, walls, or other defenses or shelter, on any part thereof, constructed by the armies formerly engaged in the battles on the lands or approaches to the park, any person so offending and found guilty thereof, before any justice of the peace of the county in which the offense may be committed, shall for each and every such offense forfeit and pay a fine, in the discretion of the justice, according to the aggravation of the offense, of not less than five nor more than fifty dollars, one-half to the use of the park and the other half to the informer, to be enforced and recovered, before such justice, in like manner as debts of like nature are now by law recoverable in the several counties where the offense may be committed.

SEC. 11. That to enable the Secretary of War to begin to carry out the purpose of this act, including the condemnation and purchase of the necessary land, marking the boundaries of the park, opening or repairing necessary roads, maps and surveys, and the pay and expenses of the commissioners and their assistant, the sum of one hundred and twenty-five thousand dollars, or such portion thereof as may be necessary, is hereby appropriated, out of any moneys in the Treasury not otherwise appropriated, and disbursements under this act shall require the approval of the Secretary of War, and he shall make annual report of the same to Congress. (16 U.S.C. §424 as amended.)

AN ACT TO PROTECT THE BIRDS AND ANIMALS IN YELLOWSTONE NATIONAL PARK, AND TO PUNISH CRIMES IN SAID PARK, AND FOR OTHER PURPOSES,

Approved May 7, 1894 (28 Stat. 73)

Be it enacted by the Senate and House of Representatives of the United States of America in Congress assembled, That the Yellowstone National Park, as its boundaries are now defined, or as they may be hereafter defined or extended, shall be under the sole and exclusive jurisdiction of the United States; and that all the laws applicable to places under the sole and exclusive jurisdiction of the United States shall have force and effect in said park: *Provided, however,* That nothing in this act shall be construed to forbid the service in the park of any civil or criminal process of any court having jurisdiction in the States of Idaho, Montana, and Wyoming. All fugitives from justice taking refuge in said park shall be subject to the same laws as refugees from justice found in the State of Wyoming. (U.S.C., title 16, sec. 24.)

SEC. 2. That said park, for all the purposes of this act, shall constitute a part of the United States judicial district of Wyoming, and the district and circuit courts of the United States in and for said district shall have jurisdiction of all offenses committed within said park.

SEC. 3. That if any offense shall be committed in said Yellowstone National Park, which offense is not prohibited or the punishment is not specially provided for by any law of the United States or by any regulation of the Secretary of the Interior, the offender shall be subject to the same punishment as the laws of the State of Wyoming in force at the time of the commission of the offense may provide for a like offense in said State; and no subsequent repeal of any such law in the State of Wyoming shall affect any prosecution for said offense committed within said park. (U.S.C., title 16, sec. 25.)

SEC. 4. That all hunting, or the killing, wounding, or capturing at any time of any bird or wild animal, except dangerous animals, when it is necessary to prevent them from destroying human life or inflicting an injury, is prohibited within the limits of said park; nor shall any fish be taken out of the waters of the park by means of seines, nets, traps, or by the use of drugs or any explosive substances or compounds, or in any other way than by hook and line, and then only at such seasons and in such times and manner as may be directed by the Secretary of the Interior. That the Secretary of the Interior shall make and publish such rules and regulations as he may deem necessary and proper for the management and care of the park and for the protection of the property therein, especially for the preservation from injury or spoilation of all timber, mineral deposits, natural curiosities, or wonderful objects within said park; and for the protection of the animals and birds in the park from capture or destruction, or to prevent their being frightened or driven from the park; and he shall make rules and regulations governing the taking of fish from the streams or lakes in the park. Possession within the said park of the dead bodies, or any part thereof, of any wild bird or animal, shall be prima facie evidence that the person or persons having the same are guilty of violating this act. Any person or persons, or stage or express company or railway company, receiving for transportation any of the said animals, birds or fish so killed, taken or caught, shall be deemed guilty of a misdemeanor, and shall be fined for every such offense not exceeding three hundred dollars. Any person found guilty of violating any of the provisions of this act or any rule or regulation that may be promulgated by the Secretary of the Interior with reference to the management and care of the park, or for the protection of the property therein, for the preservation from injury or spoilation of timber, mineral deposits, natural curiosities or wonderful objects within said park, or for the protection of the animals, birds and fish in the said park, shall be deemed guilty of a misdemeanor, and shall be subjected to a fine or not more than one thousand dollars or imprisonment not exceeding two years, or both, and be adjudged to pay all costs of the proceedings.

That all guns, traps, teams, horses, or means of transportation of every nature or description used by any person or persons within said park limits when engaged in killing, trapping, ensnaring, or capturing such wild beasts, birds, or wild animals shall be forfeited to the United States, and may be seized by the officers in said park and held pending the prosecution of any person or persons arrested under charge of violating the provisions of this act, and upon conviction under this act of such person or persons using said guns, traps, teams, horses, or other means of transportation such forfeiture shall be adjudicated as a penalty in addition to the other punishment provided in this act. Such forfeited property shall be disposed of and accounted for by and under the authority of the Secretary of the Interior. (U.S.C., title 16, sec. 26.)

SEC. 5. That the United States circuit court in said district shall appoint a commissioner, who shall reside in the park, who shall have jurisdiction to hear and act upon all complaints made, of any and all violations of the law, or of the rules and regulations made by the Secretary of the Interior for the government of the park, and for the protection of the animals, birds, and fish and objects of interest therein, and for other purposes authorized by this act. Such commissioner shall have power, upon sworn information, to issue process in the name of the United States for the arrest of any person charged with the commission of any misdemeanor, or charged with the violation of the rules and regulations, or with the violation of any provision of this act prescribed for the government of said park, and for the protection of the animals, birds, and fish in the said park, and to try the person so charged, and, if found guilty, to impose the punishment and adjudge the forfeiture prescribed. In all cases of conviction an appeal shall lie from the judgment of said commission to the United States district court for the district of Wyoming, said appeal to be governed by the laws of the State of Wyoming providing for appeals in cases of misdemeanor from justices of the peace to the district court of said State; but the United States district court in said district may prescribe rules of procedure and practice for said commissioner in the trial of cases and for appeal to said United States district court. Said commissioner shall also have power to issue process as hereinbefore provided for the arrest of any person charged with the commission of any felony within the park, and to summarily hear the evidence introduced, and, if he shall determine that probable cause is shown for holding the person so charged for trial, shall cause such person to be safely conveyed to a secure place for confinement, within the jurisdiction of the United States district court in said State of Wyoming, and shall certify a transcript of the record of his proceedings and the testimony in the case to the said court, which court shall have jurisdiction of the case: Provided, That the said commissioner shall grant bail in all cases bailable under the laws of the United States or of said State. All process issued by the commissioner shall be directed to the marshal of the United States for the district of Wyoming; but nothing herein contained shall be construed as preventing the arrest by any officer of the Government or employee of the United States in the park, without process, of any person taken in the act of violating the law or any regulation of the Secretary of the Interior: Provided, That the said commissioner shall only exercise such authority and powers as are conferred by this act. (U.S.C., title 16, sec. 27).

SEC. 6. That the marshal of the United States for the district of Wyoming may appoint one or more deputy marshals for said park, who shall reside in said park, and the said United States district and circuit courts shall hold one session of said courts annually at the town of Sheridan in the State of Wyoming, and may also hold other sessions at any other place in said State of Wyoming or in said National Park at such dates as the said courts may order. (U.S.C., title 16, sec. 28.)

SEC. 7. That the commissioner provided for in the act shall, in addition to the fees allowed by law to commissioners of the circuit courts of the United States, be paid an annual salary of one thousands dollars, payable quarterly, and the marshal of the United States and his deputies, and the attorney of the United States and his assistants in said district, shall be paid the same compensation and fees as now provided by law for like services in said district. (U.S.C., title 16, sec. 29.)

SEC. 8. That all costs and expenses arising in cases under this act, and properly chargeable to the United States, shall be certified, approved, and paid as like costs and expenses in the courts of the United States are certified, approved, and paid under the laws of the United States. (U.S.C., title 16, sec. 31.)

SEC. 9. That the Secretary of the Interior shall cause to be erected in the park a suitable building to be used as a jail, and also having in said building an office for the use of the commissioner, the cost of such building not to exceed five thousands dollars, to be paid out of any moneys in the Treasury not otherwise appropriated upon the certificate of the Secretary as a voucher therefor. (U.S.C., title 16, sec. 30.)

SEC. 10. That this act shall not be construed to repeal existing laws conferring upon the Secretary of the Interior and the Secretary of War certain powers with reference to the protection, improvement, and control of the said Yellowstone National Park.

16 U.S.C. 24–31.

AN ACT FOR THE PRESERVATION
OF AMERICAN ANTIQUITIES,

Approved June 8, 1906 (34 Stat. 225)

Be it enacted by the Senate and House of Representatives of the United States of American in Congress assembled, That any person who shall appropriate, excavate, injure, or destroy any historic or prehistoric ruin or monument, or any object of antiquity, situated on lands owned or controlled by the Government of the United States, without the permission of the Secretary of the department of the Government having jurisdiction over the lands on which said antiquities are situated, shall, upon conviction, be fined in a sum of not more than five hundred dollars or be imprisoned for a period of not more than ninety days, or shall suffer both fine and imprisonment, in the discretion of the court. (U.S.C., title 16, sec. 433.)

SEC 2. That the President of the United States is hereby authorized, in his discretion, to declare by public proclamation historic landmarks, historic and prehistoric structures, and other objects of historic or scientific interest that are situated upon the lands owned or controlled by the Government of the United States to be national monuments, and may reserve as a part thereof parcels of land, the limits of which in all cases shall be confined to the smallest area compatible with the proper care and management of the objects to the protected: *Provided,* That when such objects are situated upon a tract covered by a bona fide unperfected claim or held in private ownership, the tracts, or so much thereof as may be necessary for the proper care and management of the object, may be relinquished to the Government, and the Secretary of the Interior is hereby authorized to accept the relinquishment of such tracts in behalf of the Government of the United States. (U.S.C., title 16, sec. 431.)

SEC. 3. That permits for the examination of ruins, the excavation of archaeological sites, and the gathering of objects of antiquity upon the lands under their respective jurisdictions may be granted by the Secretaries of the Interior, Agriculture, and War to institutions which they may deem properly qualified to conduct such examination, excavation, or gathering, subject to such rules and regulations as they may prescribe: *Provided,* That the examinations, excavations, and gatherings are undertaken for the benefit of reputable museums, universities, colleges, or other recognized scientific or educational institutions, with a view to increasing the knowledge of such objects, and that the gatherings shall be made for permanent preservation in public museums. (U.S.C., title 16, sec. 432.)

SEC. 4. That the Secretaries of the departments aforesaid shall make and publish from time to time uniform rules and regulations for the purpose of carrying out the provisions of this act. (U.S.C., title 16, sec. 432.)

AUTO USE IN THE NATIONAL PARKS
PROCEEDINGS OF THE NATIONAL PARK
CONFERENCE HELD AT THE YOSEMITE
NATIONAL PARK

October 14, 15, and 16, 1912

WALTER L. FISHER [Secretary of the Interior]: Once more, are we ready for the automobile question? If we are, perhaps before starting it might be well to make a brief reference to a little discussion we had last night, which, of course, is known to the selected representatives of the automobile people who are here present, but should be fully known to all the others. It may be desirable to clear away the fog on this question as far as we can. There is said to be a tendency toward fog on certain portions of the Pacific coast, and I want to make sure none has gotten into the automobile issue. It will not be necessary to argue with the present Secretary of the Interior that the automobile is an improved means of transportation which has come to stay; it will not be necessary to argue with him that if it can be introduced into the Yellowstone Park or to the Yosemite Park or any other park, under conditions which are otherwise proper, it ought to be done. The interesting and important question is whether the conditions are proper, and upon that what I wish is constructive suggestion. It will not be necessary for any representative of any automobile concern or of any automobile organization to argue with me upon the proposition that the machines should be admitted if we can find a proper way; but they should not pass up to me the question of what that proper way is. If I knew a proper way to admit the automobiles into the Yosemite Park it would not be necessary to discuss that question at all to-day or at any other time. The difficulty is that with all the consideration and attention we have given the subject, including the examination and report of engineers, we do not know of such a way, and we want to hear the question discussed from that point of view.

Now, there are several classes of automobile, as you know, and a greater variety of automobilists. If all the automobiles were of certain types and if automobilists operated that type of machine in the way that some operate their automobiles, it would be a tame animal and we could introduce it into the parks with impunity. Unfortunately, in the process of evolution we have not got that far. It is not necessary to argue with the automobilists, if we are going to be frank with each other and talk man fashion, that there are still a great many gentlemen who buy automobiles who have not yet ceased to be peripatetic nuisances. We do know that some automobiles make a great deal of noise; that they emit very obnoxious odors; that they drop their oil and gasoline all over the face of the earth wherever they go; that those automobiles are sold by people who regard it as a hardship to be excluded from any particular road. We know much more clearly that even machines which, as machines, have reached a high degree of perfection, are operated by gentlemen who don't know how to operate them, and are operated by other gentlemen who may know how, but don't take the necessary pains to operate them properly, and by still a third class of gentlemen who are perfectly fearless themselves and, liking adventure, operate them in such a way as to create the impression on passers-by on foot or in a horse-drawn vehicle that it is very dangerous to be on the road at the same time.

The daily papers are full of reports of the results of these things, and it does very little good to demonstrate even if it could be demonstrated to the satisfaction of a court, that after all, if the driver of a horse-drawn vehicle had handled his team with proper circumspection the accident would not have occurred. It does occur every day and therefore it is very important that we do not bring about a situation where it is more likely to occur, under conditions where the Government is inviting people into a national park on the theory that it is a playground and that they can largely relax the habits they may have in crowded centers of civilization of being everlastingly on the watch unless they be run into. There are several phases of the situation as it relates to the Yosemite.

There are a number of suggestions that have reached me, and I am going to try to get rid of a few of the questions right at the start. I am in receipt, as I said yesterday, of a considerable number of telegrams brought about by the very laudable and active influence of the automobile organizations and, I should judge, of the automobile manufacturers and agents, who want to see that the machines are admitted into this park; and in this connection permit me to say that I have not the slightest objection to the automobile business as a business. It is a very excellent business, and I would like to see it succeed, and I am willing to assume that a man in that business will be very earnest in trying to extend it. I have no objection to that. I think it is his right as an American citizen to do that and he is entitled to careful consideration. Now these telegrams have reached me; but among them there is apparently not an entire unanimity. Some of the telegrams object most strenuously to the introduction of automobiles in the parks, apparently on any basis, even to the rim of the park, so there is that difference to start with among automobile

people. I have received other letters and communications with regard to the admission of automobiles on the floor of the valley, from
men who have said they would be in favor of the admission of machines
to the rim if it could be worked out, but would be radically opposed
to the introduction of those machines on the floor of the valley, and
I may say, without violating any confidences, you have among you
here in attendance, gentlemen who most heartily concur in that view.

There are men who say that the machines should not be admitted
to the floor of the valley. Some think they should be admitted to the
rim, and they disagree among themselves as to whether that should
be upon a road which is also used by horse-drawn vehicles or whether
it should be on a separate road, and some of them have suggestions
with regard to a separate road and others have suggestions with regard
to the use of a road jointly with horse-drawn vehicles, but at different hours and under regulations that would protect the two kinds of
traffic, as they think. Those are the things about which I would like
to hear from you, and if the representatives will address themselves
to those questions right at the outset, I think we will make more
progress than in any other way. . . .

Col. FORSYTH [Acting Superintendent of Yosemite National Park]:
I still think, Mr. Secretary, that there are features even that I am not
prepared to deal with—to give an opinion on. It is a question of
engineering. I think that if the roads are made safe, and the question
as to what is safe brings out such diversity of opinion, it must be
settled by the engineers. My only opposition to the automobile in
this park is the safety to human life. . . .

Now, in connection with all the national parks, the bill setting
aside that park either says so explicitly or by implication that the
park shall be a place of resort and recreation for the people, a place
of benefit and enjoyment for the people for all time. Now, when the
Government sets aside a park for that purpose, it takes on itself the
obligation of making that park accessible for all the people; that is,
possible for all time. Now, that obligation goes with the very establishment of parks, but that obligation is limited. It is overshadowed
by this other obligation on the Government to throw around the people
every reasonable safeguard to life and limb. Now, that obligation is
of greater importance than the other. It overshadows it. It is fundamental. That very same obligation in a different aspect compels our
Government to send our Army and Navy to distant lands to protect
our lives and people. That is the same obligation resting on us right
here, on the Secretary and on myself, in the protection of life and
limb here in the park.

Now, in the way of mountain roads, this park is much more dangerous than the Yellowstone Park in the main. The roads here are pretty narrow. This Big Oak Flat Road is only 8 feet wide in perhaps a hundred places in 4 miles, where a rocky cliff rises abruptly on one side and sinks down abruptly on the other. Now, no teams and motor cars can pass each other there, nor are the turnouts sufficiently numerous, so that my position on the automobile question is I want a reasonable safeguard to life and limb, and if that is provided, why nobody will welcome the automobile more than I.

The SECRETARY: You represent the Sierra Club, and I see Mr. [John] Muir is here. You have personally been over this valley a great many times. You are familiar with these roads. You say that the position of the Sierra Club is that the automobile ought to be admitted when the proper time comes. Do you think the time has come?

Mr. [William] COLBY [Sierra Club]: I think it is very close at hand. I feel, as far as the Glacier Point proposition is concerned, that automobiles should be allowed to go as far as Glacier Point with perhaps that thousand dollar expenditure, and as far as coming down into the valley is concerned, that we should rely upon engineering reports, because naturally when it comes to turnouts and the erection of barriers and so on to prevent the machines from going over, you should exercise every precaution, but with the construction of these turnouts and the construction of these walls in the most dangerous places automobiles could be safely allowed to come to this valley at the present time.

The SECRETARY: Another fundamental question is also involved. What do you think of the joint use of the roads by automobiles and horses as compared with the countersuggestions as to a separate road?

Mr. COLBY: I believe in joint use. The cost of construction of a separate road is too great, and it is an obstacle which we can not overcome. I think the testimony given here by Mr. Marshall and Mr. Curtis and by yourself regarding these difficult mountain roads over which you have ridden, and also the Kings River Road, over which an automobile stage climbs daily, illustrates it. If we take parallel conditions we don't find accidents. We must take parallel conditions. We find the same conditions on Market Street if a driver gets drunk or his machine gets wrecked as we do anywhere in the mountains. It doesn't matter where he is. I think I have about covered what I have to say—maybe a word or two more.

Washington, D.C.: GPO, 1913, 58–61, 129, 130, 139–140.

AN ACT TO ESTABLISH A NATIONAL PARK SERVICE, AND FOR OTHER PURPOSES,

Approved August 25, 1916 (39 Stat. 535)

Be it enacted by the Senate and House of Representatives of the United States of America in Congress assembled, That there is hereby created in the Department of the Interior a service to be called the National Park Service, which shall be under the charge of a director, who shall be appointed by the Secretary and who shall receive a salary of $4,500 per annum. There shall also be appointed by the Secretary the following assistants and other employees at the salaries designated: One assistant director, at $2,500 per annum; one chief clerk, at $2,000 per annum; one draftsman, at $1,800 per annum; one messenger, at $600 per annum; and, in addition thereto, such other employees as the Secretary of the Interior shall deem necessary: *Provided*, That not more than $8,100 annually shall be expended for salaries of experts, assistants, and employees within the District of Columbia not herein specifically enumerated unless previously authorized by law. The service thus established shall promote and regulate the use of the Federal areas known as national parks, monuments, and reservations hereinafter specified by such means and measures as conform to the fundamental purpose of the said parks, monuments, and reservations, which purpose is to conserve the scenery and the natural and historic objects and the wild life therein and to provide for the enjoyment of the same in such manner and by such means as will leave them unimpaired for the enjoyment of future generations. (U.S.C., title 16, sec. 1.)

SEC. 2. That the director shall, under the direction of the Secretary of the Interior, have the supervision, management, and control of the several national parks and national monuments which are now under the jurisdiction of the Department of the Interior, and of the Hot Springs Reservation in the State of Arkansas, and of such other national parks and reservations of like character as may be hereafter created by Congress: *Provided*, That in the supervision, management, and control of national monuments contiguous to national forests the Secretary of Agriculture may cooperate with said National Park Service to such extent as may be requested by the Secretary of the Interior (U.S.C., title 16, sec. 2.)

SEC. 3. That the Secretary of the Interior shall make and publish such rules and regulations as he may deem necessary or proper for the use and management of the parks, monuments, and reservations under the jurisdiction of the National Park Service, and any violations of any of the rules and regulations authorized by this Act shall be punished as provided for in section fifty of the Act entitled "An Act to codify and amend the penal laws of the United States," approved March fourth, nineteen hundred and nine, as amended by section six of the Act of June twenty-fifth, nineteen hundred and ten (Thirty-sixth United States Statutes at Large, page eight hundred and fifty-seven). He may also, upon terms and conditions to be fixed by him, sell or dispose of timber in those cases where in his judgment the cutting of such timber is required in order to control the attacks of insects or diseases or otherwise conserve the scenery or the natural or historic objects in any such park, monument, or reservation. He may also provide in his discretion for the destruction of such animals and of such plant life as may be detrimental to the use of any of said parks, monuments, or reservations. He may also grant privileges, leases, and permits for the use of land for the accommodation of visitors in the various parks, monuments, or other reservations herein provided for, but for periods not exceeding twenty years; and no natural curiosities, wonders, or objects of interest shall be leased, rented, or granted to anyone on such terms as to interfere with free access to them by the public: *Provided, however,* That the Secretary of the Interior may, under such rules and regulations and on such terms as he may prescribe, grant the privilege to graze live stock within any national park, monument, or reservation herein referred to when in his judgment such use is not detrimental to the primary purpose for which such park, monument, or reservation was created, except that this provision shall not apply to the Yellowstone National Park. (U.S.C., title 16, sec. 3.)

SEC. 4. That nothing in this Act contained shall affect or modify the provisions of the Act approved February fifteenth, nineteen hundred and one, entitled "An Act relating to rights of way through certain parks, reservations, and other public lands." (U.S.C., title 16, sec. 4.)

SECRETARY LANE'S LETTER ON
NATIONAL PARK MANAGEMENT

May 13, 1918

Mr. Stephen T. Mather
Director
National Park Service

Dear Mr. Mather:
    The National Park Service has been established as a bureau of
this Department just one year. During this period our efforts have
been chiefly directed toward the building of an effective organization while engaged in the performance of duties relating to the
administration, protection, and improvement of the national parks
and monuments, as required by law. This constructive work is now
completed. The New Service is fully organized; its personnel have
been carefully chosen; it has been conveniently and comfortably
situated in the new Interior Department Building; and it has been
splendidly equipped for the quick and effective transaction of its
business.
    For the information of the public, an outline of the administrative
policy to which the new Service will adhere may be announced. This
policy is based on three broad principles: First that the national parks
must be maintained in absolutely unimpaired form for the use of
future generations as well as those of our own time; second, that
they are set apart for the use, observation, health, and pleasure of
the people; and third, that the national interest must dictate all decisions affecting public or private enterprise in the parks.
    Every activity of the Service is subordinate to the duties imposed
upon it to faithfully preserve the parks for posterity in essentially
their natural state. The commercial use of these reservations, except
as specially authorized by law, or such as may be incidental to the
accommodation and entertainment of visitors, will not be permitted
under any circumstances.
    In all of the national parks except Yellowstone you may permit
the grazing of cattle in isolated regions not frequented by visitors,
and where no injury to the natural features of the parks may result
from such use. The grazing of sheep, however, must not be permitted in any national park.
    In leasing lands for the operation of hotels, camps, transportation

facilities, or other public service under strict Government control, concessioners should be confined to tracts no larger than absolutely necessary for the purpose of their enterprises.

You should not permit the leasing of park lands for summer homes. It is conceivable, and even exceedingly probable, that within a few years under a policy of permitting the establishment of summer homes in national parks, these reservations might become so generally settled as to exclude the public from convenient access to their streams, lakes, and other natural features, and thus destroy the very basis upon which this national playground system is being constructed.

You should not permit the cutting of trees except where timber is needed in the construction of buildings or other improvements within the park and can be removed without injury to the forests or disfigurement of the landscape, where the thinning of forests or cutting of vistas will improve the scenic features of the parks, or where their destruction is necessary to eliminate insect infestations or diseases common to forests and shrubs.

In the construction of roads, trails, buildings, and other improvements, particular attention must be devoted always to the harmonizing of these improvements with the landscape. This is a most important item in our program of development and requires the employment of trained engineers who either possess a knowledge of landscape architecture or have a proper appreciation of the esthetic value of park lands. All improvements will be carried out in accordance with a preconceived plan developed with special reference to the preservation of the landscape, and comprehensive plans for future development of the national parks on an adequate scale will be prepared as funds are available for this purpose.

Wherever the Federal Government has exclusive jurisdiction over national parks, it is clear that more effective measures for the protection of the parks can be taken. The Federal Government has exclusive jurisdiction over the national parks in the States of Arkansas, Oklahoma, Wyoming, Montana, Washington, and Oregon, and also in the Territories of Hawaii and Alaska. We should urge the cession of exclusive jurisdiction over the parks in the other States, and particularly in California and Colorado.

There are many private holdings in the national parks, and many of these seriously hamper the administration of these reservations. All of them should be eliminated as far as it is practicable to accomplish this purpose in the course of time, either through Congressional appropriation or by acceptance of donations of these lands. Isolated tracts in important scenic areas should be given first consideration, of course, in the purchase of private property.

Every opportunity should be afforded the public, wherever possible, to enjoy the national parks in the manner that best satisfies the individual taste. Automobiles and motorcycles will be permitted in all of the national parks; in fact, the parks will be kept accessible by any means practicable.

All outdoor sports which may be maintained consistently with the observation of the safeguards thrown around the national parks by law will be heartily endorsed and aided wherever possible. Mountain climbing, horseback riding, walking, motoring, swimming, boating, and fishing will ever be the favorite sports. Winter sports will be developed in the parks that are accessible throughout the year. Hunting will not be permitted in any national park.

The educational, as well as the recreational, use of the national parks should be encouraged in every practicable way. University and high-school classes in science will find special facilities for their vacation period studies. Museums containing specimens of wild flowers, shrubs, and trees and mounted animals, birds, and fish native to the parks, and other exhibits of the character, will be established as authorized.

Low-priced camps operated by concessioners should be maintained, as well as comfortable and even luxurious hotels wherever the volume of travel warrants the establishment of these classes of accommodations. In each reservation, as funds are available, a system of free camp sites will be cleared, and these grounds will be equipped with adequate water and sanitation facilities.

As concessions in the national parks represent in most instances a large investment, and as the obligation to render service satisfactory to the Department at carefully regulated rates is imposed, these enterprises must be given a large measure of protection, and generally speaking, competitive business should not be authorized where a concession is meeting our requirements, which, of course, will as nearly as possible coincide with the needs of the traveling public.

All concessions should yield revenue to the Federal Government, but the development of the revenues of the parks should not impose a burden upon the visitor.

Automobile fees in the park should be reduced as the volume of motor travel increases.

For assistance in the solution of administrative problems in the parks relating both to their protection and use, the scientific bureaus of the Government offer facilities of the highest worth and authority. In the protection of the public health, for instance, the destruction of insect pests in the forests, the care of wild animals, and the propagation and distribution of fish, you should utilize their hearty cooperation to the utmost.

You should utilize to the fullest extent the opportunity afforded by the Railroad Administration in appointing a committee of western railroads to inform the traveling public how to comfortably reach the national parks; you should diligently extend and use the splendid cooperation developed during the last three years among chambers of commerce, tourist bureaus, and automobile highway associations, for the purpose of spreading information about our national parks and facilitating their use and enjoyment; you should keep informed of park movements and park progress, municipal, county, and State, both at home and abroad, for the purpose of adapting, whenever practicable, the world's best thought to the needs of the national parks. You should encourage all movements looking to outdoor living. In particular you should maintain [a] close working relationship with the Dominion Parks Branch of the Canadian Department of the Interior, and assist in the solution of park problems of an international character.

The Department is often requested for reports on pending legislation proposing the establishment of new national parks or the addition of lands to existing parks. Complete data on such park projects should be obtained by the National Park Service and submitted to the Department in tentative form of report to Congress.

In studying new park projects, you should seek to find scenery of supreme and distinctive quality or some national feature so extraordinary or unique as to be of national interest and importance. You should seek distinguished examples of typical forms of world architecture; such, for instance, as the Grand Canyon, as exemplifying the highest accomplishment of stream erosion, and the high, rugged portion of Mount Desert Island as exemplifying the oldest rock forms in America and the luxuriance of deciduous forests.

The national park system as now constituted should not be lowered in standard, dignity, and prestige by the inclusion of areas which express in less than the highest terms the particular class or kind of exhibit which they represent.

It is not necessary that a national park should have a large area. The element of size is of no importance as long as the park is susceptible of effective administration and control.

You should study existing national parks with the idea of improving them by the addition of adjacent areas which will complete their scenic purposes or facilitate administration. The addition of the Teton Mountains to the Yellowstone National Park, for instance, will supply Yellowstone's greatest need, which is an uplift of glacier-bearing peaks; and the addition to the Sequoia National Park of the Sierra summits and slopes to the north and east, as contemplated by pending legislation, will create a reservation unique in the world, because of its gigantic trees, extraordinary canyons, and mountain masses.

In considering projects involving the establishment of new national parks or the extension of existing park areas by delimination of national forests, you should observe what effect such delimination would have on the administration of adjacent forest lands, and wherever practicable you should engage in an investigation of such park projects jointly with officers of the Forest Service, in order that questions of national park and national forest policy as they effect the lands involved may be thoroughly understood.

FRANKLIN K. LANE, *Secretary of the Interior.*

National Park Service Handbook of Administrative Policies for Natural Areas, 1968, 68–71.

2

Defining the System

1919–1932

During the period from 1919 to 1932, Director Stephen Mather, his assistant and successor, Horace Albright, and a cadre of handpicked superintendents consolidated the national park system and defined its operating policies. These "Mather men" would concentrate primarily on defining preservation priorities and molding the system according to definite ideas about its purpose. The national parks were for inspiration and education of the people, not, as many supposed, for recreation per se.

Several considerations had to be accommodated by Mather and his employees in devising these policies. First, the National Park Service was not a universally praised and supported idea nor were its parks. Many believed that the U.S. Forest Service could easily, cheaply, and more ably administer these areas of scenic beauty. For a period, the parks' rationale for existence wavered in the minds of lawmakers. Second, Mather, Albright, and the others reflected the attitudes of the time regarding public access to the parks. It was implicit that any and all efforts be made to assure maximum use of the units by the public. It was both democratic and prudent to so promote use. Third, they believed that the existing parks were all so remote and of such limited access that only large-scale concession monopolies stood a chance of financially surviving and developing them. Finally, science was too underdeveloped to provide hard evidence against many popular wildlife management practices. It would be another decade before scientists would be heeded and then only for an interim.

As the National Park Service struggled to establish and define itself the first challenge appeared. In 1920 the Federal Power Act authorized construction of dams on federal lands. After the debacle at Hetch Hetchy in Yosemite National Park, this was an ominous threat. After a flurry of action by conservationists and others, however, Congress added an amendment in 1921 forbidding dams in national parks and monuments without its specific approval.

For the next several years, as the National Park Service moved toward the end of its first decade, it further detailed its policies and philosophy. In answer to criticism about potential overdevelopment, the superintendents of the various parks passed a resolution encouraging some development but rejecting "overdevelopment." Although undefined, there is in the latter term an understanding that some limits must be contemplated for park tourism. Three years later Secretary of the Interior Hubert Work continued the tradition of secretary's letters on national park management by restating and elaborating the policies of Secretary Franklin Lane. In some cases such as grazing and summer home leases, the Work letter took a stronger tone to reject abuses of earlier orders.

Three years later, in 1928, Congress specifically defined the matter of concession operations and privileges, addressed legally only obliquely in the Organic Act. That same year the National Park Service moved to develop a rigid plan and staff to prevent or control forest fires in the parks. An address by U.S. Forest Service fire control expert Jay Price outlined the concepts of the time. Later that year the NPS appointed its first Fire Control Expert under the Branch of Forestry and devoted funds specifically to fire prevention programs.

In 1931, after a decade and a half, Director Horace Albright could reflect on many issues affecting the growing system of parks. Two issues in particular needed addressing. First, park planning, despite the adoption of a program for five-year plans in 1926, was haphazard and varied from park to park. In issuing an office order, Albright sought to streamline and make uniform the planning procedure. This represents a further attempt to cope with the parks as a system instead of as an amalgamation of independent units.

The second problem was the quality of care and preservation being devoted to the natural resources of the parks. Early anthropocentric ideas had led to predator destruction, vista clearing of forests, introduction of exotics, and use of herbicides and pesticides in the parks. Often released through articles in popular or scholarly journals, Albright's policy pronouncements sought to correct or at least rationalize these early activities. Thus, in a policy memorandum the director specified forestry policy including such matters as fire control, insect and fungal damage, hazard tree removal, and prevention of illicit cutting and pruning. In an article for the *Journal of Mammalogy*, Albright rejected predator destruction and promoted the parks as the homes of all species.

Coming on the heels of the "predator policy" letter was one of the most advanced and profound statements about the parks for this era. Wildlife scientist George Wright and a staff he hired and personally funded released a seminal work in 1932 entitled *Fauna of the National Parks*. Wright and his colleagues laid down the ground rules for scientific wildlife management and for the provision of further research to shape decision-making. It was a piece ahead of its time and for a while led to a Wildlife Division with Wright as its head. However, after his untimely death in 1936, many of his ideas were ignored until the Leopold report of 1963 resurrected them.

Although the majority of profound actions and decisions during this period reflected concern about management of the parks, the system continued to grow, and Mather and his aides gave much thought to how to shape that growth. Studies of potential new areas commenced early in the 1920s, and proposals were fielded from all over the nation. Investigation particularly centered on the eastern United States where most of the people and almost none of the parks resided. Still no significant statements or studies appeared. The only exception to this organizational inactivity came in an exchange between the War Department and Congress. In 1926, acting on congressional orders, the War Department issued a study of potential areas in the United States for inclusion in a system of national battlefields and memorials. It directed public and congressional attention to a number of these historic places and influenced the shaping of this part of the national park system.

AMENDMENT TO THE FEDERAL POWER ACT,

Approved March 3, 1921 (41 Stat. 1353)

Be it enacted by the Senate and House of Representatives of the United States of America in Congress assembled, That hereafter no permit, license, lease, or authorization for dams, conduits, reservoirs, power houses, transmission lines, or other works for storage or carriage of water, or for the development, transmission, or utilization of power, within the limits as now constituted of any national park or national monument shall be granted or made without specific authority of Congress, and so much of the Act of Congress approved June 10, 1920, entitled "An Act to create a Federal Power Commission; to provide for the improvement of navigation; the development of water power; the use of the public lands in relation thereto; and to repeal section 18 of the River and Harbor Appropriation Act, approved August 8, 1917, and for other purposes," approved June 10, 1920, as authorizes licensing such uses of existing national parks and national monuments by the Federal Power Commission is hereby repealed. (U.S.C., title 16, sec. 797.)

SUPERINTENDENTS' RESOLUTION
ON OVERDEVELOPMENT

Prepared at the National Park Service Conference
Nov. 13–17, 1922; Yosemite Park, Calif.
With Explanatory Letter.

BE IT RESOLVED by the officers and superintendents of the National Park Service, that the best interests of the nation will be served by a more adequate development of the national parks. Roads and trails should be improved and extended, ample accommodations should be provided for visitors, and other improvements carried out, so that the parks may better fulfill their mission of healthful recreation and education to a larger number of people.

It is, however, stated as a policy of the National Park Service, that over-development of any national park, or any portion of a national park, is undesirable and should be avoided. Certain areas should be reserved in each park, with a minimum amount of development, in order that animals, forest, flowers and all native life shall be preserved under natural conditions.

One of the objects of the National Park Service is to preserve some of the finest of our scenery for future generations, that they may always know the quiet dignity of our forests and the rugged grandeur of our mountains. In the development of the parks, some of these areas should be made possible of access, but they should be protected from anything that will impair them.

Plans for the development of each national park should be outlined as far ahead as possible, in order that the park may receive adequate development, without over-development.

In developing the national parks of the United States, care will be exercised to prevent the over-development of any park.

This policy was discussed at the recent conference of National Park officials, held at Yosemite, California, from November 12 to 17, 1922. A resolution was passed stating that the National Park Service favored the continued improvement and development of the national parks, but that it did not favor, and would guard against, over-development of any national park, or any portion of a national park.

The present difficulty in most of the national parks is to secure satisfactory development, maintenance and protection with the limited funds that are available, but nevertheless, future plans will be so laid out that no park, or portion of a park, will become over-developed.

It is the intention to make the chief scenic features of each park accessible to the average visitor, but to set aside certain regions of each park, which will not be traversed by automobile roads, and will have only such trails or other development as will be necessary for protection of the area.

The national parks offer to the American public not only great opportunities for healthful recreation but constantly increasing opportunities for acquiring information on many phases of natural history and science.

In our schools, the youth of the land receives, with more or less labor and reluctance, such education and instruction as can be derived from books. A vital part of the education of every individual is to acquire at least a partial understanding and appreciation of nature and scenery. This is best obtained at the fountain source, in the out-of-doors, where Nature's works are unimpaired and unrestricted by the hand of man. An education of this sort is usually achieved with the keenest interest and the most genuine pleasure. If all of our education could be obtained in such an enjoyable manner, there would be less ignorance and much more knowledge.

A child starts his school life at the age of six years, let us say, and continues his studies until the necessity, the incentive or the opportunity is withdrawn. The book study ceases. The study of nature should supplement but not replace the study of books, and it has this advantage, that it may be begun at any time, carried on without interfering with other work, and need never be stopped, for there are always new fields of knowledge stretching ahead, and all that is needed is a keen interest and appreciation.

The study of nature develops power of observation, quickens the senses, increases the usefulness of an individual in any line of work or occupation, and makes his life broader, deeper, happier.

Every boy and girl knows a few flowers, a few trees, a few birds and something about rocks and physical geography. Who is there who would not gladly know more of these interesting subjects? The story is told of a prairie boy who, when asked to name three trees, replied, "Cottonwood, willow, —sagebrush." We laugh at the originality of his answer but the heart is touched by the poverty of experience and the lack of opportunity that it suggests.

To go back to our subject, the mission of the national parks is to provide, not cheap amusement, but healthful recreation and to supplement the work of schools by opening the doors of Nature's laboratory, to awaken an interest in natural science as an adjunct to the commercial and industrial work of the world.

Another object for which the parks were created, is to set aside for future generations, certain areas that are typical of our finest scenery. They are to be held free from commercial exploitation. The standing forests will prove more valuable than the lumber they would produce, the graceful waterfall will prove more precious than the power it would yield, the unscarred beauty of the mountain is worth more than the mineral wealth that may be buried in its heart. In order that our nation may grow and prosper, forests must be cut, streams must be turned onto dry lands, cararacts must give up their power, meadows must be shorn to feed the flocks. These things are necessary. Scenery must often be destroyed by commerce, beauty must often be sacrificed to industry. But in order that we shall not squander all of our birthright, a few jewels of scenery are set aside for ourselves and for posterity to enjoy.

To obtain the greatest good from our national parks, some portions of every park must be made readily accessible by first class roads. Provision must be made for the automobile camper, and for the accommodation and transportation of other visitors. But not all of Nature's treasures are to be seen from the seat of an automobile; one does not receive at twenty miles an hour, the inspiration that results from a pilgrimage on foot; and an automobile horn is less effective than the silence of solitude, to awaken thoughts that are deep and abiding. Someone had said, "Great views make great thoughts, great thoughts make great men". The national parks should be a real factor in the building of a better, stronger race.

When a camping area becomes fully developed, other areas should be opened, and thus prevent the over-crowding of any locality. Before the travel on a road becomes excessive, other roads should be built that will divert a part of the traffic and prevent congestion.

It is not desirable to fully develop all portions of the park. Some portions should be made easily accessible to motorists, by means of good roads; other regions should attract parties on horseback, by means of good trails; still other areas should be accessible only to those who journey on foot.

The development of a park will, therefore, not be uniform throughout. Some portions will be fully developed, others partly developed, and still others will be left in their natural, wild condition. Such variety will best serve the varied needs of the different classes of visitors. Geographic conditions will play an important part, and the plans for development must be adapted to topography. For example, in a mountainous region, the principal roads may be best located in the valleys, leading to the principal scenic points. Trails may be built up the tributary streams. The greater portion of the area, including the high regions, the ridges and the peaks will be left untouched, but still accessible to those with sufficient vigor and enthusiasm.

If there were no development, no roads or trails, no hotels or camps,

a national park would be merely a wilderness, not serving the purpose for what it was set aside, not benefitting the general public. No
one is selfish enough to wish to withhold development, but many are
keenly interested in seeing development properly directed. The parks
should be popular, but never commonplace. They should accommodate crowds if necessary but without over-crowding. Animals should
be protected in their natural surroundings rather than caged in a zoo.
Outdoor recreation should supplant cheap amusements. Museums
and nature study should be offered to stimulate along educational
and beneficial lines rather than to accentuate sight-seeing of an
unintelligent order.

There will always be some parts of every national park that will
be kept in their original state of nature. These areas will offer quiet
retreats for wild life. The forests will be in less danger of destruction by the careless camper. The hiker, the mountaineer, the artist,
the student, he who wishes to leave the throng and penetrate the
unfrequented places would find delight in their sanctuaries. They
will not be visited by those, who in restless haste, see much and
appreciate but little.

Every national park now has such undeveloped areas. Yellowstone
has vast regions in the southwest and in the southeast corners, that
are rarely visited. The northern part of Glacier Park is known to but
the few who have explored on foot or by pack train. Mount Rainier
has snow fields and glaciers that are rarely trodden by the foot of
man. Crater Lake is viewed by many but explored by few. Lassen
and Mt. McKinley parks are difficult of access and almost wholly
undeveloped. Yosemite has thousands of visitors to its beautiful valley
floor, but mile after mile of quiet forest and leaping stream that are
rarely visited. The Giant Forest of Sequoia Park has been an inspiration to many though few have really explored the high Sierra country
that lies to the eastward. The Grand Canyon and Zion Park each
contain valleys and gorges into which man has never penetrated.
Mesa Verde has a hundred ruins built by a prehistoric race, whose
secrets have not been brought to light by the archaeologist's shovel.
Rocky Mountain has many a peak and pinnacle, canyon and cliff,
lovely meadow and charming lakelet that is unvisited save by those
who love the rough trail and the trackless wilds. Hawaii National
Park has, in Kilauea, a lake of fire, that has drawn visitors from all
over the world, but how many have looked into the summit crater of
Mauna Loa?

It is a conservative statement to say that ninety percent of the
visitors to any park, never get far from the automobile roads. Probably not one park can claim to have ten percent of its area fully

developed, and readily accessible. Nine tenths of the park travel is, therefore, in one tenth of the park area.

There is no sharp line between necessary, proper development and harmful over-development. The best judgment and active work of all concerned must be focused together in order to secure the best results. At present the educational and economic value of the national parks to the nation, is restricted by insufficient development. Far-sighted men, however, are making plans for years ahead, and it is to guide future protection that the National Park Service announces its stand, "For adequate development, but against over-development."

Roger W. Toll
Superintendent.
Rocky Mountain Nat'l Park.

December 1, 1922.

National Park Service Archives, Harpers Ferry, Box K5410.

STATEMENT OF NATIONAL PARK POLICY
—SECRETARY WORK, MARCH 11, 1925

Memorandum for the Director,
National Park Service.

Owing to changed conditions since the establishment in 1917 [*sic*] of the National Park Service as an independent bureau of the Department of the Interior, I find it advisable to restate the policy governing the administration of the national park system to which the Service will adhere.

This policy is based on three broad, accepted principles:

First, that the national parks and national monuments must be maintained untouched by the inroad of modern civilization in order that unspoiled bits of native America may be preserved by future generations as well as our own;

Second, that they are set apart for the use, education, health and pleasure of all the people;

Third, that the national interest must take precedence in all decisions affecting public or private enterprise in the parks and monuments.

The duty imposed upon the National Park Service in the organic act creating it to faithfully preserve the parks and monuments for posterity in essentially their natural state is paramount to every other activity.

The commercial use of these reservations, except as specially authorized by law, or such as may be incidental to the accommodation and entertainment of visitors, is not to be permitted.

In national parks where the grazing of cattle has been permitted in isolated regions not frequented by visitors, such grazing is to be gradually eliminated.

Lands leased for the operation of hotels, camps, transportation facilities, or other public service under strict Government control, should be confined to tracts no larger than absolutely necessary for the purposes of their enterprises.

The leasing of park and monument lands for summer homes will not be permitted. Under a policy of permitting the establishment of summer homes, these reservations might become so generally settled as to exclude from convenient access to their streams, lakes, or other natural features, and thus destroy the very basis upon which this national playground system is being constructed.

The cutting of trees is not to be permitted except where timber is needed in the construction of buildings or other improvements within a park or monument and only when the trees can be removed without injury to the forests or disfigurement of the landscape; where the thinning of forests or cutting of vistas will reveal the scenic features of a park or monument; or where their destruction is necessary to eliminate insect infestations or diseases common to forests and shrubs.

In the construction of roads, trails, buildings and other improvements, these should be harmonized with the landscape. This important item in our program of development requires the employment of trained engineers who either possess a knowledge of landscape architecture or have a proper appreciation of the esthetic value of parks and monuments. All improvements should be carried out in accordance with a preconceived plan developed with special reference to the preservation of the landscape. The overdevelopment of parks and monuments by the construction of roads should be zealously guarded against.

Exclusive jurisdiction over national parks and monuments is desirable as more effective measures for their protection can be taken. The Federal Government has exclusive jurisdiction over the national parks in the States of Arkansas, Oklahoma, Wyoming, Montana, Washington, and Oregon, and of three of the parks in California; also in the Territories of Hawaii and Alaska. The cession of exclusive jurisdiction over the parks in the other States, and particularly in Arizona and Colorado, is urged, as over all the national monuments.

There still remain many private holdings in the national parks, although through the generosity of public-spirited citizens many of these which seriously hampered their administration have been donated to the Federal Government. All of them should be eliminated as far as it is practicable to accomplish this purpose in the course of time, either through Congressional appropriation or by acceptance of donations of these lands. Isolated tracts in important scenic areas should be given first consideration, of course, in the purchase of private property.

The public should be afforded every opportunity to enjoy the national parks and monuments in the manner that best satisfies the individual taste. Automobiles and motorcycles operated for pleasure but not for profit, except automobiles used by transportation companies operating under Government franchise, are permitted in the national parks. The parks and monuments should be kept accessible by any means practicable.

All outdoor sports within the safeguards thrown around the national parks by law, should be heartily endorsed and aided wherever possible. Mountain climbing, horseback riding, walking, motoring, swimming, boating, and fishing will ever be the favorite sports. Winter sports are being rapidly developed in the parks and this form of recreation promises to become an important recreational use. Hunting is not permitted in any national park or monument except in Mount

McKinley National Park, Alaska, in accordance with the provisions of the organic act creating it.

The educational use of the national parks should be encouraged in every practicable way. University and high school classes in science will find special facilities for their vacation period studies. Museums containing specimens of wild flowers, shrubs, and trees, and mounted animals, birds, and fish native to the parks and monuments, and other exhibits of this character, should be established as funds are provided.

Low-priced camps operated under Government franchise are maintained, as well as comfortable and even luxurious hotels. Free camp grounds equipped with adequate water and sanitation facilities are provided in each reservation. These camp grounds should be extended as travel warrants and funds are available.

As franchises for the operation of public utilities in the national parks represent in most instances a large investment, and as the obligation to render service satisfactory to the Department at carefully regulated rates is imposed, these enterprises must be given a large measure of protection, and generally speaking competitive business is not authorized where an operator is meeting service requirements, which coincide as nearly as possible with the needs of the traveling public.

All franchises yield revenues to the Federal Government which, together with automobile license fees collected in the parks where a license fee is charged, are deposited to the credit of miscellaneous receipts in the Treasury of the United States. Due allowance is made by Congress for revenues collected in appropriating funds for the upkeep and improvement of the parks and monuments.

In the solution of administrative problems in the parks and monuments relating both to their protection and use, the scientific bureaus of the Government are called upon for assistance. For instance, in the protection of public health, the Public Health Service of the Treasury Department cooperates; in the destruction of insect pests in the forests, the Bureau of Entomology of the Department of Agriculture is called upon; and in propagation and distribution of fish, the Bureau of Fisheries of the Department of Commerce gives its hearty cooperation.

In informing the traveling public how to reach the parks and monuments comfortably, the splendid cooperation given by the railroads, automobile highway associations, chambers of commerce and tourist bureaus is acknowledged and should be furthered for the purpose of spreading information about the national parks and monuments and facilitating their use and enjoyment. Every effort should be made to keep informed of park movements and park progress, municipal, county, and State, both at home and abroad, for the purpose of adapting, whenever practicable, the world's best thought to the needs of the national park system. All movements looking to outdoor living should be encouraged. A close working relationship with the Dominion Parks Branch of the Canadian Department of the Interior should be maintained to assist in the solution of park problems of an international character.

Our existing national park system is unequaled for grandeur. Additional areas when chosen should in every respect measure up to the dignity, prestige, and standard of those already established. Proposed park projects should contain scenery of distinctive quality or some natural features so extraordinary or unique as to be of national interest and importance, such as typical forms of natural architecture as those only found in America. Areas considered for national parks should be extensive and susceptible of development so as to permit millions of visitors annually to enjoy the benefits of outdoor life and contact with nature without confusion from overcrowding.

In considering projects involving the establishment of national parks or the extension of existing park areas by transfer of lands from national forests the effect such change of status would have on the administration of adjacent forest lands should be carefully considered. It might be well to point out the basic difference between national parks and national forests. National forests are created to administer lumbering and grazing interests for the people, the trees being cut in accordance with the principles of scientific forestry, conserving the smaller trees until they grow to a certain size, thus perpetuating the forests. Grazing is permitted in national forests under governmental regulations, while in the national parks grazing is only permitted where not detrimental to the enjoyment and preservation of the scenery and may be entirely prohibited. Hunting is permitted in season in the national forests but never in the national parks, which are permanent game sanctuaries. In short, national parks unlike national forests, are not properties in a commercial sense, but natural preserves for the rest, recreation and education of the people. They remain under Nature's own chosen conditions. Therefore, in an investigation of such park projects the cooperation of officers of the Forest Service should be sought in accordance with the recommendations of the President's Committee on Outdoor Recreation in order that questions of national park and national forest policy as they affect the lands involved may be thoroughly understood.

HUBERT WORK, *Secretary of the Interior.*

National Park Service Handbook of Administrative Policies for Natural Areas, 1968, 72–75.

STUDY AND INVESTIGATION OF BATTLE FIELDS IN THE UNITED STATES FOR COMMEMORATIVE PURPOSES

May 4, 1926.—Committed to the Committee of the Whole House on the State of the Union and Ordered to be Printed

Mr. JOHNSON of Indiana, from the Committee on Military Affairs, submitted the following REPORT [To accompany H. R. 11613]

The Committee on Military Affairs, to which was referred the bill (H. R. 11613) to provide for the study and investigation of battle fields in the United States for commemorative purposes, having considered the same, report thereon with the recommendation that it do pass.

During the past several years quite a number of bills have been introduced providing for the inspection of battle fields and the establishment of national military parks. In the present session 28 bills have been introduced of which 14 provide for establishment of national military parks with appropriations authorized approximating nearly $6,000,000. The other bills provide for markers on battle fields, the inspection of sites with a view to eventual establishment of parks, etc.

Because of the number of measures introduced and the evident interest of Congress in the establishment of these military parks, your committee believes that the study and investigation called for in H. R. 11613 will be of the greatest interest and importance in determining what action should be taken by Congress. In this connection a study made in the Army War College in 1925, which was furnished to the committee at the request of the chairman, is made a part of this report for the information of the House.

THE ARMY WAR COLLEGE,
Washington Barracks, D.C., May 28, 1925.

Subject: Study of records pertaining to the battles of the United States with reference to the establishment of national military parks and national monuments.

To: The Commandant, the Army War College.

In accordance with instructions quoted in the following paragraphs, this section has made a study of records and data pertaining to certain wars in which the military forces of the United States were engaged:

"The object of this study will be the compilation of two lists of such battles, arranged in order of priority, under the following heads:

"(1) Those battles of such great importance and far-reaching effect as to warrant commemoration by the establishment of national military parks.

"(2) Those battles sufficiently important to warrant commemoration by the acquisition of some land and the placing of a limited number of markers or monuments and the designation of the sites so obtained as national monuments.

"It is the view of the War Department that national military parks should as a general thing cover a comparatively large area of ground, probably some thousands of acres, and so marked and improved as to make them into real parks available for detailed study by military authorities, the battle lines and operations being clearly indicated on the ground. The expense of maintaining such a park is so great as to indicate that the number should be kept fairly low.

"Less important and extensive engagements which have nevertheless a definite military and political effect should be listed under the second category, the idea being that limited areas of ground on the site of the battle could be purchased and appropriately marked and the whole aggregation of separate areas designated as a national monument."

From the instructions it was assumed that the lists were to comprise only those battles fought within the limits of the United States and from the beginning of the Revolutionary War to the present time. The lists would therefore be confined to battles of the Revolutionary War, the War of 1812, the Mexican War, Indian wars, and the Civil War.

I. THE ACTION OF CONGRESS

A preliminary study was made of acts of Congress providing for the establishment of national military parks and for otherwise commemorating the battles of past wars. It was found that these acts provided for the commemoration of battles in one of three ways:

1.   By the establishment of national military parks.

2.   By indicating the lines of battle by markers or monuments, or both, without establishing parks.

3.   By single monuments without otherwise marking the field.

1.   *National military parks.*—The first national military park was established by Congress in the act of April [*sic*] 19, 1890.

"That for the purpose of preserving and suitably marking for historical and professional military study the fields of some of the most remarkable maneuvers and most brilliant fighting in the War of the Rebellion * * * the following described highways in these States are hereby declared approaches to and parts of the Chickamauga and Chattanooga Military Park as established in the second section of this act."

The national military park established under this act consists of a main park of about 5,600 acres covering the battle field of Chickamauga, and detached areas in Wauhatchie Valley, on Lookout Mountain, on Missionary Ridge and elsewhere secured for the purpose of establishing the lines of battle on the battle field of Chattanooga.

The second national park established was that of Gettysburg. The marking of the battle lines on this field was begun under the sundry civil act of March 3, 1893, but the park itself was established under the act of February 11, 1895, which authorized the Secretary of War to accept for this purpose from the Gettysburg Memorial Association about 800 acres of ground. Since the establishment of the park it has been greatly enlarged and now covers an area of about 2,530 acres.

The third national military park established by Congress was that of Shiloh; it was established under the act of December 27, 1894.

"That in order that the armies of the Northwest which served in the Civil War, like their comrades of the eastern armies at Gettysburg and those of the Central West at Chickamauga, may have the history of one of their memorable battles preserved on the ground where they fought, the battle field of Shiloh, in the State of Tennessee is hereby declared a national military park * * *."

It will not be seen from the above that these three military parks were designed by Congress not only to preserve for historical and professional study the battle fields themselves, but also to serve as lasting memorials to the great armies of the war. The field of Gettysburg was to be a memorial to the Union Army of the Potomac as well as the Confederate Army of northern Virginia; the field of Chickamauga, a memorial to the Union Army of the Cumberland

and the Confederate Army of Tennessee, and the field of Shiloh a memorial to the Union Army of the Tennessee and the Confederate armies which, under various designations, opposed it in western Tennessee and in Mississippi.

As a fitting memorial to the Union Army of the Tennessee the field of Shiloh was not so appropriate as that of Vicksburg; the campaign of Vicksburg was the most brilliant operation of that army. Some years later, therefore, the Vicksburg National Military Park was established by the act of February 21, 1899. This park has an area of about 1,300 acres, covering the siege zone about the intrenched camp of 1863.

That these national military parks might be fitting memorials to the great armies engaged and be accurately marked for historical and professional military study, a commission consisting of survivors of the opposing armies was appointed for each park to fix the location of every monument and marker on the field. The legislatures of many of the States cooperated in the work by making appropriations for monuments to mark the positions of their regiments and batteries on the field and also to serve as memorials to these regiments and batteries.

These four are the only national parks that have been established to cover battle fields of the Civil War.

It is to be noted that in the four battles marked by these four existing national military parks were represented all corps of the Union Army but four and practically all the organizations of the Confederate Army. The four Union corps not represented in these battles were the Tenth and Eighteenth (consolidated in the latter part of the war to form the Twenty-fourth) which were on the Carolina coast, the Nineteenth, which was near the mouth of the Mississippi River, and the Twenty-third, organized in Kentucky in 1863 and later a part of the Army of the Ohio.

One small national military park has been established to commemorate one of the battles of the Revolutionary War. The history of its establishment is as follows:

By the act of February 13, 1911, the sum of $30,000 was appropriated "for the erection of a monument on the battle field of Guilford Court House * * * to commemorate the great victory won there on March 15, 1781, by the American forces commanded by Maj. Gen. Nathaniel Greene and in memory of Maj. Gen. Nathaniel Greene and the officers and soldiers who participated in the battle."

This was followed by the act of March 2, 1917:

"That in order to preserve for historical and professional military study one of the most memorable battles of the Revolutionary War, the battle field of Guilford Court House in the State of North Carolina is hereby declared to be a national military park whenever the title shall have been acquired by the United States.

"The Secretary of War is authorized to receive from the Guilford Battle Ground Co. a deed of conveyance to the United States of all lands belonging to said corporation embracing about 125 acres."

2. *Marking lines of battle without establishing national military parks.*—In the sundry civil act of August 19, 1890, there appeared this item:

"For the purpose of surveying, locating, and preserving the lines of battle of the Army of the Potomac and the Army of Northern Virginia at Antietam and for marking the same and for locating and marking the position of 43 different commands of the Regular Army engaged in the battle of Antietam, and for the purchase of sites for tablets for the marking of such positions * * * $15,000."

Under this and subsequent legislation of like character the lines of the battle fields of Antietam have been very satisfactorily marked without the establishment of a national military park. The lines consist of about 5 miles of improved avenues along which are (as a rule) placed the monuments and markers of the different organizations that took part in the battle. A significant indication of State policy may be seen in the fact that the State of Pennsylvania refused to erect a monument on the Antietam field for any unit already represented by a monument on the battle field of Gettysburg.

The field of Chattanooga is marked in a manner similar to that of Antietam, as the field lies outside of the main park of Chickamauga.

In this method of marking battle fields there is less latitude for locating monuments and markers than if greater areas are acquired, but it gives very satisfactory results for historical and professional military study at a much smaller expenditure of money for the purchase of land and a much smaller expenditure for maintenance.

3. *Single monuments.*—Single monuments have generally been erected to commemorate battles of the Revolutionary War, the War of 1812, and Indian wars. Some of these have been erected by appropriations made by Congress to supplement funds by States or raised by patriotic associations; others have been erected by the National Government alone. Among the former class may be cited:

Revolutionary War

Saratoga monument:
United States - $70,000
New York State -   25,000
Monument association - - - - - - - - - - - - - - - - -   10,000

Princeton battle field monument:

United States -	30,000
Monument association - - - - - - - - - - - - - - - - -	30,000

Monmouth Court House battle field monument:

United States -	20,000
New Jersey -	10,000
Monmouth County - - - - - - - - - - - - - - - - - - -	10,000

Bennington battle field monument:

United States -	40,000
Vermont -	15,000
New Hampshire -	5,000
Massachusetts -	10,000
Monument association - - - - - - - - - - - - - - - - -	10,000

Moores Creek battle field monument, to repair: United
States - 5,000

War of 1812

New Orleans battle field monument, to complete: United
States - $25,000

Indian wars

Tippecanoe battle-field monument:

United States -	$12,500
Indiana -	12,500

Among those erected by the United States alone may be cited:

Revolutionary War

Yorktown monument - - - - - - - - - - - - - - - - - - -	$100,000
Guilford Court House battle monument - - - - - - - - -	30,000
Kings Mountain battle monument - - - - - - - - - - - -	30,000

Indian wars

Point Pleasant battle monument - - - - - - - - - - - - -	$10,000
Fort Recovery monument - - - - - - - - - - - - - - - -	25,000
Horseshoe battle-field monument - - - - - - - - - - - -	5,000
Fort Phil Kearny monument - - - - - - - - - - - - - - -	500

The policy as thus outline[d] by acts of Congress gives an appropriate scheme for the commemoration of the battle in which the military forces of the United States have been engaged.

Class I. Battles worthy of commemoration by the establishment of national military parks. These should be battles of exceptional political and military importance and interest, whose effects were far-reaching, whose fields are worthy of preservation for detailed military and historical study, and which are suitable to serve as memorials to the armies engaged.

Class II. Battles of sufficient importance to warrant the designation of their sites as national monuments. The action of Congress and the great difference in the importance of these battles give reason for the subdivision under this class into:

Class IIa. Battles of such great military and historic interest as to warrant locating and indicating the battle lines of the forces engaged by a series of markers or tablets, but not necessarily by memorial monuments.

Class IIb. Battles of sufficient historic interest to be worthy of some form of monument, tablet, or marker to indicate the location of the battle field.

If it is deemed necessary for the protection of the markers and monuments already erected on any field of Class II battles, the ground acquired by the Government on this field might be declared a national monument. This has not been done heretofore, as the battle field of Antietam has a caretaker, and on fields where single monuments have been erected it has been the policy of the Government, as soon as they have been completed, to transfer them to some local association for care and maintenance.

II. LIST OF BATTLES

In making out the lists of battles as directed by the instructions it was decided to treat each war separately.

1. REVOLUTIONARY WAR

(a) Battles of Class I

(1)   Battle of Saratoga—September 19 and October 7, 1777.
(2)   Siege of Yorktown—September 28 to October 19, 1781.

The two decisive events of the Revolutionary War were the surrender of the British Army under General Burgoyne as a result of his defeat in the battles variously called Saratoga, Stillwater, Freeman's Farm or Bemis Heights, and the surrender of the British Army under General Cornwallis as a result of the siege of Yorktown. Of the importance of these engagements there can be no doubt. Saratoga brought about the definite intervention of France in favor of the American Colonies and is listed by Creasy as one of the 15 decisive battles of the world; Yorktown ended the war. The surrender of Burgoyne is commemorated by a monument on the site where the surrender took place a few miles from the battle field. The surrender of Cornwallis is commemorated by a monument erected within his lines at Yorktown. In accordance with the scheme proposed both events would be worthy of national military parks.

(b) Battles and engagements of Class IIb

From a list of about 400 battles, engagements, etc., given in Heitman's Historical Register of Officers of the Continental Army the following have been selected as of more than ordinary military and historic interest. There are few military events of the Revolutionary War that have not been commemorated by monuments or tablets through the efforts of patriotic societies.

1. Engagement at Lexington, Mass., April 19, 1775. Monument.
2. Engagement at Concord, Mass., April 19, 1775. Monument.
3. Capture of Ticonderoga, N.Y., May 10, 1775. Tablet.
4. Battle of Bunker Hill, Mass., June 17, 1775. Monument.
5. Siege of Boston, Mass., June 1775-March, 1776.
6. Battle of Moores Creek Bridge, N.C., February 27, 1776. Monument.
7. Defense of Sullivans Island, S.C., June 28, 1776. Monument.
8. Battle of Long Island, N.Y., August 27, 1776.
9. Harlem Heights, September 16, 1776.
10. Engagement at White Plains (or Chatterton's Hill), N.Y., October 28, 1776.
11. Defense of Fort Washington, N.Y., November 16, 1776.
12. Battle of Trenton, N.J., December 26, 1776. Monument.
13. Battle of Princeton, N.J., January 3, 1777. Monument.
14. Battle of Oriskany, N.Y., August 6, 1777. Monument.
15. Battle of Bennington, N.Y., August 16, 1777. Monument.
16. Battle of Brandywine, Pa., September 11, 1777.
17. Battle of Germantown, Pa., October 4, 1777.
18. Defense of Forts Clinton and Montgomery, N.Y., October 6, 1777.
19. Battle of Monmouth, N.J., June 28, 1778. Monument.
20. Engagement at Quaker Hill, R.I., August 29, 1778.
21. Capture of Vincennes, Ind., February 23, 1779. Tablet.
22. Capture of Stony Point, N.Y., July 16, 1779. State park.
23. Siege of Savannah, Ga., September 23 to October 9, 1779.
24. Defense of Charleston, S.C., March 29 to May 12, 1780.
25. Battle of Camden, S.C., August 16, 1780.
26. Battle of Kings Mountain, S.C., October 7, 1780. Monument.
27. Battle of Cowpens, S.C., January 17, 1781.
28. Battle of Hobkirk Hill, S.C., April 25, 1781.
29. Battle of Eutaw Springs, S.C., September 8, 1781.

It is believed that the events on the above list not shown as having monuments or tablets may properly be so commemorated.

2. WAR OF 1812

In the War of 1812 the principal actions, within the limits of the United States in which the military forces were engaged, were of a defensive character.

No battle of this war is placed in Class I.

(a) Battles of Class IIa

Battle of New Orleans, January 8, 1815: The Battle of New Orleans is the best known of the battles of this war and more troops were engaged on that field than on any other. It was a brilliant defense by raw troops against the attack of a much larger force of highly trained troops with war experience and led by well-known British generals. As it was fought after the treaty of peace had been signed, though not confirmed, it had no effect on the general conduct or outcome of the war or on the terms of the treaty of peace. Its immediate effect was to protect New Orleans from capture and perhaps from the experience of the National Capital.

In honor of this victory a monument has been erected; it was completed under the act of March 4, 1907, which appropriated $25,000 for this purpose. In view of the unique character of this battle it is believed that the line of defense should be located and properly marked; the battle is therefore listed in Class IIa.

(b) Battles of Class IIb

1. Battle of Black Rock, N.Y., December 30, 1813.
2. Fort Meigs, Ohio, April 28 to May 9, 1813.
3. Frenchtown, Mich., January 18, 1813, and January 22, 1813.
4. North Point, Md. (Long Log Lane), September 12, 1814.
5. Plattsburg, N.Y., September 6 to 11, 1814.
6. Sacketts Harbor, N.Y., May 29, 1813.

3. WAR WITH MEXICO

In the Mexican War there were but two battles fought within the limits of the United States. Each battle is worthy of commemoration by some form of monument.

Battles of Class IIb

Battle of Palo Alto, Tex., May 8, 1846.
Battle of Resaca de la Palma, May 9, 1846.

4. INDIAN WARS

From the beginning of the Revolutionary War to the engagement with the Chippewa Indians in October, 1898, there were innumerable encounters with Indian tribes. Until the Mexican War these encounters took place east of the Mississippi River; thereafter, with the advance of the pioneers, they were transferred to the west of that river. According to lists prepared by The Adjutant General more than 1,000 such engagements occurred between January, 1866, and January, 1891. There were comparatively few engagements in which large forces of the United States participated, but as all these encounters are more or less intimately related to the development of the Western States and the advance westward of civilization, the most important of them are worthy of commemoration. The following list is believed to contain those of greatest historical interest and importance. Monuments have been erected to commemorate some of them.

Battles of Class IIb

1. Battle of Newton (Elmira), N.Y., August 22, 1779. Defeat of Iroquois by General Sullivan.
2. Battle of Miami Village, by General Harmer, September 30, 1790.
3. Engagements near Fort Recovery, Ohio, November 4, 1791, by General St. Clair. Monument.
4. Battle of Fallen Timbers (Miami Rapids), Ohio, August 20, 1794, by General Wayne.
5. Battle of Tippecanoe, Ind., November 7, 1811, by General Harrison. Monument.
6. Massacre at Fort Mims, Creek Nation, August 30, 1813.
7. Battle of Talladega, by General Jackson, November 7, 1813.
8. Battle of Tohekeka or Horseshoe Bend, Ala., March 27, 1814, by General Jackson. Monument.
9. Battle of the Wisconsin (called also Battle of the Bad Axe and Battle of the Iowa), August 2, 1832, by General Atkinson.
10. Engagement at Tampa Bay, Fla., December 28, 1835.
11. Engagement at the Withlacoochie, Fla., December 31, 1835, and near that river on February 27, 1836.
12. Battle of Okeechobee, Fla., December 25, 1837.
13. Engagement near Fort Phil Kearney, Dakota, December 21, 1866. Monument.
14. Engagement at Prairie Dog Creek, Kans., August 21-22, 1867.
15. Engagement on the Arickaree, Kans., September 17-25, 1868. Monument.

16. Attack on Black Kettles Village on Washita River, Indian Territory, November 27, 1868.

17. Attack on Modocs in lava beds, California, January and April, 1873.

18. Battle of the Rosebud River, Mont., June 17, 1876.

19. Battle of Little Big Horn, Mont., June 25-26, 1876. Monument.

20. Engagement at Slim Buttes, S. Dak., September 9, 1876. Monument near site.

21. Engagement at Powder River, Mont., November 25-26, 1876.

22. Engagement at White Bird Canyon, Idaho, June 17, 1877.

23. Engagement at Clearwater, Idaho, July 11-12, 1877.

24. Engagement at Big Hole Basin, Mont., August 9-10, 1877. Monument.

25. Attack on Nez Perce camp at Snake Creek, near Bear Paw Mountains, Mont., September 30, 1877.

26. Engagement at Milk River, Colo., September 29-October 1, 1879.

27. Engagement at Wounded Knee Creek, S. Dak., December 29, 1890. Monument erected by Sioux.

It will be noted that a gap extending from 1837 to 1866 exists in this list. The engagements with Indians that occurred between these years have not been compiled and must be sought in the old records section of The Adjutant General's office. As this will require considerable time I shall submit later a supplementary list of engagements with Indians between these years, should any be found of sufficient importance to warrant inclusion in the list.

5. CIVIL WAR

(a) Battles of Class I

1. Battle of Gettysburg, July 1-3, 1863.

2. Battles and siege of Vicksburg, May 1 to July 4, 1863.

3. Battles of Chickamauga and Chattanooga, September 19-20, and November 23-25, 1863.

The following considerations governed the selection of the battles named as being entitled to be placed in Class I.

The year of 1863 may well be considered the critical or decisive year of the war and the battles listed the decisive events of that year.

The year 1862 opened very auspiciously for the Union armies, and by May the Army of the Potomac was within striking distance of Richmond, and the Armies of the Ohio and the Tennessee had compelled the Confederate armies to evacuate the States of Kentucky and Tennessee. Then, however, the Union advance was checked, and in midsummer the Confederate armies made a counter movement which carried the Army of Northern Virginia into Maryland and the armies of Tennessee and East Tennessee almost to the Ohio River. The Confederates were unable to maintain their advanced positions, retired slowly, and the year closed with the disastrous Union assaults at Fredericksburg and Chickasaw Bluff and the practically drawn battle of Murfreesboro or Stone River. At the end of 1862 the outcome of the war was still uncertain.

The year of 1863 opened auspiciously for the Confederate armies by the defeat of the Army of the Potomac at Chancellorsville in the east. In the west the Union Army of the Tennessee was meeting with one obstacle after another in its attempt to open the Mississippi River. However, in July the high hopes awakened in the Confederate States by Chancellorsville were cast down by the defeat of Gettysburg and the loss of Vicksburg and its garrison. The hopes of the Confederacy were slightly raised by the victory of their central army at Chickamauga in September, but its fate was sealed when this army, too, was defeated in the Battle of Chattanooga and forced to retreat. At the end of 1863 the outcome of the war was no longer doubtful.

The struggle was prolonged through the year 1864 and into 1865, but at no time could the advance of the Union armies be checked. When attempts were made to divert the Union commanders from their objectives they ended only in disaster.

In addition to those named Shiloh has also been placed in this class by an act of Congress.

Since each of the three great Union armies, the Army of the Potomac, the Army of the Middle West (known in succession as the Army of the Ohio, Army of the Cumberland, and Army of Georgia) and the Army of Tennessee, has its national military park on the site of its most famous battle, all the other important battles of these armies are placed in Class II.

(b) Battles of Class IIa

In this class are placed battles of far-reaching importance, in which the numbers engaged and the losses sustained, or the resultant military or political effects, were so great as to warrant their inclusion. While the greater portion of these fields lies in the State of Virginia, the States of Tennessee, Georgia, and North Carolina are represented in the list. Should it be deemed important to preserve any one of these fields for professional military and historical study, it would be sufficient to mark the battle lines as on the field at Antietam, otherwise the battle might be commemorated as an important historical event by the erection of a single monument.

These battles are listed in chronological order, as it has been found impracticable to arrange battles fought by different armies, in different theaters, with different objectives, in a satisfactory order of importance.

1.   Battle of Bull Run, Va. July 21, 1861.

2.   Fort Donelson, Tenn. February 14-15, 1862, Army of the Tennessee.

3.   Battles around Richmond, Va. June 26-July 1, 1862, Army of the Potomac.

4.   Second Manassas or Groveton, August 30, 1862, Army of Virginia and Army of the Potomac.

5.   Fredericksburg, Va. December 13, 1862, Army of the Potomac.

6.   Murfreesboro, Tenn. December 31, 1862, Army of the Cumberland.

7.   Chancellorsville, Va. May 1-4, 1863, Army of the Potomac.

8.   The Wilderness, Va. May 5-9, 1864, Army of the Potomac.

9.   Spottsylvania, Va. May 8-18, 1864, Army of the Potomac.

10.   Cold Harbor, Va. June 1-12, 1864, Army of the Potomac.

11.   Battles around Atlanta, Ga. July 20-September 1, 1864, Armies of the Cumberland, Tennessee, and Ohio.

12.   Battles around Petersburg, Va. June 15, 1864, to April 2, 1865, Army of the Potomac and The James.

13.   Battle of the Opequan (or Winchester), Va. September 19, 1864, Army of the Shenandoah.

14.   Nashville, Tenn. December 15-16, 1864, Corps of the Armies of the Cumberland, the Tennessee, and the Ohio.

15.   Bentonville, N.C. March 19-21, 1865, Army of Georgia.

(c) Battles of Class IIb

In a war covering a period of four years, fought over an extensive territory, in which there occurred over 2,000 listed battles, engagements, and sieges wherein organizations of various sizes participated, it is very difficult, if not impossible, to make a satisfactory list of all the battles and engagements that might be considered worthy of some form of monument as a memorial to the organizations and to the men who took part. It is believed, however, that a single monument should suffice to commemorate any battle or engagement not listed in Class IIb, since none can be more important in our history than some of the battles of the Revolutionary War thus commemorated, even though in the Civil War battles the forces engaged and the losses suffered were greater. Distinctions within this class—i.e., between important battles such as Franklin, Cedar Creek, Kenesaw Mountain, Champions Hill, Perryville, Pea Ridge, and smaller engagements, such as the Monocacy, Brandy Station, etc.—might fittingly be indicated by the size of the monument.

C. A. BACH
Lieutenant Colonel, Cavalry,
Chief Historical Section, A.W.C.

Approved, June 16, 1925.

DWIGHT F. DAVIS, *Acting Secretary of War.*

H.R. 1071, 69th Congress, 1st Session.

EXCERPT FROM "AN ACT MAKING APPROPRIATIONS FOR THE DEPARTMENT OF THE INTERIOR FOR THE FISCAL YEAR ENDING JUNE 30, 1929, AND FOR OTHER PURPOSES,"

(Act Amending Concessions Policy)
Approved March 7, 1928 (45 Stat. 235)

Section 3 of the Act of August 25, 1916 (39 Stat. 535), entitled "An Act to establish a National Park Service, and for other purposes," is hereby amended by adding the following thereto: "*And provided further,* That the Secretary of the Interior may grant said privileges, leases, and permits and enter into contracts relating to the same with responsible persons, firms, or corporations without advertising and without securing competitive bids: *And provided further,* That no contract, lease, permit, or privilege granted shall be assigned or transferred by such grantees, permittees, or licenses, without the approval of the Secretary of the Interior first obtained in writing: *And provided further,* That the Secretary may, in his discretion, authorize such grantees, permittees, or licensees to execute mortgages and issue bonds, shares of stock, and other evidences of interest in or indebtedness upon their rights, properties, and franchises, for the purposes of installing, enlarging, or improving plant and equipment and extending facilities for the accommodation of the public within such national parks and monuments." (U.S.C., 6th supp., title 16, sec. 3.)

FIRE PREVENTION PLAN FOR THE NATIONAL PARKS 10TH NATIONAL PARK CONFERENCE, FEBRUARY 15–21, 1928

MR. JAY H. PRICE INTRODUCED

MR. HALL: We have with us today another former member of the Forest Service who is now District Fire Inspector under the Clarke McNary Act. At this time I will introduce Mr. Jay H. Price, who will speak on fire prevention.

PAPER BY MR. PRICE

MR. PRICE read the following paper on "Fire Plans":

"In considering the subject of fire plans, the first question that naturally arises is, what is a fire plan? In formulating a definition the first thing to remember is that a fire plan is a means to an end. In this discussion we shall assume that the desired end is adequate fire control, although, of course, fire control itself is but a means to an end. The ultimate goal is the conservation of all forest resources for the benefit of man. Now unless a fire plan, however elaborately prepared, serves this end, it is absolutely worthless unless, perchance, your filing cabinets are in need of ballast. A plan that rests securely and continually in the files is, by that very fact, branded as worthless.

"Going further, we may define a fire plan as:

(1) A clear portrayal of the fire control problem, based upon the best data available.

(2) A plan of action designed to solve the problem; and

(3) Detailed instructions governing the execution of the plan of action, in which responsibility for each duty is clearly defined.

"To get a clear picture of the fire control problem on an administrative unit, we must study all the factors that govern the origin and spread of fires. For the sake of clarity I'll refer to these factors under four main headings, namely, risks, hazards, weather and topography. As a start, let us consider those factors that govern the likelihood of fire starting. First, there are the risks, which may be defined as those agencies capable of starting fires. Lightning is the most important natural risk; man and his engines supply the artificial risks. The latter, for the sake of study, are often divided into many and varied classifications, such as smokers, brush burners, locomotives, incinerators, etc. Second, there are the hazards. I am here using this term to designate the material and substance on the ground that can be ignited and burned. Common forest hazards are dry grass, rotten wood, brush, slash, duff, etc. Third, there is weather which governs the start of fires through its influence on both risks and hazards. For example, the presence of a strong wind may cause a living spark from a flue to travel a much greater distance that it would do otherwise. Thus the risk is influenced by the weather. Ground cover and forest litter respond very quickly to atmospheric changes. For example, a hazard such as dry grass reacts much more surely to the discarded burning pipe heel during periods of low humidity.

"Before a fire can start, it is plain that both a risk and hazard must be present. A thousand burning cigarette butts tossed on a paved avenue will cause no fires simply because there is no inflammable material on the pavement. Likewise a field of ripe grain is quite safe when all igniting agencies are absent. Obviously the prevention problem is either to do away with the hazard or to remove or sterilize the risk.

"The spread of fires, as distinguished from the origin, is governed by the following factors:

(1) Hazard, that is, the amount and nature of the material on the ground.

(2) Weather, directly by wind conditions, and indirectly through its influence on the inflammability of the cover; and

(3) Topographic conditions. For example, other conditions being equal, a fire will spread much more rapidly up a steep slope than across a plain. Also topographic features often have a marked effect on local wind directions and velocities.

"Perhaps we can learn more readily what a fire plan should contain by attempting to build one for a typical area. Let us consider, as an example, a relatively small watershed traversed by a mountain road. There are at least two tasks before us; one is the intelligent study of the fire history of the area, and the other is a careful survey of the conditions on the ground. Suppose that the fire history of the area shows that there have been no fires of record before July 1 and after September 30, but that fires are frequent between these dates; that all of the fires have started along the highway and that they have been traced with a fair degree of certainty to burning tobacco and unextinguished camp fires. Furthermore, we find that several of them have attained large size and have done much damage before control was possible. From these considerations our fire problem is already beginning to take shape. We make our field survey and find that the road is located on a grassy bench, sparsely wooded with oak trees and flanked on the upper side by a steep slope covered with dense brush. This brush slope rises to a rolling plateau covered with a fine stand of pine trees of all ages. We find that the general exposure is toward the south, and as a result the grass and brush are very dry by early July. We find that travel on the road is very heavy during the summer and that on account of the low embankments many tourists draw to one side for picnic lunches or for over-night camps.

"With the foregoing picture of the conditions on the ground we can well understand the fire history of the unit. Our portrayal of the fire problem, part one of our fire plan, is reasonably complete.

"Our next job is to construct part two of our fire plan; that is, to develop a plan of action.

"To prevent the fires, which is certainly the most desirable form of attack, we have seen that we must remove or render innocuous either the risks or the hazards. As a matter of safety we may well work on both factors. For the risks on this particular area there are several courses of action to consider, among which are:

(1) We may forbid all travel during the dangerous part of the season.

(2) We may forbid smoking and camp fires, either completely or everywhere except at designated safe places, such as prepared campgrounds.

(3) Or we may impose no restrictions but concentrate our prevention effort in educating the travelers to be careful with matches, burning tobacco and camp fires.

"The first plan is often impossible and is always undesirable; the last is generally the best if it can be made effective. Our survey has shown that the dry grass is the hazard responsible for the origin of the fires. This danger may be removed or reduced in various ways, such as by roadside burning or plowing. From all of these plans we must select those that appear most practical. Our selection will necessarily be influenced by outside considerations, such as the use policy for the area, the disfigurement that may result from hazard reduction, etc. Whatever the plan selected, probably some prevention patrol will be necessary.

"Having determined our plan of action as far as prevention is concerned, and realizing that few such plans prove to be 100 per cent effective, we must organize for the occasional fire that will start in spite of our preventive efforts. Both the fire history and the field survey have shown that rapid spread is inevitable. This calls for immediate detection and short elapsed time between discovery and attack. It may be that our prevention patrolman can handle the problem. If not, we must provide a detection system that will have the area under observation at all times. We must also provide a fireman and station him so that a minimum of time will be consumed by travel.

"Even with this suppression plan built up, we know there is a chance of a fire getting away during days of bad fire weather or during periods of extremely heavy travel, such as week-end holidays. We determine that in such periods two men, or even more, rather than one, should be held ready for instant dispatch. Perhaps there is a possibility of smoke pouring over the area due to fires burning outside the administrative unit. The result may be interference with the visibility of our lookout, and we must plan to supplement our detection service with ground or air patrol.

"So far we have planned either to prevent the fires or to suppress them before they become large. But we know from experience that plans sometimes fail, no matter how carefully made and executed. There is always the possibility of a fire getting away and becoming large, due perhaps to unforeseen weather conditions or to a failure on the part of the protection force. We know that such a fire in this area will spread rapidly and by nightfall may reach several hundred acres. From our study on the ground, taking into consideration cover, slope, and the natural barriers if any, we know that a large crew will be necessary to control the fire during the night and early morning. We know also that employment of such a crew demands a certain amount of qualified overhead, sufficient supplies of food and camp gear, a supply of tools suited to the work at hand, special transportation, and perhaps an emergency communication system.

"Despite all of these suppression efforts, a fire may evade control even after an all night fight, and the winds of [the] next day may develop it into a major conflagration. If such an event is at all possible we must have an emergency plan that will cover the procuring of great numbers of firefighters, and vast quantities of supplies and equipment from the large purchasing and labor centers, qualified overhead and special equipment from neighboring units, and such specialized services as for example, airplane scouting.

"Thus we have attempted to build up a plan of action for the particular area under consideration. Similarly we must construct a plan for all areas in the administrative unit, taking into consideration the nature of the factors affecting the fire control problem. When this is done there remains the job of correlating the whole and evolving a final plan of action for the entire unit. Here the attempt must be to fill each need but at the same time avoid unnecessary duplication. And, of course, we must keep in mind the financial limitations under which we are working.

"We have now pictured our fire problem and have built up a comprehensive plan of action aimed to solve it. The third and very important task is to work out the detailed arrangements and instructions, and to definitely fix responsibility for action. I cannot attempt to go into this task completely but I do wish to mention a few of the main considerations.

"The protective organization must have an executive head. Preferably this executive head should be free to cover the field at all times even during large fires. In general, the best brains should be on the fire line during suppression emergencies. This means that a dispatching system must be arranged so that all the fire boss will have to do is press the button and the sinews of war will be at his command. To make this possible, detailed arrangements must be made in advance for supplies, equipment, man power, etc., and transportation for the same. In short, all materials that may be needed must be listed by location, amounts and ownership, and the means of communication and transportation between the supply points and the forest areas must be arranged for in advance and described in the written plan.

"The fire plan should show the opening dates for lookouts, patrols, and firemen, and their stations or patrol routes. Written instructions must be provided for each man, covering his specific duties. The reinforcements of the regular protective force during periods of emergency must be arranged for in advance and recorded in the written plan.

"In considering our sample area, we have seen that we shall need a large crew of men to handle the occasional fire that gets away. It is not enough to know that a certain road crew will be available, and that a certain number of firefighters may be recruited in a near-by town. We must make definite arrangements with the foreman of the crew and with a labor agent in the town, and record them in the fire plan, or else there will be misunderstandings and delays in getting the men when we want them.

"No protection plan would be complete without provision for training. Such training should cover both prevention and suppression activities. Even with training, man is prone to error, so fire control work must be given frequent and constructive inspection. In big suppression emergencies inspection is particularly important. We must plan to send two men to do one man's job.

"Roads and trails have an important place in the fire plan, both in determining the plan of action and in carrying it out.

"Before closing I want to refer once more to the subject of fire data. We have seen that a study of the fire history of an area is very helpful in constructing a plan of action. It follows that the more nearly complete the record is, the more serviceable it will be. The individual fire report is the only means yet developed of getting fire history onto paper, of building up a reservoir of basic data.

"We have seen that there is nothing mysterious about constructing a fire plan. It merely calls for an intelligent analysis of the fire problem and a carefully thought-out plan aimed to solve it. It should be as simple as possible but should not omit essential detail. It must be subject to modification whenever conditions on the ground change, and whenever we learn something new and better to apply in meeting the fire problem. There is much still to learn, therefore the best of present fire plans are far from perfect."

Following Mr. Price's talk, which ended at about 5.00 P.M., the fire alarm sounded and all present left the building to witness an outdoor demonstration of recently developed fire-fighting equipment, tools, fire pumps, etc., under the supervision of the U.S. Forest Service.

Minutes of the Tenth National Park Conference, 131–136. National Park Service Archives, Harpers Ferry, Box A40.

THE NATIONAL PARK SERVICE'S POLICY
ON PREDATORY MAMMALS, 1931

The National Park Service is attempting to put the parks to their highest use. Every policy developed is an attempt to meet the purposes for which the parks were formed; First, the national parks must be maintained in absolutely unimpaired form for the use of future generations as well as those of our own time; second, they are set apart for the use, observation, health, pleasure, and inspiration of the people; and third, the national interest must dictate all decisions affecting public or private enterprise in the parks.

Certainly, one of the great contributions to the welfare of the Nation that national parks may make is that of wild life protection. It is one of the understood functions of the parks to give total protection to animal life. A definite policy of wild life protection is being developed with the result that fine herds of game are furnished as a spectacle for the benefit of the public, and those same herds furnish the best of opportunity for scientific study. Many disappearing species are to be found within park areas, so that in some instances we may speak of the parks as providing "last stands."

Of late there has been much discussion by the American Mammalogical Society and other scientific organizations relative to predatory animals and their control. The inroads of the fur trapper and widespread campaigns of destruction have caused the great reduction of some and the near disappearance of several American carnivores. The question naturally arises as to whether there is any place where they may be expected to survive and be available for scientific study in the future.

The National Park Service believes that predatory animals have a real place in nature, and that all animal life should be kept inviolate within the parks. As a consequence, the general policies relative to predatory animals are as follows:

1. Predatory animals are to be considered an integral part of the wild life protected within national parks and no widespread campaigns of destruction are to be countenanced. The only control practiced is that of shooting of coyotes or other predators when they are actually found making serious inroads upon herds of game or other mammals needing special protection.

2. No permits for trapping within the borders of a park are allowed. A resolution opposing the use of steel traps within a park was passed several years ago by the superintendents at their annual meeting.

3. Poison is believed to be a non-selective form of control and is banned from the national parks except where used by Park Ser-

vice officials in warfare against rodents in settled portions of a park, or in case of emergency.

Though provision is made for the handling of special problems which may arise, it is the intention of the Service to hold definitely to these general policies. It can be seen, therefore, that within the national park system definite attention is given to that group of animals which elsewhere are not tolerated. It is the duty of the National Park Service to maintain examples of the various interesting North American mammals under natural conditions for the pleasure and education of the visitors and for the purpose of scientific study, and to this task it pledges itself.

HORACE M. ALBRIGHT,
Director, National Park Service

Reprinted by permission of *The Journal of Mammalogy* (12, 2, 1931, 185–186).

A FORESTRY POLICY FOR
THE NATIONAL PARKS

Approved May 6, 1931
Horace M. Albright, Director

FOREST PROTECTION, GENERAL. Until such time as forest investigations disclose special methods of handling, necessary in maintaining wilderness areas under natural conditions, the parks will be as completely protected as possible against damage by vandalism, fire, insects (except as modified in subsequent statement dealing with insect control), fungi, mechanical injury, and grazing by domestic animals. This protection will be extended to all park areas, including land covered by forests, brush, grass or other cover.

FIRE PROTECTION

1. Fire Protection Plan. The protection of each park from fire will be based upon a written FIRE PROTECTION PLAN covering the prevention, detection and suppression of all fires on or threatening the park area. Where no Fire Protection Plan is yet available, a preliminary plan will be prepared and put into operation by the superintendent. Technical assistance in the preparation of more complete Fire Protection Plans will be provided by the Fire Control Expert of the Forestry Division, who will visit the individual parks to study conditions at first hand and prepare, in cooperation with the superintendents, fire chiefs and other members of the local organizations, detailed Fire Protection Plans. These will be revised from year to year as conditions necessitate, the revision being submitted through Field Headquarters (Forestry Division, Berkeley) for approval by the Director before being put into effect.

2. Training of Fire Control Personnel, for forest fire protection, is exceedingly important. Wherever practicable, superintendents will, at the beginning of the fire season, arrange for a local conference of all men taking part in fire protection activities in that particular park. The purpose of this meeting will be:

(a) To give each individual a definite idea of the fire protection system of the park as a whole, and of his place and work in that organization.

(b) To instruct each man in the use and repair of fire suppression tools and equipment; installation of emergency telephone lines and instruments; etc.

(c) To see that each member of the fire suppression organization understands the proper technique for fire suppression in the various types of cover with which he will have to deal, and under varying conditions of slope, wind, humidity, etc.

(d) To instruct each man in the essentials for finding and preserving evidence in the case of man-caused fires, and securing statements and affidavits in connection with fire law enforcement.

Wherever possible, specialists in fire protection should be called in to meet the men in these park fire conferences. This would include the N.P.S. Fire Control Expert, the District Fire Weather Forecaster, and officers in control of adjacent Federal, State, or private areas under fire protection.

3. <u>Cooperation with Other Protective Agencies</u>. Wherever there is an area under fire protection adjacent to a national park, efforts will immediately be made to coordinate the Fire Protection Plans for those two areas. This plan for cooperation should ordinarily be reduced to a written agreement, or at least to memorandum form.

In case of fire close to the park boundary, communication should immediately be established and maintained with the fire chief of the other agency or agencies involved and all control measures shall be coordinated as closely as possible.

4. <u>Hazard Reduction</u>. Provision will be made to fireproof so far as possible all camp grounds and roadsides where the annual vegetation creates an especially high fire hazard, by such means as local conditions necessitate, due regard for the preservation of scenic factors and wild flowers being given every possible consideration consistent with proper and adequate protection. Clean-up of dead and down timber and brush along roads and trails will also be attended to as far as money for such purpose can be secured.

All hazard reduction projects should be graphically shown on a hazard reduction map, one copy of which will be kept at the headquarters of the individual park, a copy submitted to the Forestry Division of Field Headquarters, and a copy to the Director.

5. <u>Fire Equipment</u>. The purchase of fire equipment should be planned in cooperation with the Fire Control Expert in order to insure the type of equipment best suited for local conditions.

Fire Protection Equipment of any character purchased from fire protection funds (which includes both the Fire Prevention Appropriation and the Emergency Reconstruction and Fighting Forest Fires Fund) shall be held exclusively for forest protection purposes. Such equipment must be available for fire suppression during the entire fire season, but outside that period may be utilized in connection with insect control and tree disease control work, but shall at no time be assigned to use or requisition for general park operations or for activities outside those financed from forest protection funds.

The only exception to the foregoing will be where light-delivery patrol trucks purchased for fire protection work are needed for the general transportation pool outside the period of the fire season and new trucks or those in equally as fit condition will be furnished for protection purposes at the beginning of the next fire season to re-place them. In any event the replacement of the light-delivery fire protection trucks shall be financed from other funds to the propor-tionate extent to which they were used for other than fire protection purposes, and if their use for other than fire protection purposes will result in impaired efficiency for fire protection use, then their use for other activities is prohibited.

The above exception for light-delivery fire patrol trucks does not apply at any time to the larger special trucks purchased for transpor-tation of firefighters and fire suppression equipment nor to tank fire trucks. Such trucks are to be held exclusively for fire suppression purposes at all times until they are replaced or condemned for fire suppression use. Trucks purchased for fire use exclusively shall be painted red and shall have thereon the name of the park, the words Fire Truck and the number under which the truck is recorded.

6. Law Enforcement. An effort will be made to ascertain the exact cause of each fire, and where such a fire proves to be man-caused every effort will be made to apprehend the person or persons responsible for the starting of the fire. Whether the starting of the fire was intentional or accidental, the party or parties responsible will, if possible, be made to pay (a) the cost of fire suppression, and (b) the commercial value of the forest and improvements destroyed. In case legal action is necessary, especially in the case of large fires, the services of the Fire Control Expert will be available for assis-tance in obtaining affidavits and in the preparation of the case for legal procedure. Every effort should be made by the superintendent and fire chief, however, to secure legal depositions from witnesses, and other data, at the time of the fire.

7. Fire Reports.

(a) Telegraphic Reports. Each serious fire or one which will apparently cover 200 or more acres, or cost $500 or more for suppression will be immediately reported by wire both to the Director and to Field Headquarters, with the request to the latter, if necessary, for the advisory services of the Fire Control Expert.

(b) Individual Fire Reports will be prepared in duplicate as soon as possible after the fire is out. One copy of the report will be retained at park headquarters and one copy sent to the Forestry Division of Field Headquarters.

[Note: page missing from original document]

(1) Volume of work maps, showing:
    (a) Starting point of fire separately by A's, B's and C's.
    (b) Lightning zones.
    (c) Camper zones.
    (d) Other, if needed.
(2) Hazard map: showing all fires over 40 acres in actual area by years.
(3) Visibility maps: showing visibility from lookouts and classifying visibility of areas covered into
    (a) Direct
    (b) Indirect
    (c) Blind.
(4) Hour control map; to assist in making and checking distribution of fire guards.
(5) Physical improvement map: diagram of communication and transportation systems; locations of fire protection buildings, pastures, etc.
(6) Statistical Records:
    (a) Summary of individual fire reports.
    (b) Cost of fires by classes.
    (c) Analysis of man-caused fires.
    (d) Detection record.
    (e) Elapsed time record summary.
    (f) Other records as found necessary.
(7) Type cover map.

The Fire Atlas will be prepared and kept up to date in triplicate; one copy at the headquarters of each park; one copy for Field Headquarters (Forestry Division, Berkeley); and one copy for the Director.

9. Fire Reviews. For the purpose of determining and developing the best practice and technique in fire suppression work, a review of large or important fires or of the fire control work for the season on one or more parks, following the close of the fire season, is a highly desirable practice. A board consisting of the Chief Forester, Fire Control Expert, and a representative of the field force designated by the Director, will cooperate with the park personnel in making such a review, unless circumstances or the exigencies of the situation should make it desirable for the review to be performed simply by the Fire Control Expert in cooperation with the local organization.

The purpose of such review will be to examine in detail the history of each fire studied with a view to discovering principles and techniques which can be effectively applied in fire control activities on that and other parks, and also to indicate the ways in which fire control in the individual parks can be strengthened and made most effective.

Such reviews are also highly valuable in connection with important fires in the suppression of which other agencies have participated, in which case a joint board of review embracing all organizations concerned can be established. Such reviews should be for the purpose of perfecting cooperation so as to effect the closest coordination of protection work by all the cooperating agencies.

10. Fire Research. The data submitted to Field Headquarters (Forestry Division, Berkeley) on individual fire reports, on fire atlas sheets, etc., will be used in a careful study by the Fire Control Expert of fire conditions in the individual parks. This officer will also correlate the available data from the Weather Bureau, the Forest Service, Forest Experiment Stations, and other agencies involved in research on fire control subjects, so that this information can be used in preparing or revising the Fire Protection Plans for the individual parks.

INSECT CONTROL

Cooperation by the Bureau of Entomology. The functions of the Bureau of Entomology in respect to forest insects are three-fold. These are (1) research or investigations; (2) cooperative service in control; and (3) educational service. These activities are centered in the Division of Forest Insect Investigations.

The handling of the preliminary surveys, necessary for keeping the Bureau of Entomology acquainted with the conditions within each park, is an obligation of the Park Service itself, as are also the actual control operations. The Bureau of Entomology, however, is the clearing house on all technical matters pertaining to the technique and advisability of control. Where control operations are advisable the Bureau of Entomology will detail a member of its organization to assist in and direct the more detailed survey necessary in the development of the program for insect control operations, and in the case of the larger control projects will provide technical supervision for a period of time varying from a few days to the entire period of control, depending on the technicalities involved.

Insect Control Policy. It will be the policy to secure and maintain, so far as practicable, full protection from insect epidemics in areas of the following character within the national parks and monuments.

(1)   Areas of intensive use, such as camp grounds.
(2)   Areas of important scenic or esthetic attraction (unless the partial loss of the tree species attached within a mixed stand

will not materially affect the general appearance of the stand and its scenic or esthetic value, nor materially add to the fire hazard).

(3)   Areas of prospective intensive use within the next ten-year period.

(4)   Areas within the national park threatening protected areas either within or outside the national park.

(5)   Areas of unusual fire hazard.

(6)   Areas set aside for study and research (unless natural agencies are to be left undisturbed).

Complete protection in the sense here used would call for removal of light endemic infestation in areas of intensive use.

With such insects as the mountain pine beetle in lodgepole pine and the Black Hills beetle in yellow pine, there can be no question but that every outbreak should be immediately controlled before it develops into a widespread epidemic costing often thousands of dollars.

Quite a different example is presented in case of the western pine beetle in Oregon and California. This beetle takes annually a small percentage of the stand and at intervals of some years a considerably larger percentage. The main objective in controlling the depredations of this beetle would be to prevent the peaks of this type of infestation developing and thus prolong the life of the existing stand over a longer rotation of gradual replacement; in other words, the objective would be to carry on a certain amount of maintenance control from year to year in an effort to keep the losses at the lowest possible status all the time.

With defoliating insects, it is possible to readily control them where the trees are accessible to high-powered pumping equipment such as along main highways. Within a few years it may be practical to use airplanes for dusting some of these infestations.

Under the above policy, remote areas of no special scenic value and not of high fire hazard, little used or seen by the public and not planned for intensive use within a reasonable period of years, may be omitted from insect control plans if they will not endanger control in adjacent areas, unless there are other special factors which make their protection from insects important.

Maintenance Control must be carried on during subsequent years to follow up initial control work in any area. It is an essential part of our insect control policy that such maintenance control shall be systematically planned for in the annual estimates in order that the good accomplished by initial control shall not be lost or dissipated.

Right of Way Clearing. In the preparation of instructions for disposal of timber cut in clearing rights of way for roads, trails, telephone lines, power lines and other clearings, provision shall be made to insure satisfactory measures to avoid the development of any insect epidemic. The advice of the appropriate field station of the Bureau of Entomology should be secured to insure the incorporation of the proper provisions in this regard.

Handling Insect Control Work. On each timbered park the super-intendent will designate some member of his organization to give special attention to insect matters within that park. This man should keep himself thoroughly informed at all times regarding the condition of the forest stand, and in case of any insect infestation shall make such preliminary survey as is necessary to furnish intelligent information regarding conditions for transmittal to the field station of the Bureau of Entomology in order to secure action by that Bureau. The man assigned to this work should also accompany the field representative of the Bureau of Entomology in making more detailed field surveys and in planning insect control activities, and is the logical man to supervise the actual control operations.

In selecting the man to be assigned to insect work, it is highly desirable that a man trained and experienced in this line of work should be selected, if available. If no one with these qualifications is available within the park in question, possibly a transfer of a suitable man elsewhere within the Service can be arranged. If not, it will be necessary to provide training and experience for the best available man as early as possible on those parks where this line of work is of importance. In this connection, other qualities being equal, a training in entomology, biology or forestry will furnish a better background for intelligent understanding of the problems involved.

The man assigned to insect work must be given sufficient time to devote to a survey of the forest stands each year to insure that no epidemics become established because of a lack of information of the presence of insect infestation. Our aim must be to discover infestation promptly so that insect control can be maintained at the minimum expense and under conditions which will give the greatest promise of success. It is important also that all the rangers be given instruction in recognizing the signs of insect infestation, either by the man detailed to insect work or by field representatives of the Bureau of Entomology, so that they may participate in the discovery and report of incipient infestations within their respective districts.

An Annual Report of Insect Conditions within each forested park will be prepared and submitted not later than November 1 to the Forestry Division on the forms furnished for this purpose. Such reports should indicate where important problems exist, and the assistance of the Bureau of Entomology will be requested where the problems are of sufficient importance. In case of a serious situation at any time, it should be immediately reported to the Forestry Division and to the Bureau of Entomology for action without delaying report until the regular annual report date.

TREE DISEASE CONTROL

The Bureau of Plant Industry is the agency designated to furnish technical advice relative to fungus diseases and pathological conditions affecting the park trees and forests. Dr. E.P. Meinecke, Principal Pathologist, Bureau of Plant Industry, Ferry Building, San Francisco, California, has been designated as general advisor to the National Park Service in matters of forest pathology.

In the control of blister-rust the Office of Blister-Rust Control of the Bureau of Plant Industry will act in the capacity of technical adviser and supervisor of control operations in a manner similar to that of the Bureau of Entomology in insect control. For this work in the West the Office of Blister-Rust Control, Senior Pathologist S.N. Wyckoff in Charge, 618 Realty Building, Spokane, Washington, will furnish men for field examination and preparation of estimates for control work, and for technical supervision of control operations. In the East, this work will be handled through the Office of Blister-Rust Control of the Bureau of Plant Industry, Department of Agriculture, Washington, D.C.

In order to permit complete and accurate surveys by members of the Office of Blister-Rust Control, it is essential that type maps be completed at the earliest possible date showing the distribution of all species of pine belonging to the white pine group susceptible to blister-rust damage, in order that such surveys by the representatives of the Office of Blister-Rust Control may cover the entire field in which protection might be desirable without having to spend time in determining the distribution of these species.

It is further essential that in any eradication of currant and gooseberry bushes (Ribes) for the prevention of the spread of blister-rust to the white pine species, consideration should be given to all the values at stake and the local importance of Ribes as a part of the landscape as contrasted with scattered white pines. The viewpoint of the Landscape Division and of the Branch of Research and Education should be obtained in connection with any contemplated Ribes eradication in order that all factors shall be given proper consideration in the determination of the most advisable program.

CAMP GROUND PROTECTION

Where camp grounds are located within timber, intensive use year after year results in removal of the humus, packing of the surface layers of the ground where the feeder roots are located, and in mechanical injury to the trees themselves where they are unfenced, resulting in conditions unfavorable to tree growth which may eventually so disturb the vitality of the trees that they will become easy prey to insect attacks and tree diseases. In some instances this has already resulted in the gradual loss of shade trees from year to year to such an extent that new camp grounds or the fencing off of the remaining trees constitute the only remedy.

In solving this important problem of camp ground protection the advice of the forest pathologists of the Bureau of Plant Industry and of the entomologists of the Bureau of Entomology is extremely desirable in determining the protective measures to be applied.

LAND AND TIMBER EXCHANGE

In case of any land and timber exchange, all cutting will be done under conservative forestry regulations, which must be submitted to Field Headquarters (Forestry Division, Berkeley) for examination as to their sufficiency before incorporation in the contract.

Cutting regulations in exchange contracts shall be strictly enforced by the superintendents and no changes shall be effected without prior approval of the Director through Field Headquarters (Forestry Division, Berkeley).

FOREST PRODUCTS

1.   Clean-up. Dead and down timber constitutes a fire menace and should be removed from the forest, especially along roads, whenever practicable. In the case of large burns this may sometimes be effected by the sale of burned material or by the granting of it free to the park operator or others who will guarantee to take all of it from a specified area, thus reducing the fire hazard. Clean-up work along roadsides, lake shores, etc., must be planned in cooperation with Field Headquarters (Landscape Division), and when in progress will be under the supervision of the landscape architect assigned to the park.

2.   Firewood. Firewood will, so far as practicable, be secured from dead timber and from clearing operations on roads and other projects. If it becomes necessary to cut live standing trees for firewood purposes solely, the marking of such trees for cutting will be done under the supervision of a representative of Field Headquarters (Forestry Division, Berkeley).

GRAZING

It is the policy of the Park Service to eliminate the grazing of domestic livestock on national park ranges as soon as practicable.

Local regulations relative to the grazing of saddle and pack animals will be made and enforced by the superintendent, with approval by the Director.

National Archives, Record Group 79, Entry 18, Records of Arno Cammerer; also available from the files of the Natural Resources Division, National Park Service, Washington, D.C.

OFFICE ORDER NO. 228
PARK PLANNING

April 3, 1931

There is inclosed [*sic*] a copy of the Employment Stabilization Act of 1931, Public No. 616, approved February 10, 1931. Its purpose is to provide for the advance planning and regulated construction of public works, for the stabilization of industry, and for aiding in the prevention of unemployment during periods of business depression.

This requirement of law for advance planning makes necessary the restatement of certain fundamental facts and policies. The national parks and national monuments have been set aside for the enjoyment and the benefit of the people, and the National Park Service is enjoined by organic law creating the Service to administer them in such manner and by such means as will leave them unimpaired for the enjoyment of future generations. Therefore, in all planning and construction of physical improvements, to permit their present use by visitors and at the same time not impair the areas so as to interfere with their enjoyment by future generations, the strictest landscape control and supervision will be required.

The Superintendent is responsible for the proper development of his park or monument. He initiates projects, is responsible for estimates of appropriations and the expenditure of construction funds, and for the proper completion of the work within the limitation of such funds.

The professional services of Field Headquarters are available to assist him in the planning, the designing and the construction of his park development work, and are to be fully and completely utilized.

THE PARK DEVELOPMENT PLAN

In order to properly coordinate all phases of planning, the park development outline, or five-year program, which was begun in 1926 shall be brought to date as of October 1, 1931, and hereafter shall be known as the Park Development Plan. This plan will give the general picture of the park or monument showing the circulation system (roads and trails), the communication systems (telephone and telegraph), wilderness areas and developed areas. More detailed plans of developed areas (villages, tourist centers, etc.) will be required to properly portray these special features. These plans being general guides will naturally be constantly in a state of development and shall hereafter be brought up-to-date and made a matter of record annually. Their success depends upon the proper collaboration of study and effect on the part of the Park Superintendent, the Chief Landscape Architect, the Chief Engineer, and the Sanitary Engineer. The resulting plan will not be the work of any one but will include the work of all. Since park development is primarily a Landscape development, it shall be the duty of the Landscape Division to collaborate with the Park Superintendent in the preparation of the Park Development Plan. Exact copies must be made for the park or monument, Field Headquarters and the Washington Office. Revised copies shall be made and furnished each office annually.

Accompanying each Park Development Plan there shall be prepared a written outline of the items deemed necessary in the development of the park or monument. Exact copies must be made for the park or monument, Field Headquarters and the Washington Office. Revised copies shall be made and furnished each office annually. As this outline furnishes information for the preparation of the Park Development Plan, it too will be constantly in a state of progression and will be the clearing house of proposed items. It classifies the development of the park or monument as to areas and these areas into units according to use.

A general outline or classification to be used as a guide is given below. It will naturally vary in detail in different parks and monuments.

DEVELOPMENT OUTLINE OR CLASSIFICATION

1. Circulation—
    a. Road System:
        1. General Road System Plan. (Outlined on Park Topographic Map.)
        2. Project Plans. (Plan for each unit in road system.)
    b. Trail System:
        1. General Trail System. (Outlined on Park Topographic Map.)
        2. Project. (A plan or field report for each unit.)
2. Wilderness (Sacred Areas) Areas. (Outlined on Park Topographic Map or Park General Plan.)
    a. Wilderness Areas—large areas to be generally protected as undeveloped wilderness areas.
    b. Sacred Areas—small areas to be protected against all development for the protection of a special natural feature—i.e., 1/8 mile radius around Old Faithful Geyser. Similar areas around important water falls—a special group of trees or geological feature, etc.
    c. Research Reserves.
3. Developed Areas. Including Building Group units such as villag-

es or tourist centers. Each should have all or part of the following according to the use of the area:

a. Circulation System:
1. Roadways.
2. Parking Areas.
3. Bridle Paths.
4. Footpaths.

b. Public Utilities. (General Layouts by Sanitary Engr. or Civil Engr.)
1. Water System.
2. Sewerage System.
3. Garbage Disposal.
4. Telephone System.
5. Power System.
6. Other Utilities (each should have a number).

c. Government Building Units:
1. Administrative Group.
Administration Building, Museum, Post Office, etc.
2. Residential Group.
All Employee Housing.
3. Utility Group.
Shops, Equipment Housing Barns, etc., possibly Laborers Mess and Bunkhouses.

d. Tourist Facilities:
1. Hotel Areas.
2. Lodge Areas.
3. Housekeeping Camp Areas.
4. Government Auto Camp Areas.
5. Retail Areas (only in larger parks).

e. Public Utility Operators Non-Tourist Units:
1. Administration Area (often in hotel and not a distinct unit).
2. Residential Area (Residences and Dormitories).
3. Utility Area (Warehouses, Shops, etc.)
4. Transportation System Area (usually is part of Utility Area).

f. Outlying Units (minor developed areas):
1. Ranger Station.
2. Road Camp.
3. Other facilities (give separate number to each).

The Park Development Plan and written Development Outline or Classification are the working tools to coordinate the thought and effort of the various offices engaged on planning and construction work. They are general guides and are not binding as to details. The Park Superintendent will find them of value in selecting items for the estimates, being able to select items from a list that is constantly being studied instead of making a list entirely from memory at the time he is asked for estimates.

The Chief Landscape Architect can obtain information as to size, purposes, and the relation of the unit to other facilities in preparing sketches and working drawings. The Chief Engineer and Sanitary Engineer likewise can get data and information as to the relation of their projects to other work.

The Bureau of Public Roads work can also be related to other work.

The Public Utility Operators' developments can likewise be co-ordinated with all Government activities.

SIX YEAR ADVANCE PLAN

The estimates for 1933 will be submitted in their usual form and constitute the first year of the six year program. The attached sample form will be used in submitting the items for the subsequent five years. The law requires that these programs be resubmitted annually, therefore when the 1933 appropriations have become law and the 1934 estimates are submitted, items will be selected from the five year advance program for the current year list and new items will be added to the five year list. The revised six year program shall be submitted each year with the final estimates of appropriations.

Sketches and estimates should be prepared in advance as rapidly as it is feasible to accomplish this and ultimately it is hoped that sketch plans and estimates will have been prepared for all items in the six year program. As a development plan is studied, the necessary revisions can be noted from time to time for the annual revision. The Park Development Plan and the written Outline or Classification are important, as they allow the proper method of studying the full development of the entire park and each developed area within the park. This is the best point of view to approach the study of a development.

The selection of items for the current year's estimate[s] and the five year list is a recapitulation of the same items in the order of their priority and it is a relatively simple problem to select this priority list from the items in the park development outline which are arranged according to locality.

In the 1934 estimates, the superintendents will be required to have a clearance from the Chief Landscape Architect on all items that are included in his estimates. The study of these items should begin immediately and be carried on through the construction season. It is best that the items to go in the coming year's estimates be approved during the construction season while the Landscape Architects are in the field.

EXPLANATION OF FORM
(EMPLOYMENT STABILIZATION)

The attached sample form is to be followed in supplying the information required, except that the "Dates" for approval of plans, shown in upper right-hand margin, are not to be inserted.*

Each project deemed necessary in properly administering and developing the park (or monument) as contemplated in this act should be listed in the order of its priority, and the project number should be preceded by the letters E. S. (Employment Stabilization). The E. S. project number is assigned temporarily to identify it until the project is authorized for construction at which time it will be given a project number in the customary manner. The name assigned should be descriptive of the project. The nature of the work to be performed should be set forth on separate letter-size sheets, writing lengthwise across the paper. While the descriptive matter should be concisely stated, do not sacrifice needed information. If a unit of road, give the type, length, width, cost, months during which construction can be carried on, and the same character of information for bridges, retaining walls, tunnels, and other prominent features.

Under Section 8 of the Act (c) and (e) it will later on be necessary to:

Prepare a year in advance plans for prompt commencement and carrying out an expanded program at any time, which is to include expansion in organization and detailed construction plans as well as plans for the acquisition of sites, all of which information is to be assembled and made available to the Employment Stabilization Board and the Bureau of the Budget, upon request.

Section 8 (d) provides that such programs, plans, and estimates for the six-year period shall be submitted to the Board and to the Director of the Bureau of the Budget. Therefore, the data required on the form (Employment Stabilization) is to be submitted annually to the Washington Office in quadruplicate.

PLEASE ACKNOWLEDGE RECEIPT OF THIS ORDER.

HORACE M. ALBRIGHT,
Director.

Historic Files, Office of Park Planning and Special Studies, National Park Service, Washington, D.C.

* Ed. Note: For reasons of space, the form has not been included.

FAUNA OF THE NATIONAL PARKS
OF THE UNITED STATES
A PRELIMINARY SURVEY OF
FAUNAL RELATIONS IN NATIONAL PARKS

By George M. Wright, Joseph S. Dixon, Ben H. Thompson
Contribution of Wild Life Survey
Fauna Series No. 1, May 1932

APPROACH TO WILD-LIFE ADMINISTRATION

The national parks of the United States began by rescuing from the immediate dangers of private exploitation certain areas which were climax examples of Nature's scenic achievements.

With rapid expansion of frontiers to the end that European culture not only replaced that of the red man but actually altered the physical appearance of his environment, the national parks were quickly projected into a larger sphere of purpose. This involved a magnificent conception. The American people intrusted the National Park Service with the preservation of characteristic portions of our country as it was seen by Boone and La Salle, by Coronado, and by Lewis and Clark. This was primitive America, and it was to be kept for the observation of the recreation-seeking public and scientists of to-day, and their descendants in the generations of to-morrow.

PLACE OF WILD LIFE IN AMERICA'S CIVILIZATION

Throughout the new land, the wild-life resources were a vital necessity to the explorer. He penetrated the wilderness for fur. This was the first great crop to be harvested. Wherever he went, he depended upon the game for his very existence. So it was inevitable that the history and tradition of our national life should be replete with references to the animals that occurred so profusely. Emblematically, we live among these birds and animals to-day. Witness the eagle and the buffalo of our coins, the antlered emblems of fraternal orders, the wild creatures depicted on State flags and seals. Daily speech bristles with descriptive words based on concepts of these animals and their habits.

Yet many species continue to exist in the living state only as small remnants hidden away in the wildest corners of the country, remote from the perils of human contact. Others have persisted through their ability to escape greedy human eyes. With the alteration of their habits to meet the rigors of civilization, many of them are no longer to be observed in their primitive state.

To all of this the national parks present one of the outstanding exceptions. In them, the carnivores classed as predatory find their only sure haven. Fur-bearers and game have benefit of partial protection elsewhere, but in the parks alone are they given opportunity to forget that man is the implacable enemy of their kind, so that they lose their fear and submit to close scrutiny.

The national parks owe much of their unique charm to the unusual opportunities they afford for observing animals amid the intimacies of wild settings in which even the observers feel themselves a part. It is one of the causes contributing to their constantly increasing popularity. The thrill of being in the same meadow with an elk, no fence or bar between, reaches everyone, young or old. Without the scurry and scratch of a chipmunk along the bark or the call of a jay and the flash of its blue, the high mountain and the deep gorge would be cold, dead indeed. The visitor would not linger after his first comprehensive gaze at awesome scenery if the vista did not include the intimate details of those living things, the plants, the animals that live on them, and the animals that live on those animals.

Appreciation of the importance that the wild life commands among the resources of the national parks rests upon comprehension of the important points developed above.

In logical sequence, these points are:

1.   That the wild life of America exists in the consciousness of the people as a vital part of their national heritage.

2.   That in its appointed task of preserving characteristic examples of primitive America, the National Park Service faces an especially important responsibility for the conservation of wild life. This is emphasized by the wholesale destruction which has decimated the fauna in nearly every part of the land outside of the park areas.

3.   That the observation of animals in the wild state contributes so much to the enjoyment derived by visitors that this is becoming a park attraction of steadily increasing rank.

CONSERVATION OF THE WILD-LIFE
RESOURCES OF THE PARKS

Recognition of these factors by those entrusted with the care of the parks led to intensification of the protective function until vandalism was wiped out, while poaching has been reduced to a minimum in all but a few parks where it, too, will be eliminated as conditions grow more favorable. But this part of conservation has not been enough. The need to supplement protection with more constructive wild-life management has become manifest with a steady increase of problems both as to number and intensity.

An early example of classic note is the bison of Yellowstone. The necessity of saving the bison left no alternative to intensive management. No one questioned the sacrifice of policy involved in the maintenance of the herd in a state that was only semiwild. It was either that or lose the great buffalo to this country except as Exhibit A in a zoological garden. Still, this case was an exception, and no one even for a moment considered that it established a precedent for dealing with other species.

The policy of noninterference with wild life became more and more deeply intrenched. Protection would do the rest. Nevertheless, time proved that management of some sort would have to be invoked to save certain situations, especially as the parks were opened to thousands of visitors, causing a flood of fresh complications.

The conclusion was unavoidable. Protection, far from being the magic touch which healed all wounds, was unconsciously just the first step on a long road winding through years of endeavor toward a goal too far to reach, yet always shining ahead as a magnificent ideal. This objective is to restore and perpetuate the fauna in its pristine state by combating the harmful effects of human influence.

The park faunas face immediate danger of losing their original character and composition unless the tide can be turned. The vital significance of wild life to the whole national-park idea emphasizes the necessity for prompt action. The logical course is a program of complete investigation, to be followed by appropriate administrative action.

The unique feature of the case is that perpetuation of national conditions will have to be forever reconciled with the presence of large numbers of people on the scene, a seeming anomaly. A situation of parallel circumstance has never existed before. Therefore, the solution can not be sought in precedent. It will challenge the conscientious and patient determination of biological engineers. And because of the nature of the task, it is inherently an inside job. Constancy to the objective can be made a certainty only by employment of a staff whose members are of the Service, conversant with its policies, and imbued with a devotion to its ideals.

HISTORY AND PROGRAM OF THE SURVEY

During service with the educational department in Yosemite National Park in 1928 and 1929, thoughts of this trend led one of the writers to the hope that something might be done towards concentrating greater interest on the fundamental aspects of wild-life administration throughout the national-park system. In collaboration with another of the co-authors, the general outline for a preliminary investigation was developed. Having received the sanction of the director, the idea then assumed concrete form under his guidance, becoming at once the Preliminary Wild Life Survey, with headquarters at Berkeley, California.

Personnel included Joseph S. Dixon, economic mammalogist; George M. Wright, scientific aide in the National Park Service; Ben H. Thompson, research associate; and Mrs. George Pease, secretary. All expense of the survey, inclusive of office, field, and salaries, was met with private funds until July 1, 1931. Since then it has been supported about equally from public appropriation and private contribution.

Stated objectives were:

1. To focus attention upon the need for a well-defined wild-life policy of the National Park Service, including the extension of the protective function to embody a definitely constructive program.

2. To assist park superintendents in dealing with the urgent animal problems immediately confronting them.

3. To present a report which would delineate the existing status of wild life in the parks, analyze unsatisfactory conditions, and outline a proposed plan for the orderly development of wild-life management.

The present paper is intended to fulfill the third objective. In the meantime certain results have been attained toward the accomplishment of the first and second purposes and may be mentioned briefly here.

Through the existence of a group actively functioning in the field, the director was provided with arguments and data and with a living organization for which he could solicit support to assure perpetuation. This has facilitated the securing of one permanent position of field naturalist and appropriations to cover the field expenses of the staff. Office quarters were provided in conjunction with field offices of the Branch of Research and Education of the National Park Service in Hilgard Hall, University of California, Berkeley, Calif. Thus the foundations of a wild-life organization as a continuing function of the Service have already been laid.

The photographic collection, numbering 2,523 negatives and accompanied by prints all filed and indexed in readily available form, is proving useful to the educational program. Though its prime purpose is for scientific record, the service rendered in this other field increases its value measurably.

Field notes on 279 species of birds and mammals have been systematically recorded. In part they have provided data for the report, but the whole of them remains an important reservoir of classified information for the intensive biological studies which should follow the first general investigation.

Nevertheless, throughout the preliminary survey, fixity to the main purpose of obtaining a perspective of the problem in its entirety has been the paramount consideration. Consequently, the search focused on the general trends in the status of animal life, with particular regard to the motivating factors. If a finger can be placed on the mainsprings of disorder, there is hope of discovering solutions that will be adequate in result. Meeting existing difficulties with superficial cures might be temporarily expedient and, in cases of emergency, necessary but if continued would build up a costly patchwork that must eventually give out. It would be analogous to placing a catch-basin under a gradually growing leak in a trough and then trying to keep the trough replenished by pouring the water back in. The task mounts constantly and failure is the inevitable outcome. The only hope rests in restoration of the original vessel to wholeness. And so it is with the wild life of the parks. Unless the sources of disruption can be traced and eradicated, the wild life will ebb away to the level occupied by the fauna of the country at large. Admitting the magnitude of the task, it still seems worth the undertaking, for failure here means failure to maintain a characteristic of the national parks that must continue to exist if they are to preserve their distinguishing attribute. Such failure would be a blow injuring the very heart of the national-park system.

The field studies were conducted in accordance with this point of view. Findings as presented in this report are calculated to lay a foundation of approach and practice useful in dealing with wild-life problems of all categories wherever or whenever they may occur, rather than to stand as an enumeration of a lengthy list of individual problems without correlation. However, numerous examples have been given detailed treatment in the development of the arguments, and in a separate section all problems met with in each area have been enumerated in order to record for future reference the situations obtaining in the parks at the time of the investigation, as seen by the field party.

In the following pages the technique developed for the preliminary survey is detailed for the benefit of others who may undertake similar projects and for the usefulness it will have as a skeletal outline for the intensive studies in each park later on. Further, this account of the methods employed will facilitate a critical evaluation of the results.

SUGGESTED NATIONAL-PARK POLICY
FOR THE VERTEBRATES

Every tenet covering the vertebrate life in particular must be governed by the same creed which underlies administration of wild life in general throughout the national parks system, namely:

That one function of the national parks shall be to preserve the flora and fauna in the primitive state and, at the same time, to provide the people with maximum opportunity for the observation thereof.

In the present state of knowledge, and until further investigations make revision advisable, it is believed that the following policies will best serve this dual objective as applied to the vertebrate land fauna. Without further comment, inasmuch as the supporting reasons have been developed in preceding sections, it is proposed:

Relative to areas and boundaries—

1.   That each park shall contain within itself the year-round habitats of all species belonging to the native resident fauna.

2.   That each park shall include sufficient areas in all these required habitats to maintain at least the minimum population of each species necessary to insure its perpetuation.

3.   That park boundaries shall be drafted to follow natural faunal barriers, the limiting faunal zone, where possible.

4.   That a complete report upon a new park project shall include a survey of the fauna as a critical factor in determining area and boundaries.

Relative to management—

5.   That no management measure or other interference with biotic relationships shall be undertaken prior to properly conducted investigation.

6.   That every species shall be left to carry on its struggle for existence unaided, as being to its greatest ultimate good, unless there is real cause to believe that it will perish if unassisted.

7.   That, where artificial feeding, control of natural enemies, or other protective measures, are necessary to save a species that is unable to cope with civilization's influence, every effort shall be made to place that species on a self-sustaining basis once more; whence these artificial aids, which themselves have unfortunate consequences, will no longer be needed.

8.   That the rare predators shall be considered charges of the national parks in proportion that they are persecuted everywhere else.

9.   That no native predator shall be destroyed on account of its normal utilization of any other park animal, excepting if that animal is in immediate danger of extermination, and then only if the predator is not itself a vanishing form.

10.   That species predatory upon fish shall be allowed to continue in normal numbers and to share normally in the benefits of fish culture.

11.   That the numbers of native ungulates occupying a deteri-

orated range shall not be permitted to exceed its reduced carrying capacity and, preferably, shall be kept below the carrying capacity at every step until the range can be brought back to original productiveness.

12. That any native species which has been exterminated from the park area shall be brought back if this can be done, but if said species has become extinct no related form shall be considered as a candidate for reintroduction in its place.

13. That any exotic species which has already become established in a park shall be either eliminated or held to a minimum provided complete eradication is not feasible.

14. That the threatening invasion of the parks by other exotics shall be anticipated; and to this end, since it is more than a local problem, encouragement shall be given for national and State cooperation in the creation of a board which will regulate the transplanting of all wild species.

Relative relations between animals and visitors—

15. That presentation of the animal life of the parks to the public shall be a wholly natural one.

16. That no animal shall be encouraged to become dependent upon man for its support.

17. That problems of injury to the persons of visitors or to their property or to the special interests of man in the park, shall be solved by methods other than those involving the killing of the animals or interfering with their normal relationships, where this is at all practicable.

Relative faunal investigations—

18. That a complete faunal investigation, including the four steps of determining the primitive faunal picture, tracing the history of human influences, making a thorough survey and formulating a wild-life administrative plan shall be made in each park at the earliest possible date.

19. That the local park museum in each case shall be repository for a complete study skin collection of the area and for accumulated evidence attesting to original wild-life conditions.

20. That each park shall develop within the ranger department a personnel of one or more men trained in the handling of wild-life problems, and who will be assisted by the field staff appointed to carry out the faunal program of the Service.

Washington, D.C.: GPO, 1933, 1–7, 147–148.

3

The New Deal Years

1933–1941

Through the 1920s, the young National Park Service cemented its support among the public, defined its management priorities, and established the philosophy that was to guide the agency for the next four decades. However, the steady sequence of development and expansion was shattered by the stock market crash of 1929 and ensuing Depression. Ironically, although it did have a negative effect on visitation, the economic crisis actually spawned the greatest booms in construction of visitor facilities, road and trail development, park planning, identification of new areas, and new initiatives for expansion of the system to ever occur.

With the arrival of the new administration under Franklin Roosevelt came a whirlwind of government changes. In that first year, 1933, two pieces of legislation profoundly impacted the parks. First came an act to relieve unemployment among the nation's young men. This act led to the Civilian Conservation Corps (CCC), a military-style program of public works conducted in large measure in and around national and state parks and forests. Over the next nine years the CCC constructed more infrastructure in the parks than in the entire history of the system to that time. Five-year plans were completed in a season and park superintendents, in charge of choosing jobs for these brigades of laborers, were able to achieve development only dreamed of a few years earlier.

The second action came with congressional attempts to streamline and reduce the costs of the federal government. The executive branch reorganized its departments and their duties. One result was that the National Park Service received all the national monuments, many from the U.S. Forest Service, and all the battlefields and

111

memorials from the War Department. Thus the system with which we are now familiar began operation under a single agency.

Over the remainder of the Depression years through 1941, three major processes characterized the national park system—the continued definition of natural resources policy, the development of infrastructure and resulting challenges to perceived overdevelopment, and the accelerated expansion of the system into new areas of preservation. In 1933 the spate of policy statements provided by Horace Albright a couple of years earlier rounded out with an article by the director on the importance of research in the national parks. It was a tacit admission of the inadequacy of scientific data for management purposes and the first of a long string of calls for more systematic research in park policy-making.

Three years later a loophole in the philosophical framework of fauna management was closed with issuance of Office Order No. 323 defining the fish policies of the National Park Service. Director Arno Cammerer, following the lead of his Wildlife Division, ordered the exclusion or removal of exotic species and the encouragement of native ones as his predecessor had done in the case of mammals.

While natural resource policies matured and their scientific basis was at least temporarily encouraged, the construction of visitor infrastructure continued apace with CCC labor. Hundreds of miles of roads and trails opened hitherto wild backcountry and thousands of structures, from museums to employee houses and campground comfort stations, appeared across the system. By the middle of the decade conservationists and others were registering alarm at the scale and pace of development. The Emergency Conservation Committee released in 1936 a pamphlet entitled "Roads and More Roads in the National Parks and National Forests," in which this progress was challenged on the basis of its destruction of pristine wilderness. The author of the pamphlet and driving force behind the committee was Rosalie Edge. Her views were repeated in the publications of the Sierra Club, the National Parks Association, and other conservation groups. Clearly the time had come to challenge the Park Service's priorities, and increasing numbers of observers were ready to do so.

Within the agency as well, a variety of opinions existed on appropriate development. However, most NPS personnel practiced a philosophy which might best be called "atmosphere preservation." Spawned and maintained by landscape architects, it was an immature holistic approach to the environment that considered entire visual and experiential scenes and the inspiration they provided as the highest preservation targets. This philosophy and its attendant reaction to the extraordinary development under the CCC was best expressed by Superintendent John White in a 1936 speech to his fellow super-intendents entitled "Atmosphere in the National Parks."

The final process of the period concerned the expansion of the park system. With reorganization of the executive branch, the National Park Service became much more focused on historic preservation. At the same time concern about preserving the nation's historic structures and sites led Congress to enact the Historic Sites Act of 1935. This far-reaching act ordered the NPS to survey historic and archaeologic sites, buildings, and objects, and take steps to identify and protect them or assist others in protecting them.

The following year, a decade of discussion and concern about the availability of recreation resources for the nation's burgeoning population surfaced in the form of the Park, Parkway, and Recreation Area Study Act of 1936. This charged the Park Service to conduct wide-ranging studies of potential additions to the park system specifically to provide recreation opportunities. Given the philosophy of the time on park purpose, this was a departure into new areas of operation for the agency. Much of the ensuing study concentrated on coastal areas and resulted the following year in the authorization of Cape Hatteras National Seashore. In 1941 the Park Service released its report on the park and recreation problem of the United States. Although the report strongly recommended that state and local governments bear the brunt of responsibility for recreation, reflecting the Mather philosophy, many of the current national seashores, lakeshores, parkways, and recreation areas were studied and proposed as a result of this 1936 act.

AN ACT FOR THE RELIEF OF UNEMPLOYMENT THROUGH THE PERFORMANCE OF USEFUL PUBLIC WORK, AND FOR OTHER PURPOSES,

Approved March 31, 1933 (48 Stat. 22)

Be it enacted by the Senate and House of Representatives of the United States of America in Congress assembled, That for the purpose of relieving the acute condition of widespread distress and unemployment now existing in the United States, and in order to provide for the restoration of the country's depleted natural resources and the advancement of an orderly program of useful public works, the President is authorized, under such rules and regulations as he may prescribe and by utilizing such existing departments or agencies as he may designate, to provide for employing citizens of the United States who are unemployed, in the construction, maintenance and carrying on of works of a public nature in connection with the forestation of lands belonging to the United States or to the several States which are suitable for timber production, the prevention of forest fires, floods and soil erosion, plant pest and disease control, the construction, maintenance or repair of paths, trails and firelanes in the national parks and national forests, and other work on the public domain, national and State, and Government reservations incidental to or necessary in connection with any projects of the character enumerated, as the President may determine to be desirable: *Provided,* That the President may in his discretion extend the provisions of this Act to lands owned by counties and municipalities and lands in private ownership, but only for the purpose of doing thereon such kinds of cooperative work as are now provided for by Acts of Congress in preventing and controlling forest fires and the attacks of forest tree pests and diseases and such work as is necessary in the public interest to control floods. The President is further authorized, by regulation, to provide for housing the persons so employed and for furnishing them with such subsistence, clothing, medical attendance and hospitalization, and cash allowance, as may be necessary, during the period they are so employed, and, in his discretion, to provide for the transportation of such persons to and from the places of employment. That in employing citizens for the purposes of this Act no discrimination shall be made on account of race, color, or creed; and no person under conviction for crime and serving sentence therefor shall be employed under the provisions of this Act. The President is further authorized to allocate funds available for the purposes of this Act, for forest research, including forest products investigations, by the Forest Products Laboratory.

SEC. 2. For the purpose of carrying out the provisions of this Act the President is authorized to enter into such contracts or agreements with States as may be necessary, including provisions for utilization of existing State administrative agencies, and the President, or the head of any department or agency authorized by him to construct any project or to carry on any such public works, shall be authorized to acquire real property by purchase, donation, condemnation, or otherwise, but the provisions of section 355 of the Revised Statutes shall not apply to any property so acquired.

SEC. 3. Insofar as applicable, the benefits of the Act entitled "An Act to provide compensation for employees of the United States suffering injuries while in the performance of their duties, and for other purposes," approved September 7, 1916, as amended, shall extend to persons given employment under the provisions of this Act.

SEC. 4. For the purpose of carrying out the provisions of this Act, there is hereby authorized to be expended, under the direction of the President, out of any unobligated moneys heretofore appropriated for public works (except for projects on which actual construction has been commenced or may be commenced within ninety days, and except maintenance funds for river and harbor improvements already allocated), such sums as may be necessary; and an amount equal to the amount so expended is hereby authorized to be appropriated for the same purposes for which such moneys were originally appropriated.

SEC. 5. That the unexpended and unallotted balance of the sum of $300,000,000 made available under the terms and conditions of the Act approved July 21, 1932, entitled "An Act to relieve destitution," and so forth, may be made available, or any portion thereof, to any State or Territory or States or Territories without regard to the limitation of 15 per centum or other limitations as to per centum.

SEC. 6. The authority of the President under this Act shall continue for the period of two years next after the date of the passage hereof and no longer.

Approved, March 31st 1933.

16 U.S.C. 585–590.

EXCERPTS FROM EXECUTIVE ORDER NO. 6166
OF JUNE 10, 1933 AND EXECUTIVE ORDER
NO. 6228 OF JULY 28, 1933 (5 U.S.C. SECS. 124–132)

Executive Order No. 6166:

ORGANIZATION OF EXECUTIVE AGENCIES

WHEREAS section 16 of the act of March 3, 1933 (Public, No. 428. 47 Stat. 1517), provides for reorganizations within the executive branch of the Government; requires the President to investigate and determine what reorganizations are necessary to effectuate the purposes of the statute; and authorizes the President to make such reorganizations by Executive order; and

WHEREAS I have investigated the organization of all executive and administrative agencies of the Government and have determined that certain regroupings, consolidations, transfers, and abolitions of executive agencies and functions thereof are necessary to accomplish the purposes of section 16;

NOW, THEREFORE, by virtue of the aforesaid authority, I do hereby order that:

SECTION 2.—*National Parks, Buildings, and Reservations*

All functions of administration of public buildings, reservations, national parks, national monuments, and national cemeteries are consolidated in an Office of National Parks, Buildings, and Reservations in the Department of the Interior, at the head of which shall be a Director of National Parks, Buildings, and Reservations; except that where deemed desirable there may be excluded from this provision any public building or reservation which is chiefly employed as a facility in the work of a particular agency. This transfer and consolidation of functions shall include, among others, those of the National Park Service of the Department of the Interior and the National Cemeteries and Parks of the War Department which are located within the continental limits of the United States. National cemeteries located in foreign countries shall be transferred to the Department of State, and those located in insular possessions under the jurisdiction of the War Department shall be administered by the Bureau of Insular Affairs of the War Department.

The functions of the following agencies are transferred to the Office of National Parks, Buildings, and Reservations of the Department of the Interior, and the agencies are abolished:

Arlington Memorial Bridge Commission
Public Buildings Commission
Public Buildings and Public Parks of the National Capital
National Memorial Commission
Rock Creek and Potomac Parkway Commission
Expenditures by the Federal Government for the purposes of the Commission of Fine Arts, the George Rogers Clark Sesquicentennial Commission, and the Rushmore National Commission shall be administered by the Department of the Interior.

SECTION 19.—*General Provisions*

Each agency, all the functions of which are transferred to or consolidated with another agency, is abolished.

The records pertaining to an abolished agency or a function disposed of, disposition of which is not elsewhere herein provided for, shall be transferred to the successor. If there be no successor agency, and such abolished agency be within a department, said records shall be disposed of as the head of such department may direct.

The property, facilities, equipment, and supplies employed in the work of an abolished agency or the exercise of a function disposed of, disposition of which is not elsewhere herein provided for, shall, to the extent required, be transferred to the successor agency. Other such property, facilities, equipment, and supplies shall be transferred to the Procurement Division.

All personnel employed in connection with the work of an abolished agency or function disposed of shall be separated from the service of the United States, except that the head of any successor agency, subject to my approval, may, within a period of four months after transfer or consolidation, reappoint any of such personnel required for the work of the successor agency without reexamination or loss of civil-service status.

SECTION 20.—*Appropriations*

Such portions of the unexpended balances of appropriations for any abolished agency or function disposed of shall be transferred to the successor agency as the Director of the Budget shall deem necessary.

Unexpended balances of appropriations for an abolished agency or function disposed of, not so transferred by the Director of the Budget, shall, in accordance with law, be impounded and returned to the Treasury.

SECTION 21.—*Definitions*

As used in this order—

"Agency" means any commission, independent establishment, board, bureau, division, service, or office in the executive branch of the Government.

"Abolished agency" means any agency which is abolished, transferred, or consolidated.

"Successor agency" means any agency to which is transferred some other agency or function, or which results from the consolidation of other agencies or functions.

"Function disposed of" means any function eliminated or transferred.

SECTION 22.—*Effective Date*

In accordance with law, this order shall become effective 61 days from this date; *Provided,* That in case it shall appear to the President that the interests of economy require that any transfer, consolidation, or elimination be delayed beyond the date this order becomes effective, he may, in his discretion, fix a later date therefor, and he may for like cause further defer such date from time to time.

FRANKLIN D. ROOSEVELT.

THE WHITE HOUSE,
June 10, 1933.

Executive Order No. 6228:

ORGANIZATION OF EXECUTIVE AGENCIES

WHEREAS executive order No. 6166 dated June 10, 1933, issued pursuant to the authority of Section 16 of the Act of March 3, 1933 (Public No. 428—47 Stat. 1517) provides in Section 2 as follows:

"All functions of administration of public buildings, reservations, national parks, national monuments, and national cemeteries are consolidated in an office of National Parks, Buildings, and Reservations in the Department of the Interior, at the head of which shall be a Director of National Parks, Buildings, and Reservations; except that where deemed desirable there may be excluded from this provision any public building or reservation which is chiefly employed as a facility in the work of a particular agency. This transfer and consolidation of functions shall include, among others, those of the

National Park Service of the Department of the Interior and the National Cemeteries and Parks of the War Department which are located within the continental limits of the United States. National Cemeteries located in foreign countries shall be transferred to the Department of State, and those located in insular possessions under the jurisdiction of the War Department shall be administered by the Bureau of Insular Affairs of the War Department."
and;

WHEREAS to facilitate and expedite the transfer and consolidation of certain units and agencies contemplated thereby, it is desirable to make more explicit said Section 2 of the aforesaid executive order of June 10, 1933, insofar as the same relates to the transfer of agencies now administered by the War Department:

NOW, THEREFORE, said executive order No. 6166, dated June 10, 1933, is hereby interpreted as follows:

1. The cemeteries and parks of the War Department transferred to the Interior Department are as follows:

NATIONAL MILITARY PARKS

Chickamauga and Chattanooga National Military Park, Georgia and
   Tennessee.
Fort Donelson National Military Park, Tennessee.
Fredericksburg and Spotsylvania County Battle Fields Memorial,
   Virginia.
Gettysburg National Military Park, Pennsylvania.
Guilford Courthouse National Military Park, North Carolina.
Kings Mountain National Military Park, South Carolina.
Moores Creek National Military Park, North Carolina.
Petersburg National Military Park, Virginia.
Shiloh National Military Park, Tennessee.
Stones River National Military Park, Tennessee.
Vicksburg National Military Park, Mississippi.

NATIONAL PARKS

Abraham Lincoln National Park, Kentucky.
Fort McHenry National Park, Maryland.

BATTLEFIELD SITES

Antietam Battlefield, Maryland.
Appomattox, Virginia.
Brices Cross Roads, Mississippi.
Chalmette Monument and Grounds, Louisiana.

Cowpens, South Carolina.
Fort Necessity, Wharton County, Pennsylvania.
Kenesaw Mountain, Georgia.
Monocacy, Maryland.
Tupelo, Mississippi.
White Plains, New York.

NATIONAL MONUMENTS

Big Hole Battlefield, Beaverhead County, Montana.
Cabrillo Monument, Ft. Rosecrans, California.
Castle Pinckney, Charleston, South Carolina.
Father Millet Cross, Fort Niagara, New York.
Fort Marion, St. Augustine, Florida.
Fort Matanzas, Florida.
Fort Pulaski, Georgia.
Meriwether Lewis, Hardin County, Tennessee.
Mound City Group, Chillicothe, Ohio.
Statue of Liberty, Fort Wood, New York.

MISCELLANEOUS MEMORIALS

Camp Blount Tablets, Lincoln County, Tennessee.
Kill Devil Hill Monument, Kitty Hawk, North Carolina.
New Echota Marker, Georgia.
Lee Mansion, Arlington National Cemetery, Virginia.
Battleground, District of Columbia.
Antietam, (Sharpsburg) Maryland.
Vicksburg, Mississippi.
Gettysburg, Pennsylvania.
Chattanooga, Tennessee.
Fort Donelson, (Dover) Tennessee.
Shiloh, (Pittsburg Landing) Tennessee.
Stones River, (Murfreesboro) Tennessee.
Fredericksburg, Virginia.
Poplar Grove, (Petersburg) Virginia.
Yorktown, Virginia.

    2.   Pursuant to Section 22 of said executive order it is hereby ordered that the transfer from the War Department of national cemeteries other than those named above be, and the same is hereby postponed until further order.

    3.   Also pursuant to Section 22 of said executive order it is hereby ordered that the transfer of national cemeteries located in foreign

countries from the War Department to the Department of State and the transfer of those located in insular possessions under the jurisdiction of the War Department to the Bureau of Insular Affairs of said Department be, and the same are hereby postponed until further order.

FRANKLIN D. ROOSEVELT.

THE WHITE HOUSE,
July 28, 1933.

5 U.S.C. 124–132.

RESEARCH IN THE NATIONAL PARKS

By Horace M. Albright
June 1933

Being equipped by nature with the most complete and magnificent laboratories imaginable, it was inevitable that scientific research should become an important and popular activity of the National Park Service. Nevertheless, it is one of the newest developments in national park work, which is primarily of a human welfare nature.

National parks began back in 1872, with the establishment of the Yellowstone National Park, the first reservation of its kind to be established anywhere. At the time of its establishment, of course, no thought was given to the scientific aspects of the geysers and other natural phenomena, yet it was because of their presence that the explorers of the Washburn-Langford-Doane party conceived the idea of a national park and gained the support to put this idea through Congress.

The organic act establishing the park provides that the area be "set apart as a public park or pleasuring ground for the benefit of people," and further that regulations be enacted by the Secretary of the Interior "for the preservation from injury or spoilation of all timber, mineral deposits, natural curiosities or wonders within the park, and their retention in their natural condition."

On this foundation has grown up the great national park and monument system that to-day contains 22 national parks and 40 national monuments under the jurisdiction of the National Park Service of the Department of the Interior.

As park after park succeeded the Yellowstone, each was founded upon principles of human welfare, upon the idea of public ownership in and enjoyment of the parks. Yet the underlying motive in establishing each park for the benefit of the people was to preserve something precious from a special standpoint which, when analyzed, proved to be based upon some natural phenomenon or other object of interest to scientists or historians.

Thus the Yosemite, paradise of beauty, also is a geologists' paradise. Some of the Big Trees, to preserve which Sequoia and General Grant National Parks were set aside, were young in the days of the Pharaohs. Mount Rainier, the next to be established as a national park, contains the greatest single peak glacier system in the United States—in addition to exquisite wild flower fields and other features of impressive beauty. So throughout the system.

When the nineteenth century closed, these five national parks constituted the national park system.

The first decade of the twentieth century brought more of Nature's interesting laboratories into the system. There came Crater Lake, a lake of exquisite blue in the crater of a volcano that collapsed or blew its head to bits sometime in the misty past; Platt, with hot springs possessing healing properties; Wind Cave, with unusual natural decorations, and a strangely acting wind which blew in or out, apparently without rhyme or reason; Mesa Verde, ancient home of Basket-maker, cliff dweller, and Pueblo Indians, with a mysterious past that stirs the imagination of the ethnologist and archeologist, as well as that of the average layman; and Glacier National Park, that upturned section of the Rocky Mountains where ancient sedimentary rocks rest upon much younger strata, carved and scarified by great ice sheets and still holding in its mountain fastness the remains of sixty small glaciers.

The same year that Mesa Verde National Park was created Congress broadened the system by passing what is known as the "Antiquities Act." This legislation provided for the establishment by Presidential proclamation, of national monuments of areas containing objects of historic, prehistoric or scientific interest.

As the national value of these parks and monuments became more apparent, the system grew steadily. The creation of Rocky Mountain National Park, including a typical area of the Rocky Mountains, was followed by two volcanic areas, showing a spectacular form of plastic surgery on the face of Old Mother Nature. One of these, Lassen Volcanic, contains our most recently active volcano on the mainland, and the Hawaii National Park, in addition to its vast dormant crater large enough to hold a modern city, also has two living volcanoes that periodically provide breathtaking displays of great beauty and sublimity.

Another far-away park, Mount McKinley in Alaska, contains the highest mountain on North America, snow-shrouded throughout the year. It affords remarkable opportunities for study of glaciers.

Three superb canyon parks in the Southwest, the Grand Canyon, Zion, and Bryce Canyon, show the wearing, tearing effects of water. Great granite mountains, glacier-laden, are the contribution of the Grand Teton National Park to the system. The huge chambers of Carlsbad Caverns National Park also attest to the dissolving, sculpturing powers of water, and the Hot Springs National Park also owes its place in the system to water—but in this case to medicinal waters, believed, ever since the days of the early Indians, to have definite healing powers.

Formerly a western institution, of recent years the National Park System has moved to the East. The Acadia National Park on the Maine Coast has ancient granite mountains that were old when the West was young; and the Great Smoky Mountains National Park, in addition to its hoary peaks, the highest mountain massing in the East, is famous for the variety and luxuriance of its flora.

The national monuments under the National Park System, established under the authority of the Antiquities Act, cover a wide range of objects. There are fossil plants, petrified trees, and the bones of the long-extinct dinosaur; cliff-dweller ruins and surface pueblos of long-vanished peoples; places connected with the lives of the first white men to settle in America and of early Colonial life; Revolutionary War Shrines; ruined churches erected by the padres who accompanied the gaily adventuring Spanish cavaliers to the New World; a fort built by the serious, patient Mormons—a wealth of areas of such scientific and human interest that their preservation is important to the advancement of our national culture.

Naturally, when the system was young, its first needs, like those of the young human, were protection and proper direction—or administration in the case of the parks. Protective or police organizations were first needed, then means to house the protective force and to care for the physical needs of the visiting public, in reality the parks' non-resident owners.

Once these elementary matters were well taken care of, the National Park Service turned to the aesthetic, or "higher educational" side of the parks.

Interpretation of their natural features followed. Why geysers "gyze" is perhaps the question asked most in the Yellowstone. In Yosemite, upon seeing Half Dome, visitors want to know what became of the other half—and with the opportunity thus afforded for tactfully imparting a little scientific information the educational work goes on apace, apparently casually, but always based upon careful research.

The guided field trips and popularly worded lectures of today, as well as the museum service, grew out of this demand of the visitors to know the "why" of the interesting phenomena—although most of them do not call it that—encountered along the way.

Museum work, in the parks particularly, is quite specialized. The museums are so arrayed as to give the observer a glimpse of the interesting things to be found out in the parks themselves—to interest him to see for himself what the museum suggests. In other words, the museum exhibits are only the indices to the real museum, which is the park.

In the historic and prehistoric members of the system, of course, the museums serve a different purpose. There they actually display relics of human lives—in the former, of our pioneer forbears; in the latter, of a vanished, almost unknown race. A prehistoric burial place yields a skeleton and a few trinkets; a plastered-up cache high in a cliff, when opened is found to contain pottery or basketry; here there is a grinding stone and there a weapon of the chase. These all are studied and gradually some idea of the lives of the prehistoric peoples takes shape. This is one of the most fascinating phases of the research work of the National Park Service.

Research is necessary not only to the preparation of interesting material to serve as a basis of the naturalist and historical service, but it also is fundamental to the actual protection of the natural features of the parks, as enjoined in the acts establishing the parks and in the act of August 25, 1916, creating the National Park Service. The latter act contains the following clause:

> The service thus established shall promote and regulate the use of the Federal areas known as national parks, monuments, and reservations hereinafter specified by such means and measures as conform to the fundamental purpose of the said parks, monuments, and reservations, which purpose is to conserve the scenery and the natural and historic objects and the wild life therein and to provide for the enjoyment of the same in such manner and by such means as will leave them unimpaired for the enjoyment of future generations.

There are a great variety of natural objects in the national parks. There are the wild animals, objects of intense interest to visitors, who can not see elsewhere such a wide variety of species and numbers as in these areas, since only in the national parks and national monuments are they given complete protection. The plant life, both tree and wild flower, also makes a tremendous appeal to the average visitor, for one can imagine nothing lovelier than the fields of wild flowers that carpet most of the parks during the spring and early summer. Then there are the natural scenic, scientific, and historic features, the main object of the parks' establishment. All of these natural objects need protection, and in many cases research is necessary to determine the cause of some suddenly-appearing adverse condition.

Formerly protection of the wild life was primarily a protective function, involving long ski patrols in the winter to afford protection against poachers, both hunters and trappers, and the occasional supplying of food in emergencies. Also in the Yellowstone there has been for a number of years the winter care of the buffalo herd, numbering over a thousand.

Following fundamental protection came the restocking of certain depleted natural ranges. Before going farther with this particular subject, it is important to emphasize that the policy of the National Park Service is unalterably against the introduction of exotic species of animals or plants in the national parks or national monuments, except for the occasional stocking of an otherwise barren body of water with some species of game fish for the enjoyment of lovers of the Waltonian sport. Wherever animals are introduced, it is to restock a natural range which has become depleted because of some unnatural condition or series of conditions.

Prominent among the restocking experiments are those of the bison—more generally called buffalo—in the Yellowstone, and the antelope at the Grand Canyon. Yellowstone National Park, one of the great areas ranged by buffalo in their wild state, suffered from a depletion in the herds of these animals almost to the point of extinction. A few new animals, specially selected from Texas and Montana herds, were introduced into the park, intensive management undertaken, and today the bison herd numbers over a thousand and could be much larger were the range sufficient to support a greater number.

In the Grand Canyon an interesting restocking effort with antelope is just passing out of the experimental stage. At one time these plains antelope were plentiful at the Canyon but changing conditions—possibly caused largely by the wide spreads of the descendants of hardy burros left in the Canyon by prospectors lured to other fields—brought about their disappearance from their former range. In 1924, twelve antelope kids, six bucks and six does were taken to the Canyon, fed and kept under close observation for some time, then released on the Tonto Platform, where it was hoped they would thrive and multiply. For several years prospects looked bad for the survival of these antelope, as they did not easily adapt themselves to new conditions and, possibly because of their careful raising by hand, easily became a prey to predatory animals. After five years of fighting against odds, by the end of 1929 the herd included only nine animals, four of them kids. During the past year conditions have materially improved, however, and there are twenty animals in the herd. Of ten kids born last spring, eight have survived. The outlook now is favorable for the building up of a large herd which it is hoped can be drawn upon a few years hence for the stocking of other natural antelope ranges in the national park and monument system.

Another interesting experiment at the Grand Canyon has been the transportation of deer from the North Rim across to the South Rim. At first these animals were transported across the Canyon by truck over a long detour covering a distance of 240 miles and requiring from twenty-four to thirty hours to make the trip. Later, for several years, young fawns were transported by a combination airplane and truck trip which was made in three hours. Introduction of these animals to the South Rim, and enlargement of the semi-tame herd over a period of five or six years, has been the means of presenting the public with a highly interesting feature of wild life. In addition, this herd has attracted other deer from regions adjacent to the park, thus increasing the herd to an estimated total of 1,200 head.

Of recent years it has become evident that ranger protection and restocking are not sufficient for the complete preservation of the wild animals. While in the parks it is true that the animals live as nearly as possible under primitive conditions, civilization comes close to the park boundaries, modifying the wilderness conditions; the animals wander back and forth across the boundaries, often coming in contact with domesticated animals, and thus meet vastly different conditions to those experienced by their ancestors back in the middle nineteenth century. Because of this, many situations have arisen necessitating scientific study.

Again, to mention the Yellowstone buffalo, an epidemic broke out several years ago which threatened the decimation of the herd. Experts of the Bureau of Animal Industry were called upon and studies made of the disease. It was diagnosed as hemorrhagic septicemia, and the buffalo were vaccinated against it—not an easy task, as any one who has ever seem these enormous animals in stampede will realize. Today the herd is thriving—so much so that animals have to be given away each year to keep the number of bison in the park down to the number the range can support satisfactorily.

So one study has followed another, to relieve emergency conditions, and in all such cases the National Park Service has received the unstinted cooperation of the Biological Survey and the Bureau of Animal Industry.

It has become increasingly evident, however, that a management plan of some sort must be inaugurated by the National Park Service, in order to restore and keep the park wild life in its primitive state despite the effects of human influence. This necessitates, first of all, complete investigation.

Realizing the need of this, an outline of wild life studies was prepared and work along this line undertaken in 1929 with funds made available by George M. Wright, who had become interested in the problem while serving with the educational department in Yosemite National Park. Joseph S. Dixon, economic mammalogist and scientist connected with the Museum of Vertebrate Zoology, was persuaded to assist in this work and he, Mr. Wright, and Ben H. Thompson have carried on the studies with increasing interest and vigor. Since 1931 the National Park Service fortunately has been able to assist in financing this work, and during the coming fiscal year will take over practically all the expense.

In their first printed report on the result of their studies, the members of the Wild Life Studies Group report as follows:

> . . . throughout the preliminary survey, fixity to the main purpose of obtaining a perspective of the problem in its entirety has been the paramount consideration. Consequently, the search focused on the general trends in the status of animal life, with particular regard to the motivating factors. If a finger can be placed on the mainsprings of disorder, there is hope of discovering solutions that will be adequate in result. Meeting existing difficulties with superficial cures might be temporarily expedient and, in cases of emergency, necessary, but if continued would build up a costly patchwork that must eventually give out. It would be analogous to placing a catch-basin under a gradually growing leak in a trough and then trying to keep the trough replenished by pouring the water back in. The task mounts constantly and failure is the inevitable outcome. The only hope rests in restoration of the original vessel to wholeness. And so it is with the wild life of the parks. Unless the sources of disruption can be traced and eradicated, the wild life will ebb away to the level occupied by the fauna of the country at large. Admitting the magnitude of the task, it still seems worth the undertaking, for failure here means failure to maintain a characteristic of the national parks that must continue to exist if they are to preserve their distinguishing attribute. Such failure would be a blow injuring the very heart of the national-park system.

On of the most interesting studies undertaken by this group is in connection with the trumpeter swan, one of the birds of present-day America that appears to be fast approaching the end of its journey to join the dodo in the limbo of forgotten things. It has been found that in the Yellowstone region these birds are making a last stand, and the Wild Life Division, with the cooperation of Yellowstone National Park officials, is bending every effort toward affording the necessary conditions in the park to permit the rehabilitation, if one may call it that, of this magnificent species of bird. Reports now indicate that the possibility is good for giving this species a new lease on life, just as was done in the case of the buffalo.

Typical of wild-life research of a cooperative nature has been the study of the Yellowstone elk by William Rush, an investigation initiated by the writer when superintendent of Yellowstone National Park, and later supported jointly by the Forest Service, the Biological Survey, the Montana Fish and Game Commission, and the National Park Service.

Plant life problems, while perhaps not as pressing as those pertaining to the wild animals, are equally important. Forest fires present a constant potential menace to the trees, but improvement methods of fire prevention and combat are handling this problem excellently.

Other enemies of park forests are insect infestations and tree diseases. Just as in the case of the wild animals, changing conditions outside park boundaries affect the trees inside. Insect devastations generally start outside the parks, from there encroaching on the trees inside.

Recent surveys show several serious forest situations prevailing in the national parks. One of the worst occurs in the Yellowstone, where the mountain pine beetle threatens the destruction of the lodgepole pine that constitutes about eighty per cent of the park forest. This epidemic has been carefully studied by experts of the Bureau of Entomology, as well as by Park Service men and officials of the adjoining national forests. It appears that there would be perhaps a fifty-fifty chance of saving these lodgepole pines if a five-year program of control could be undertaken immediately, at a probable cost of from $3,000,000 to $5,000,000.

This matter was discussed with the Appropriations Committee of the House of Representatives over a year ago. At that time it was decided that such an enormous expenditure was not justified, particularly as there is no definite assurance that even with such appropriations could the ravages of the infestation be stopped.

Another menace to park forests exists in the white-pine blister rust, which first appeared in the East, moved across Canada, and is now coming down into western United States through Washington, Oregon and Idaho. Blister rust control measures have been carried on successfully for several years in Acadia National Park in Maine, and it is believed that by the end of this year the white pine of Acadia will be out of danger through the eradication of host plants. In Mount Rainier Park, in the State of Washington, control measures were inaugurated last year to save a few selected stands of white pine. This work will have to be followed up intensively for a year or two, however, if any worthwhile areas of white pine are to survive in that park.

Blister rust is a fungus, its alternate host plants being the currant and gooseberry. It has been discovered that the fungus can move only a small distance from host to pine, but that after reaching the pine it can move a long distance to other host plants. So the method of control is to eliminate the host plants within the necessary radius. Present indications are that in the West control measures can be taken effectively.

If this is not done, experts of the Bureau of Plant Industry state that the resultant damage to the five-leafed pine forests of the West will be a national calamity.

A tree problem in Sequoia and Yosemite National Parks was involved in the use of the Big Tree areas by the visiting public. It was found that the constant tramping of feet around several of the oldest and largest of these trees was wearing away and packing down the ground cover to an extent that was threatening the very life of the trees. Careful studies of existing conditions led to the protection of the tree-root areas from encroachment, and the soil, which had been heavily compacted, was brought back to normal by covering the ground with forest litter and the planting of native shrubs. Dr. E.P. Meinecke, general adviser to the National Park Service on matters of forest pathology, reported recently that the oldest of the Big Trees in Yosemite, the Grizzly Giant "is in decidedly better condition now than it was six years ago. The little branchlets no longer droop as they did a few year ago, but have come back a normal bright green." This means that this old tree, estimated to be about four thousand years old, has been brought back to health, and may watch the generations come and go for a few more thousand years.

So it goes on, through a long list. As one floral or faunal problem is solved, another is presented. And the inorganic features also have their problems, often requiring a great deal of research before a solution is reached. Volcanoes are studied, investigations are made of the geyser fields, where activity in one place ceases, only to break out in another. The effects of glaciers and of running water on granite and other rocks are given attention by one group of scientists, while another is interested in the formation of great colorful canyons by the effects of wind and rain.

An especially interesting discovery at the Grand Canyon, made possible through the cooperation of the Carnegie Institution of Washington and the National Academy of Sciences, was of fossil plants and the traces of many extinct animals. Both plants and animals dated back to the age of coal plants. In the Algonkian rocks, the strata which represents one of the earliest periods from which remains of life have been obtained, were found fossils of algae, or very low types of plant life.

More than twenty distinct forms of hitherto unknown animals were discovered, not from petrifications or fossilized bones, but from footprints made by these creatures in soft, probably moist sands. Some quick covering of the sands hardened and preserved the footprints, to the end that ages later some of them might be unearthed as workmen split the rock in building a new trail, to become part of the educational program at the Grand Canyon National Park.

Increasingly experts of the National Park Service are making studies along various specialized lines, while at the same time the Service welcomes the many investigations inaugurated and carried through by organizations and individual scientists.

While perhaps not strictly in line with the general trend of this article, which has referred to research primarily from the standpoint of education, some mention should be made of the valuable research being done along landscape and sanitary lines, the former by landscape architects and architects of the National Park Service, and the latter by sanitary engineers of the Public Health Service in cooperation with the Park Service.

Again from the educational standpoint—the incalculable value of the national parks and national monuments as research laboratories has been recognized by a number of schools, including important universities, and many field classes are held therein, particularly in ecology, geology and archeology.

There is no doubt but that this use of the parks as field schools will increase in the future, side by side with the growth in tourist travel. Thus the parks have an important destiny in the future of our national life, from the standpoints of educational, spiritual and recreational values.

The Scientific Monthly, 36, 1933, 483–501. Copyright 1933 and permission given by the American Association for the Advancement of Science.

AN ACT TO PROVIDE FOR THE PRESERVATION OF HISTORIC AMERICAN SITES, BUILDINGS, OBJECTS, AND ANTIQUITIES OF NATIONAL SIGNIFICANCE, AND FOR OTHER PURPOSES,

Approved August 21, 1935 (49 Stat. 666)

Be it enacted by the Senate and House of Representatives of the United States of America in Congress assembled, That it is hereby declared that it is a national policy to preserve for public use historic sites, buildings and objects of national significance for the inspiration and benefit of the people of the United States. (16 U.S.C. sec 461.)

SEC. 2. The Secretary of the Interior (hereinafter referred to as the Secretary), through the National Park Service, for the purpose of effectuating the policy expressed in section 1 hereof, shall have the following powers and perform the following duties and functions:

(a) Secure, collate, and preserve drawings, plans, photographs, and other data of historic and archaeologic sites, buildings, and objects.

(b) Make a survey of historic and archaeologic sites, buildings, and objects for the purpose of determining which possess exceptional value as commemorating or illustrating the history of the United States.

(c) Make necessary investigations and researches in the United States relating to particular sites, buildings, or objects to obtain true and accurate historical and archaeological facts and information concerning the same.

(d) For the purpose of this Act, acquire in the name of the United States by gift, purchase, or otherwise any property, personal or real, or any interest or estate therein, title to any real property to be satisfactory to the Secretary: *Provided,* That no such property which is owned by any religious or educational institution, or which is owned or administered for the benefit of the public shall be so acquired without the consent of the owners: *Provided further,* That no such property shall be acquired or contract or agreement for the acquisition thereof made which will obligate the general fund of the Treasury for the payment of such property, unless or until Congress has appropriated money which is available for that purpose.

(e) Contract and make cooperative agreements with States, municipal subdivisions, corporations, associations, or individuals, with proper bond where deemed advisable, to protect, preserve, maintain, or operate any historic or archaeologic building, site, object, or property used in connection therewith for public use, regardless as to whether the title thereto is in the United States: *Provided,* That no contract or cooperative agreement shall be made or entered into which will obligate the general fund of the Treasury unless or until Congress has appropriated money for such purpose.

(f)   Restore, reconstruct, rehabilitate, preserve, and maintain historic or prehistoric sites, buildings, objects, and properties of national historical or archaeological significance and where deemed desirable establish and maintain museums in connection therewith.

(g)   Erect and maintain tablets to mark or commemorate historic or prehistoric places and events of national historical or archaeological significance.

(h)   Operate and manage historic and archaeologic sites, buildings, and properties acquired under the provisions of this Act together with lands and subordinate buildings for the benefit of the public, such authority to include the power to charge reasonable visitation fees and grant concessions, leases, or permits for the use of land, building space, roads, or trails when necessary or desirable either to accommodate the public or to facilitate administration: *Provided,* That such concessions, leases, or permits, shall be let at competitive bidding, to the person making the highest and best bid.

(i)   When the Secretary determines that it would be administratively burdensome to restore, reconstruct, operate, or maintain any particular historic or archaeologic site, building, or property donated to the United States through the National Park Service, he may cause the same to be done by organizing a corporation for that purpose under the laws of the District of Columbia or any State.

(j)   Develop an educational program and service for the purpose of making available to the public facts and information pertaining to American historic and archaeologic sites, buildings, and properties of national significance. Reasonable charges may be made for the dissemination of any such facts or information.

(k)   Perform any and all acts, and make such rules and regulations not inconsistent with this Act as may be necessary and proper to carry out the provisions thereof. Any person violating any of the rules and regulations authorized by this Act shall be punished by a fine of not more than $500 and be adjudged to pay all cost of the proceedings. (16 U.S.C. sec. 462.)

SEC. 3. A general advisory board to be known as the "Advisory Board on National Parks, Historic Sites, Buildings, and Monuments" is hereby established, to be composed of not to exceed eleven persons, citizens of the United States, to include representatives competent in the fields of history, archaeology, architecture, and human geography, who shall be appointed by the Secretary and serve at his pleasure. The members of such board shall receive no salary but may be paid expenses incidental to travel when engaged in discharging their duties as such members.

It shall be the duty of such board to advise on any matters relating to national parks and to the administration of this Act submitted to it for consideration by the Secretary. It may also recommend policies to the Secretary from time to time pertaining to national parks and to the restoration, reconstruction, conservation, and general administration of historic and archaeologic sites, buildings, and properties. (16 U.S.C. sec. 463.)

SEC. 4. The Secretary, in administering this Act, is authorized to cooperate with and may seek and accept the assistance of any Federal, State, or municipal department or agency, or any educational or scientific institution, or any patriotic association, or any individual.

(b) When deemed necessary, technical advisory committees may be established to act in an advisory capacity in connection with the restoration or reconstruction of any historic or prehistoric building or structure.

(c) Such professional and technical assistance may be employed without regard to the civil-service laws, and such service may be established as may be required to accomplish the purposes of this Act and for which money may be appropriated by Congress or made available by gifts for such purpose. (15 U.S.C. sec. 464.)

SEC. 5. Nothing in this Act shall be held to deprive any State, or political subdivision thereof, of its civil and criminal jurisdiction in and over lands acquired by the United States under this Act. (16 U.S.C. sec. 465.)

SEC. 6. There is authorized to be appropriated for carrying out the purposes of this Act such sums as the Congress may from time to time determine. (16 U.S.C. sec. 466.)

SEC. 7. The provisions of this Act shall control if any of them are in conflict with any other Act or Acts relating to the same subject matter. (16 U.S.C. sec. 467.)

AN ACT TO AUTHORIZE A STUDY OF THE PARK, PARKWAY, AND RECREATIONAL AREA PROGRAMS IN THE UNITED STATES, AND FOR OTHER PURPOSES,

Approved June 23, 1936 (49 Stat. 1894)

Be it enacted by the Senate and House of Representatives of the United States of America in Congress assembled, That the Secretary of the Interior (hereinafter referred to as the "Secretary") is authorized and directed to cause the National Park Service to make a comprehensive study, other than on lands under the jurisdiction of the Department of Agriculture, of the public park, parkway, and recreational-area programs of the United States, and of the several States and political subdivisions thereof, and of the lands throughout the United States which are or may be chiefly valuable as such areas, but no such study shall be made in any State without the consent and approval of the State officials, boards, or departments having jurisdiction over such lands and park areas. The said study shall be such as, in the judgment of the Secretary, will provide data helpful in developing a plan for coordinated and adequate public park, parkway, and recreational-area facilities for the people of the United States. In making the said study and in accomplishing any of the purposes of this Act, the Secretary is authorized and directed, through the National Park Service, to seek and accept the cooperation and assistance of Federal departments or agencies having jurisdiction of lands belonging to the United States, and may cooperate and make agreements with and seek and accept the assistance of other Federal agencies and instrumentalities, and of States and political subdivisions thereof and the agencies and instrumentalities of either of them. (16 U.S.C. sec. 17k.)

SEC. 2. For the purpose of developing coordinated and adequate public park, parkway, and recreational-area facilities for the people of the United States, the Secretary is authorized to aid the several States and political subdivisions thereof in planning such areas therein, and in cooperating with one another to accomplish these ends. Such aid shall be made available through the National Park Service acting in cooperation with such State agencies or agencies of political subdivisions of States as the Secretary deems best. (16 U.S.C. sec. 17*l*.)

SEC. 3. The consent of Congress is hereby given to any two or more States to negotiate and enter into compacts or agreements with one another with reference to planning, establishing, developing, improving, and maintaining any park, parkway, or recreational area. No such compact or agreement shall be effective until approved by the legislature of the several States which are parties thereto and by the Congress of the United States. (16 U.S.C. sec. 17m.)

SEC. 4. As used in sections 1 and 2 of this Act the term "State" shall be deemed to include Hawaii, Alaska, Puerto Rico, the Virgin Islands, and the District of Columbia. (16 U.S.C. sec. 17n.)

ROADS AND MORE ROADS IN THE
NATIONAL PARKS AND NATIONAL FORESTS

By Rosalie Edge
1936

INTRODUCTION

"Build a road!" Apparently this is the first idea that occurs to those who formulate projects for the unemployed. In consequence, a superfluity of four-width boulevards, with the verdure cut back for many feet on either side, goes slashing into our countrysides, without regard for the destruction of vegetation, and, too often without consideration of whether the road is needed at all. The motoring public always travels by the new road, and those who dwell along such highways, and have chosen their homes from a preference for seclusion, find themselves parked beside arteries of ceaseless traffic. No provision is made for pedestrians; and a man takes his life in his hands if he ventures on foot to call on his next door neighbor. The city dweller is forced to go far afield if he is to see aught besides asphalt, or to breathe air not polluted with carbon monoxide gas.

The work of relief employment is not based primarily, as it should be, on the usefulness and desirability of a project; such aims are, (of necessity it would seem) too often subordinated to the imperative need to put to work immediately thousands of men registered for relief through one agency or another. Vast sums are appropriated for work relief; and to use this money justly and usefully is a problem indeed. As a nation, we have adjusted our ethics to the pork-barrel; and each state, each county, city and village, loudly and insistently demands a share of the spoils.

A project which is useful, or only mildly harmful, in one county is too often repeated merely to allay jealousy in another county, where the same project may be positively detrimental. Mosquito control may be quoted as one example. The dwellers in the thickly populated suburban districts of Long Island demand that mosquitos be controlled on the surrounding great areas of salt marsh. Thousands of dollars are appropriated for this work, and armies of C.C.C. men begin to dig ditches in every direction. Then, as soon as it becomes known that Long Island townships have much money to spend on mosquito control, an outcry arises from upland communities, insisting that their mosquitoes also be destroyed. The situation in the uplands is entirely different from that of the seaside; in one the marshes are salt and in the other they are fresh. The various species of mosquitoes are not the same. While the mosquitoes of the salt marsh easily fly twenty-five miles, or more, the fresh marsh mosquito does not travel more than a mile away from its breeding-place. What do county officials or project makers care for such elementary facts? Fresh water marshes, miles distant from any town, are drained without regard to their importance as breeding places of valuable birds and fur-bearers. Thus, in order to grasp at the money which they see passed so easily from hand to hand, do the upland communities destroy sources of recreation and profit on which they might rely year by year.

So it is with roads. Through the medium of road-building, money may be buttered evenly over the whole country. There is a fixed idea in the American mind, inherited from a pioneer ancestry which suffered from having no roads at all, that any additional road must be good and that one cannot have too much of a good thing. Consequently, there have already been built with federal funds more roads than can possibly be kept in repair by state and local communities—roads parallel, roads crisscross, roads elevated, roads depressed, roads circular and roads in the shape of four-leaf clovers; a madness of roads, too many of which will be left untended to fall into disrepair and disrepute.

ROADS IN THE NATIONAL PARKS

Turning to government-owned lands, we find that work relief has entered our National Parks and Forests in force. Each one of these has its C.C.C. camps; and road-building is again the chief employment of the hundreds of men thus introduced into the wilderness. Can anyone suppose that a wilderness and a C.C.C. camp can exist side by side? And can a wilderness contain a highway?

It is conceded that the National Parks must have roads. The Parks are recreational and educational centres for all the people; and admirably do they fulfill these functions. On the other hand, no one who knows the National Parks is so naive as to believe them to be wilderness areas. They have within their borders great hotels and acres of well-equipped camps. The crowds that visit them are splendidly handled; but the management of thousands of visitors makes it necessary to have offices and living quarters for a large personnel, besides stores, parking houses, docks, corrals, and garages; all of which encroach upon the wilderness. Virgin timber has been felled to build hotels, and valuable trees are cut each year for firewood. In the past, grazing has injured both the forests and meadows; and logging operations have been extensive within the Park boundaries. Some primitive areas, however, still exist in almost all the Parks. These should be guarded as the nation's greatest treasure; and no roads should be permitted to deface their beauty.

The Park Service is eager to prevent repetition of the vandalism that has ruined Park areas in the past; but great pressure is brought to bear by commercial interests that press to have new areas opened in order to obtain new concessions. In addition, there is thrust upon the Park Superintendents the necessity to employ C.C.C. men, whether or not their services are needed; and the wilderness goes down before these conquerors. The support of the public at large must be added to the efforts of the Park Service in order to save the most beautiful of the wild places. The situation is well told in an editorial from *Glacial Drift,* the organ of Glacier National Park, as follows:

"Let those who clamor for the opening of the last primitive valleys of the park . . . remember that the charm of many places rests in their solitude and inaccessibility. Let those who consider accessibility and ease alone, weigh carefully which gives more enduring recollection, the dash over Logan Pass or the horseback or foot trip over Indian Pass, and learn that one appreciates in more lasting measure those things which one must gain through the expenditure of effort. Let those who urge more roads bear in mind that the marring of countryside does not end with the construction of a broad, two-lane, highway, absolutely safe when driven at a sane speed commensurate with the full enjoyment of a National Park, but that even the gentlest curves must be eliminated, the width ever increased, each reopening a wound to leave a more gaping scar; with no more turns with delightful surprises beyond, for there are to be no turns; only greater speed and safety, though we may well note the irony of the latter in mountainous regions where improvement always has resulted in more fatalities. Let us recall the hundreds who dash daily over Logan Pass, without so much as a stop, or the great number who, like the camper from the Atlantic seaboard, boasted he had just been in three National Parks on that day and would be in Mt. Ranier on the morrow!"

In the Parks we find hotels and other buildings in a style according, as much as possible, with the surroundings—how shocked we should be to find a skyscraper in a National Park! We need to develop roads that shall be suited to Park purposes and not to bring into their solitudes the great boulevards that are appropriate only where the population is densely crowded. Engineers are not trained in esthetic values; and when producing a triumph of their profession they give small heed to the beauty of the flora, or the interest of the other features of the landscape on which they lay their heavy hands. In the Yellowstone Park a road was last summer, quite needlessly, carried over a thermal spring. What is one less hot spring to a road-engineer? The Yellowstone Park has many hot springs—but now it has one less. A road, suitable for the transport of great loads, is not needed in the Parks; but around the camp fires any evening one may hear the boast: "We drove all the way up without changing gear," or "We never dropped below forty." Our Parks should not be desecrated for the whims of such drivers; obstacles might well be put in the way of fast driving in order to induce the tourists to contemplate the wonders of the forests and mountains spread out before them. Why cut away the crest of each rise, leaving ugly cuts with sides so steep that they cannot support plant life? A continuous easy grade is not essential for driving which is almost entirely recreational; and much primitive beauty is lost through exalting every valley and bringing every mountain low. Even the wilderness not traversed by roads is not safe from the despoiler. High up on Ptarmigan Pass in Glacier Park we met a tractor widening a so-called trail to the width of a wagon road, and watched the C.C.C. men stoop and pick out small stones with their hands. They were making a Rotten Row of a trail across what is still happily a great wilderness of virgin forest.

Last summer we stood at the top of Logan Pass and watched the cars come sweeping to the summit. They might pause for five minutes in the great parking place, decorated with landscaped beds of shrubs bordered with stone copings, which belittle what was once one of the most glorious points of the Rocky Mountains. Many people did not leave their cars, others stepped down for a few minutes to look, and to wonder that such height could be reached without a heated engine. A ranger invited and even pleaded with the sightseers to go with him on a short walk to see the secluded wonder of Hidden Lake. "You can have no idea standing here," he said, "what a wonderful thing it is to go there . . . a very little way. . . ." While he spoke, his voice was drowned in the whirr of the self-starters. The little group of nature-lovers who followed him discovered the loveliness of the lake and saw, besides, Rosy Finches and White-tailed Ptarmigan. They did not miss the company of the motorists who were by that time far in the valley below, rushing on in their enjoyment of perpetual motion.

For roads more appropriate to the National Parks, we offer two suggestions which we believe to be practical:

1.   ONE-WAY ROADS. One-way roads could be narrow and so more easily follow the grade and contour of the land. Roads, roughly paralleled, leading in opposite directions, might be separated by a strip of woodland, as has been done in some of the parkways around great cities, preserving the illusion of wilderness and reducing the great scars that wide roads make on mountain sides. The cost might be increased but the project would have the advantage of providing work for an additional number of men. With no danger from cars

coming in the opposite direction, it would not be necessary for a driver to see so far ahead as on a two-way road, and trees and shrubs could be permitted to grow close on either side. At convenient intervals the road should be widened so that a car may draw aside and stop to permit the occupants to enjoy the distant view or the nearby beauty. It is often argued that roads in the National Parks are needed for the aged and for those unable to take the horse and foot trails— but is it, indeed, fair to these people to be forced to drive along roads so wide that the flowers and shrubs and even the trees are too far distant to the right and left to be enjoyed, and which are lined on both sides with further bare spaces? What chance has anyone to identify a bird? May not a car sometimes be permitted to saunter, to linger, and even to pause? We believe one-way roads would increase enjoyment, and insure greater safety.

2.   PREVENTIVE LANDSCAPING. We recommend that the landscape gardeners, who do what they can to patch and cover up the wounds made by the engineers, precede, as well as follow, the road-builders. The ground-cover might often be put aside, and replaced after the road is completed, or used elsewhere. Last September we traversed the road that crosses the Great Smoky Mountains from Cherokee, North Carolina, to Gatlinburg, Tennessee. Two years ago this was a fine road, and adequately graded for all recreational purposes. Now, further smoothed out and straightened by C.C.C. labor, the raw, steeply-cut sides are artificial and ugly to a degree. The summit, like that of Logan Pass, is laid out in the manner of suburbs and cemeteries. This cicatrice across the heart of the lovliest of forests is a sin against Nature. The C.C.C. men were heaping and burning fifteen and twenty-foot high rhododendrons, which had been roughly uprooted, with no regard to their value. The ground cover of small, woodsy wild things had been dug under. In some places landscaping had veiled the raw earth but with man-made art. We hope that this amelioration of unsightliness may be continued; but no art can replace Nature's treasures; these must be saved before they are ground beneath the road-makers' ruthless heel. Such destruction of native grasses, plants and shrubs along our roads may be observed in every state.

Pamphlet No. 54, Emergency Conservation Committee. Library, Sequoia National Park, Ash Mountain, 1–6 (for pages included in this volume).

ATMOSPHERE IN THE NATIONAL PARKS

By Superintendent John R. White,
Sequoia National Park
1936

We have devoted a lot of time and thought in our meetings to attempts to standardize and improve our automobile regulations, to busses, to landscape preservation, to operator's accommodations and rates and so forth: all very necessary studies but not comparable in importance with the biggest problem of all, yet perhaps the most intangible; namely the preservation or the infusion of the right atmosphere.

Atmosphere, as I understand it, means the pervading influence which governs our national park development policies. It concerns the intellectual or moral, even in some aspects, the physical environment which we want in the parks. In language perhaps more specific, I take the atmosphere of a national park to depend upon what we permit in the way of public use, and equally what we do not permit.

Perhaps the public uses which affect the atmosphere of a national park are those connected with the recreation and entertainment of visitors.

Some of the questions that arise pertain to campfire entertainments, quiet camps and the use of curfew, entrance hours, radios in automobiles or operators' buildings, dances, tennis courts, golf courses, artificial swimming pools, bands, loudspeaker public announcers, electric lighting, winter sports, and so on.

Then in connection with the business of public operators we must consider the soliciting of conventions to fill up the hotels and camps, particularly at the slack opening and closing of the summer season; the publicity issued by the operators or their agents, whether by folders or in magazines or newspapers, and also by word of mouth at their own entertainments within their leased areas or on their busses.

The atmosphere of a park is also affected by the educational programs and building construction advocated by sincere friends of the parks or by technicians interested vastly in their own work, but lacking the experience or the ability to understand the administrative policy with its permanent policies and objectives.

Let us consider at least some of these matters affecting park atmosphere above outlined. It will be necessary of course for me to refer chiefly to conditions and treatments in the Sequoia National Park, while fully understanding that other superintendents have had variations of the same problems and may have solved them better than we have done. It is only by pooling all our experiences that we may be able to make visible some of those aspects of park atmosphere which are so difficult to define. If we can put them into form, we may be able better to settle upon some principles and policies which affect park atmosphere.

I would like the following observations to be considered merely as a humble contribution to thought upon a foremost and all-embracing park problem. It is so important for present and future park welfare that all of us in the Service must help in its solution.

There are four parties who have an interest in a national park; (1) future generations of Americans who have prospective inheritance in the country's natural resources and unspoiled scenery; (2) the people who visit the park and those who pay taxes to support it; (3) the Government, represented by the superintendent and his associates; and (4) the public operators who have invested their money and must have both security for the investment and fair returns from it.

Under the general policies of the Department of the Interior and the National Park Service, it is the superintendent who must play fair with all four parties and must harmonize the inevitable conflicts between the different interests.

The interest of future Americans in the park is bound up with the pride and responsibility that park officials, public operators and many visitors must feel in the share they have in preserving such choice parts of America to assist in the perpetuation of what is worth while in America. No man can live and work amid the fine permanent things of nature in the national parks and feel concerned only with his little affairs of today. Indeed, there is reason to believe that any interest that does not take the larger view cannot permanently succeed in national park business or administration. Fate may well curse those who are false to such a noble heritage.

The interest of the people is perhaps the most difficult of all to define in these perplexing days of social and economical changes. Visitors demand opportunity for physical and aesthetic enjoyment of park scenery, but also for amusement, entertainment, and instruction. There is a natural and steady pressure to place amusement and entertainment above other requirements. In many national parks the interests of local visitors conflict with those of national visitors and with the preservation of the park for the future.

The interest of the Government is in the general condition of the park, its good administration, protection and maintenance; the curtailment of expenditures, actual and prospective, to come within a reasonable budget; the orderly development of plans and their restrictions to the absolutely essential; and the morale of the park organization.

The interest of the public operators, while primarily the protection of, and return on, investment, must extend to many aspects of the interest of the other three parties, because unless the Government administration is good and fair to all, unless the visitors are satisfied and the park preserved, more indirect harm will be done to the operators' investment than by almost any direct regulation of schedules and prices.

Campfire entertainments and educational work:
There should be no paid entertainers unless they deal in the natural sciences, or in approved park diversions such as mountaineering, riding, hiking or winter sports.

The entertainment should be refined in its best sense, and so far as possible along natural history lines and explanatory of the park or other parks. This is a tremendous field, barely touched.

The audiences should not be too large; 500 is about our limit except on peak holidays. When the crowds grow we start new campfires. Of course, our policy of not to exceed about 300 camps in any one area helps. Participation by visitors is the keynote. There should be more funds to pay short-time assistants and provisions of accommodations for them.

No permanent platforms, stages or sounding boards should be permitted. If possible, an actual campfire should be lit and the entertainment center around it.

Just a warning about the scheduled, out-of-doors Peripatetic lectures, hikes and automobile caravans; they may grow to be too large or unwieldy or they may lack the best objectives, all in the endeavor to make a big showing in numbers served in the educational program.

Park Entrance Hours and Quiet Camps:
We admit travel up to 9 p.m. weekdays, 11 p.m. on Saturdays and days preceding a holiday. We open at 5 a.m. the year round. The complaints against these hours are fewer than formerly. We are reasonable about special permits and conditions, but not liberal.

We discourage rather forcibly unnecessary noise in public camps and all public areas at all times; and we impose a virtual curfew after 11 p.m. No automobiles are permitted to roam around after that hour except by special permit. If there are complaints, the visitor might be referred to the practices in many city parks which are closed by 8 or 9 p.m or at dusk.

Radios and Loudspeakers:
We do not permit any use which makes a loudspeaker publicly

available or may annoy others. We do not permit them in hotels or operators' units. But we realize that probably we must come to permitting a radio in some places in the operators' unit. Then its use must be carefully regulated to avoid jazz programs and any blaring forth that is improper.

In California the Standard Oil Company and others have loud-speaker announcer automobiles which they are glad to send to public gatherings out-of-doors. A trained announcer accompanies and humorously helps the programs along. They have twice sent such an outfit into Sequoia for the winter carnival. Last month it came in only because I was away in Death Valley while arrangements were being made.

Dances:

Dances are one of the most difficult things to regulate. We dance only from 9 to 11 p.m. None on Sundays. I wish we could abolish them, but don't think it possible until we build up more of the right kind of entertainment.

Tennis Courts and Golf Courses:

We have avoided both. Formerly there was more pressure for them than there is now.

Swimming Pools:

We have no artificial or commercial pool and don't want one. We have worked out and improved two or three river pools, and might even consider warming them if the apparatus could be hidden.

Bands and Music:

Years ago we used to have bands for the Fourth of July and other occasions, but we are gradually getting away from them.

The violin and flute with perhaps the cornet are the best instruments for campfire use, with, of course, the piano.

Just now we have a proposal to install an organ at our Giant Forest amphitheater in the heart of the Big Tree area. It sounds properly atmospheric; but we are only considering it and our thumbs are inclining downwards.

Electric Lighting:

Electric lighting is such an accepted utility that at first it seems necessary everywhere in public or operator areas. Yet nothing conduces so much to a quiet park atmosphere as general darkness except in and near buildings. We have only small Kohler plants and no general street lighting. We may install small Diesel or hydro plants in certain localities, but we are against street or highway lighting.

Operators' cabins are lit by kerosene hand lamps and candles. Many visitors like it. Few complain. Some are loud in approval. I think that with a little pressure we could have had a $100,000 electric light layout at Giant Forest a few years ago; but we are now glad that the pressure was not exercised.

Motion Pictures:

We have no commercial pictures and believe that all should be educational with perhaps newsreel added. Certainly most careful supervision is necessary. There are special problems in parks far from towns and with large permanent communities at park headquarters. Yet even in those parks the distances to moving pictures outside the parks have been so reduced by good roads and automobiles that the problem is not as serious as formerly.

The filming by commercial companies in the parks needs most careful thought and supervision, and only certain types of pictures should be permitted. A standardized form of permit and a bond are needed. The personal conduct of employees must be carefully supervised. Secretary Ickes has much improved conditions; but more thought must be given to distinguishing between the good and bad producers.

Winter Sports:

Emphasis should be placed on opportunities for everyone to take part in free sports rather than on featured performances and competitions.

Skating rink, toboggan slide, and ski-runs should be as natural as possible and with little or no artificial construction. No charge should be made for their use. No attempt should be made to rival professional winter sport areas. Winter sports should be incidental to winter use of the park, not entirely dominate it.

Any mechanical aid to winter sports such as a ski-elevator or a toboggan elevator is out of place. Improvement of facilities should be limited.

In an attempt to excel and to build up operators' winter accommodations, there is a danger of commercializing winter sports and finally of injuring atmosphere and even scenery. If operators make considerable financial investment in winter sports facilities, equipment, buildings, and so forth, there is danger that winter sports will dominate the picture, be improperly commercialized, and make a hurly-burly of the park in winter.

Public Operators:

In their desire, indeed their necessity, to make proper returns on their investments, the operators may easily damage park atmosphere to their own financial disadvantage. Their publicity should be carefully supervised to prevent extravagant and inaccurate statements or anything that will cheapen the park. The operators' entertainments are likely to degenerate into cheap vaudeville shows designed merely to entertain the average visitor unappreciative of nature.

There is danger in the conventions which are so eagerly sought to fill accommodations at [the] beginning and end of summer season. Some types of conventions are opportunities largely for dissipation rather than enjoyment of nature. A man does not need the larger part of a bottle of booze to enjoy our park scenery.

The operator, too, often wants to operate a type of large bus and ballyhoo sightseeing service which much detracts from park atmosphere. We have a strict 14-passenger limit on our busses and hope to maintain it. Perhaps it is too large now.

Fortunate indeed is the superintendent who deals with a park operator who can see or feel beyond the end of his nose; and who can realize that the proper park atmosphere will in the long run attract more of the right kind of people to patronize his facilities than will any cheapening policy which brings in crowds of the unappreciative.

It should be clearly understood that the park is not in competition with other resorts; that the favorable provisions of the operators' contract fully offset the prohibitions on certain activities common to resorts outside the parks. The park atmosphere is the operators' best friend; but he may not recognize it.

Of course, it is hardly necessary to mention certain things that operators should not install, such as slot machines of various kinds, nickel or dime in the slot telescopes, or field glasses, and so forth. Perhaps the most difficult thing to regulate is curio sales. We must confess that much sold in Sequoia is atrocious. We should try to lead our visitors generally to better things.

CCC Camps:

The thing that perturbs me about the presence of the CCC as at present organized in the parks is the possibility that park standards will be lowered because of the lack of complete control by the superintendent over camp areas. Of course, we have kept dogs out of the camps which are located within the park, but the type of entertainments, such as boxing and vaudeville shows and certain motion pictures, lowers the park standards. Park visitors attend such shows. Much can be done by reasoning with the Camp Commanders, but several hundred youths cannot be either "sissified" or immediately raised to the aesthetic level which we at least set as our goal.

We are much less anxious about the damage that may be done by CCC work to park physical values than the damage that may be done to moral values. It is easy to protect park trees or rocks; it is less easy to protect park atmosphere.

In conclusion: to preserve the national park atmosphere we must curb the human desire to develop the parks quickly to compete in popularity with other resorts, or even State or other parks or national forest areas. When a new project is proposed, the first question should be, "how will it affect the park atmosphere which we desire to maintain or restore?"

We have made a fine beginning in our educational work in the parks; and we should see to it that not only the ranger-naturalists (I still dislike the hybrid word) but all rangers and employees assist in maintaining park atmosphere and educating the park visitors to better and different things in the national parks. Often the things which adversely affect the atmosphere of a park have small and apparently innocent beginnings; but they may grow along unsuspected lines; so that constant vigilance is needed when any innovation is proposed either by park men or park operators.

We should boldly ask ourselves whether we want the national parks to duplicate the features and entertainments of other resorts, or whether we want them to stand for something distinct, and we hope better, in our national life.

We are a restless people, mechanically minded, and proud of doing constructive work. Our factories, railroads, roads, and buildings are admired by the world. We have in the parks a host of technicians, each anxious to leave his mark. But in all this energy and ambition there is danger unless all plans are subordinated to that atmosphere which though unseen, is no less surely felt by all who visit those eternal masterpieces of the Great Architect which we little men are temporarily protecting.

Address to Special Superintendents' Meeting, Washington, D.C., February 10, 1936.
Transcript in Library, Sequoia National Park, Ash Mountain.

OFFICE ORDER NO. 323
FISH POLICY

April 13, 1936

To bring all fish cultural activities in the national parks and monuments within the general policies applying to all other forms of animal life, the following policy affecting fish planting and distribution shall be followed:

No introductions of exotic species of fish shall be made in national park or monument waters now containing only native species.

In waters where native and exotic species now exist, the native species shall be definitely encouraged.

In waters where exotic species are best suited to the environment and have proven of higher value for fishing purposes than native species, plantings of exotics may be continued with the approval of the Director and the superintendent of the park in which such waters are located.

It is the definite purpose of this policy to prohibit the wider distribution of exotic species of fish within the national parks and monuments, and to encourage a thorough study of the various park waters to the end that a more definite policy of fish planting may be reached.

In waters where the introduction of exotic species threatens extinction of native species in an entire national park or monument area, such plantings should be discontinued and every effort made to restore the native species to its normal status.

The number of any species of native non-game fish should not be reduced even where such reduction may be in the interest of better fishing.

All forms of artificial stream improvement which would change natural conditions should be avoided, but the restoration of streams or lakes to their natural condition is permissible where thorough investigation indicates the desirability of such action.

There should be no effort to introduce exotic fish or other exotic aquatic life for the purpose of increasing the supply of fish food.

In cases where a lake or stream is of greater value without the presence of fishermen, there should be no stocking of such waters.

In national parks and monuments where there still remain certain lakes which do not contain fish, permission of the Director must be secured before stocking.

PLEASE ACKNOWLEDGE RECEIPT OF THIS ORDER.

(Sgd) Arno B. Cammerer
Director.

National Park Service Archives, Harpers Ferry, Box N16, "Management Biology Through 1939."

A STUDY OF THE PARK AND RECREATION
PROBLEM OF THE UNITED STATES

UNITED STATES DEPARTMENT OF THE INTERIOR
NATIONAL PARK SERVICE, 1941

A PARK AND RECREATIONAL LAND PLAN FOR
THE UNITED STATES

THE PLAN. In this section we propose to indicate what provision of lands and waters is required in order to meet adequately the recreational requirements of the American people—a provision which at the same time will give full and fair consideration to all other land-use needs. It sets forth a plan, the consummation of which, under the American form of government, cannot rest upon a single agency but instead will require the combined and coordinated efforts of all agencies, at all levels of government, which deal with any phase of recreation. It includes a brief discussion of the part the several agencies and levels of government may properly be expected to play in effectuating this plan.

1.   Every element and all types of population need areas and facilities for outdoor recreation. There is a paramount need, however, for public recreational areas, of all obtainable types and providing for a wide range of beneficial activities, within easy reach of all urban populations. Theirs is a daily life, lived among man-reared walls, man-built streets, objectionable noises and smells, and, for a large percentage of them, in poor housing. To provide what they need for frequent use requires:

(a)   Neighborhood playgrounds within easy walking distance (not more than a quarter of a mile) of all children.

(b)   Playfields and neighborhood centers within half a mile to a mile of all citizens.

(c)   Parks, or other areas characterized by natural or man-made beauty, sufficient in extent so that wear and tear will not be such as to render the cost of maintenance of their attractive features prohibitive, and sufficient in number and so distributed that all citizens, no matter how poor, may enjoy them at least occasionally. The limit of distance for areas of this type probably should not exceed two miles.

(d)   Protection of urban and suburban streams and other waters from pollution and "uglifying" uses; provision of points of access and facilities for use in such places and of such extent as prospective use will justify.

(e)   Parkways along waterways and to connect major units of the recreational-area system.

2.   For holiday and week-end use by city folk and by those who live in thickly populated or intensively cultivated rural sections, there is need of public recreational areas where picnicking, water sports, day and overnight camping, hiking, and other related activities may be enjoyed, and which are sufficiently large to provide those who use them a sense of freedom and of separation from crowds. At least one such area should be within 25 miles of those for whom they are chiefly provided. A distance even less than that will assure greater use by those for whom transportation is a difficult problem.

3.   Because of the highly artificial conditions under which city dwellers are forced to live, the pressure for adequate recreational facilities is exerted primarily in their behalf. Yet there can be no denying that the 43.8 percent of the population of the United States which the census of 1930 classed as rural have recreational needs which must be met, at least partly, by land and facilities publicly owned or freely usable by the public.

National and State parks and other related types of recreational areas are intended to attract and provide equally for all elements of the population. What the rural resident chiefly needs is something roughly equivalent to the parks and playgrounds established in and near the city for the urban dweller. Generally speaking, such areas as are acquired and developed for his needs will be smaller than the general run of State parks; they will include provision for many of the activities that characterize the city playground, such as baseball, softball, tennis, volleyball, badminton, horseshoes, and swimming. Of equal importance with these will be provision for group activities—picnics, musical and dramatic presentations, competitive sports, festivals, pageants, dancing—the means by which the dweller in the country may satisfy those gregarious instincts and the urge for self-expression which are given so little rein in the daily lives of millions of rural people.

4.   For vacation use, by all the population, wherever resident, there are needed—

(a)   Extensive public holdings in all those parts of the country characterized by forests, rugged terrain, lakes and streams, or any combination of these characteristics. In the western half of the United States, where the Federal Government's policy of retaining its remaining forest properties was adopted in time to keep much of the most spectacular mountain area from passing into private hands, most of the land required for extensive vacation uses is already in public ownership, and the problem is largely one of adjustment of use and administration so that these lands may yield the largest possible return to society. In the eastern half of the United States, which contains 70 percent of the total population, the need is very much greater, while the means of supplying it through lands now publicly owned is very much less.

Public ownership of lands, particularly by State and Federal agencies, has been rapidly extended throughout most of the region east of the Rockies during the present decade. In the more rugged sections within this region, the United States Forest Service has now acquired several million acres of land, much of it suitable for development for vacation use, and much of it developed, or under development, for such use. Most of the lands acquired by the Forest Service, however, are at such a distance from the larger cities that the mere cost of getting to them is so great as to make their use difficult or impossible for more than half of the people of the cities. Part of the rest can make fairly frequent, and the remainder at least annual, use of them, and their potential contribution to this kind of public enjoyment is considerable. Except for coastal frontage, it appears likely that in most of the States in the eastern half of the United States lands now owned by or included in the purchase program of the Forest Service, plus the holdings and proposed acquisitions of the various State conservation, park and forestry agencies, will come close to providing the lands needed for this type of use. On such properties there is need for public campgrounds, cabin colonies, group camps, and the various kinds of public resorts, allocated as to kind, extent, and location in proportion to prospective public use. In this phase of the total effort to provide adequately for desirable public recreation, it is urgently necessary to coordinate planning among the several Federal agencies and the several State agencies, in order to provide what is needed and at the same time to avoid such multiplication of locations, or facilities, or both, and duplication of planning and administrative organizations as will lay a needlessly large administrative and maintenance cost upon either the user or the general public. For any region there should be a coordinated recreational plan in which all agencies and all lands shall have their proper place as well as the individual plans prepared by each agency.

(b)   As indicated, the mountainous and other wilderness and semi-wilderness properties of the Federal and State agencies referred to are usually so located that more than half the population are unable to use them. Furthermore, the urban populations of much of Texas, Oklahoma, Iowa, Nebraska, the Dakotas, and Illinois are even less advantageously located with respect to possible use of such publically owned areas. Yet a vacation in the out-of-doors in attractive natural or naturalistic surroundings, where a reasonable variety of recreational occupation may be obtained, is desirable for all and particularly for those whose limited means normally provide only limited recreational opportunities. And while they are desirable, neither rugged terrain nor spectacular scenery is essential to the enjoyment of an outdoor vacation experience. It appears, therefore, that the most pressing problem, if this desirable and beneficial experience is to be brought within reasonably easy reach of the whole population, is to provide vacation areas close to the persons who need them and who cannot afford to seek them at a distance.

Areas of the kind needed are pretty well typified by many of the recreational demonstration areas administered by the National Park Service. These possess, generally, an interesting and occasionally exciting terrain, fair forest cover that will steadily improve under proper protection and management and flowing waters valuable for a variety of recreational uses. They are located, as a rule, sufficiently close to urban populations so that they may be used for short or long periods by persons for whom a trip of as much as 200 miles from home represents a heavy financial outlay; they are close enough, too so that subsidy of transportation for underprivileged individuals or groups, if provided, will represent a comparatively small expense for each individual served.

It is in areas such as these, close to using populations, that provision should be made for organized group camping, as well as for such other vacation facilities as inexpensive cabins and tent camp sites, supplemented by development of those recreational facilities and services which are needed to make a vacation an interesting, enjoyable, and beneficial experience.

It will doubtless be found possible frequently to provide on a single area not only vacation use but also the day, week-end, and holiday use for which provision is urged in Sections 1 and 2.

5.   Due to the extraordinary recreational value of ocean, lake, and river frontage, both because of the opportunities it offers for a variety of physical activities and because of the "refreshment of mind and spirit" provided by combinations of land and water, the necessity of providing an adequate quantity and quality of such properties for public use cannot be over-emphasized. Adequate quantity may be defined as sufficient in extent to permit all necessary development without undesirable crowding, or encroachment by private enterprise. Since the wildlife associated with water margins is often of exceptional interest to the public, ample area in which to permit its protection and display will add greatly to the public value of any such holdings. Frequently beach areas are associated with a hinterland of marsh and swamp of low monetary value, but which provides the favorite habitat to a large and interesting variety of bird and animal life. Whenever possible such hinterlands should be acquired with the frontage and full protection accorded them.

It is believed that acquisition of at least ten per cent of the shore line of the Atlantic and Pacific Oceans, the Gulf of Mexico, the Great Lakes, and of major lakes and streams would be justified by the public benefits that would result therefrom. This percentage applied to ocean and gulf frontage would provide approximately a mile of frontage for each 120,000 persons resident within 100 miles of salt water; distribution of areas should, of course, be related as directly as possible to the distribution of this tributary population, though experience has shown that ocean beaches attract heavy patronage from much greater distances than 100 miles.

6.   It would appear to be generally agreed that those areas containing scenery of such distinction as does, or is certain to, attract users from considerable distances and in fair numbers, should be in public possession. "They should be characterized by scenic and recreational resources of kinds that are unlikely to be reasonably well conserved and made available for public enjoyment under private ownership, or which under private ownership are likely to be so far monopolized as to make it seriously difficult for the ordinary citizen to secure enjoyment of them, except at a cost in time and money disproportionate to the cost of providing that enjoyment" through areas publicly owned.

The inspirational values of exceptionally striking and beautiful scenery are great—even though difficult or impossible to calculate. For that reason selection of areas that contain such scenery is justified, even though prospective attendance might, in some cases, be comparatively light.

7.   On the same plane with the scenic areas are areas or structures of historic, prehistoric, or scientific significance. There is need for continued and energetic effort on the part of public agencies to assure preservation of America's heritage of historic and archeological sites and structures for the enlightenment and inspiration of her people. As is recognized by the Historic Sites Act of 1935, such action does not necessarily involve immediate or even eventual public ownership, but may be accomplished in many cases without disturbing private or semipublic owners.

8.   For special recreational uses, six additional types of development suggest themselves as important in rounding out and completing our State and Federal recreational systems. These are:

(a)   Parkways. The modern parkway idea was represented earliest by county and other undertakings of the character of the Westchester County Parkway system in New York and the Mount Vernon Memorial Highway from Washington to Mount Vernon. Valuable precedents in the extension of the concept were established in New York State by the construction of the Long Island State Parkways and the Eastern State Parkway which is now under construction leading from

New York City to Albany. The Blue Ridge Parkway, extending from Shenandoah National Park to the Great Smoky Mountains National Park, now approaching completion; the Natchez Trace Parkway, following the course of an historic travel route, and now under construction; and the proposed Mississippi River Parkway—all these represent still further extensions of the original concept. They point the way to the utilization of the parkway on a national scale to supplement and complement both the Federal Park system and the Federal-Aid Highway system.

Travel along trunk highways has today become largely a chore, something to be accomplished in the shortest possible time. The pleasure of motion and viewing the countryside has been seriously lessened by the mixing of all kinds of traffic; by the frequency of intersections; and by the uncontrolled use of adjacent lands which, by means familiar to all of us, has made most of the roadsides of our heavily traveled roads an eyesore and an abomination.

Since the parkway is, in its essence, an elongated park traversed by a road, the abutting property owners have no right of access, light, or air. Traffic other than by passenger cars is excluded; and the traffic that is permitted gains access only at safe points separated by considerable distances. The properly planned parkway possesses beauty. Structures such as gasoline stations, restaurants, and overnight accommodations are there to serve the traveler and only so many of them are permitted as he really needs. Increasingly, as on any good highway where traffic warrants it, the opposing lines of traffic are separated by a strip of varying width.

Even should parkway construction proceed for some time without any national parkway plan, there is little likelihood that any will be constructed where it is not justified. Yet the time is surely ripe to begin the planning of a national parkway system, so that it may develop in an orderly fashion; so that, so far as possible, the most pressing needs may be met first; and so that their designation and construction may be integrated with both the Federal and State park and highway systems.

The move toward a national parkway system is a move toward restoration, to the car owner, of those returns in pleasant driving, to which his payment of a variety of special and general taxes fully entitles him.

(b)   Trails and Trailways. Millions of Americans have forgotten, or have never known, the pleasure of walking. On the one hand, the city never has made adequate provision for it, since pavements are not particularly attractive to the person who walks for the pleasure of walking. And the country road of 30 years ago has graduated into a modern highway where no provision has been made for the pedestrian, and where, even if such provision were made, the pleasure of walking would be largely destroyed by the close-by presence of noisy and odoriferous traffic.

Fine and useful as are the Appalachian Trail along the Appalachians from Maine to Georgia, the Long Trail through the Green Mountains, the John Muir Trail along the Sierras, and other trails of this type, they serve only a handful of persons. If walking, one of the best forms of recreation, is to regain the place to which it is entitled in the recreational scheme, the facilities for its enjoyment must be tremendously expanded. Men and women and children who live in cities should be able to go to the end of the bus or trolley line and, only a short distance from there, to exchange the hard pavement for the resilient surface of a footpath leading across fields and through woods, "up hill and down dale." Successfully undertaken in scattered regions by trail clubs and walking clubs, the provision of walking facilities by such groups needs encouragement and support, as a means of supplementing the wilderness trails and the trails and paths that are developed in parks and forests.

State trails were authorized as a proper function of the State by Massachusetts as long ago as 1924, but with little result in actual accomplishment there or by extension to other States. Established largely by right-of-way easement on private lands rather than by means of public ownership, and maintained by a public agency, public trails should be given recognition for their importance in any adequate recreational plan, and their use encouraged in every possible way. Desirable footpath or trail locations, in longer or shorter loops, or as connections between low-fare public transportation lines, can be found in abundance. These, when and if developed, can wisely be supplemented by the longer cross-country trails or paths, such as are being extended and mapped in several regions under the aegis of Youth Hostels. Such trails as these require, for their fullest usefulness, low-cost overnight housing such as that organization has arranged. To the walker on such trails belongs, not the narrow strip of land on which the trail happens to lie, but all that he can see, of meadow and woodland, hill and stream, to the farthest horizon.

(c) Routes of Water Travel. The commercial usefulness of the canals, constructed by the hundreds of miles before and at the time of the advent of the railroad, has now almost completely disappeared; recognition of their potential usefulness as recreational routes has barely begun. Development of facilities for recreational use of the Illinois and Michigan Canal in Illinois was started with the beginning of the Civilian Conservation Corps in 1933. Similar development of the old Chesapeake & Ohio Canal along the Potomac above the Nation's Capital is now well under way. These two examples of long-neglected assets lying close to very large populations may be expected to point the way to similar undertakings elsewhere. The canals, offering excellent recreational experiences both to the walker and the canoeist, possess in addition unusual historical interest; they are reminders of a picturesque and often exciting phase in the history of American transportation; in many places, the canals and their appurtenances, such as the lock tenders' houses along the Chesapeake & Ohio Canal, possess unusual charm that adds enjoyment to their use.

Related in character to these is the Intracoastal Waterway, extending from Massachusetts to Florida, active commercially—and none the less interesting for it—but used extensively also as a route for pleasure craft. It can logically be further developed as a recreational waterway by control of the use of its banks, provision of public areas for camping and picnicking, preservation of the many historic features along its course, and added facilities for servicing of small craft.

(d)   Waysides. This term is used to designate all those provisions of land, really integral parts of the complete highway, needed as stopping places to enable the traveler to derive greater enjoyment from use of the highway itself. It may designate simply a place off the traveled surface of the road where the motorist may pause to enjoy a pleasant view, or to read an historical marker, without endangering himself or others. It may mean a few shady acres where he may stop for a picnic lunch or just for brief relief, in pleasant surroundings, from sitting in an automobile. Some separation from the sight, sound, and smell of traffic is at least highly desirable.

As in the case of several other types of recreational areas, it is impossible to formulate any absolute quantitative recommendations as to extent or frequency of location for waysides. Attractive views, or points of historic interest, may be numerous or rare, depending on the place. Rest and picnic places might well be provided ultimately along well-traveled trunk highways at intervals of 15 to 20 miles, with somewhat longer intervals on highways less traveled.

(e)   Control of Outdoor Advertising. The blight which has been laid on the American landscape by outdoor advertising of all kinds, from the snipe sign on a tree to the modern, illuminated "landscaped" billboard which so often is planted in the foreground of the loveliest views from the highway, is one to cause any American to blush with shame. Outdoor advertising, designed to attract the motorist's attention, which, to just the degree that it is successful in accomplishing its purpose is a menace to safety on the highway, has no legitimate place in the rural or wilderness American scene. It can be eliminated or controlled, and by legal means.

The zoning of outdoor advertising out of residence and other locations, by city or county zoning agencies is recognized as a constitutional use of the police power. The State, from which cities and counties derive their zoning power, can exercise in its own behalf any power it can delegate to any other governmental agency. Exercise of that power, by passage of suitable State legislation, should be urged upon State legislatures by all those agencies and individuals interested in preserving and restoring the amenities of the American countryside.

(f)   Preservation of Roadside Beauty. Acquisition of attractive highway border strips, or use of scenic easements to obtain assurance of the preservation of natural features which contribute to the enjoyment of driving, offers an opportunity further to preserve highway recreational values for the user of the highway. Michigan and Minnesota, among the States, and the Society for the Protection of New Hampshire Forests, among the private organizations, have done notable work along these lines worthy of emulation throughout the entire country.

RESPONSIBILITY FOR EXECUTION. This chapter, thus far, represents an endeavor to indicate what is needed, if the recreational requirements of our people are to be met. Such a plan possesses little significance unless it is possible at the same time to indicate, on a logical basis, the part which the various types of agencies, Federal, State, and local, should play in providing and administering lands and facilities.

At the two ends of the scale—the Federal Government and the municipality—there appears to be general agreement as to certain responsibilities. That the Federal Government should own and administer areas which qualify as national parks and monuments, few will dispute. It is wholly logical that, within the limits of needs that may be met at reasonable cost, and subject to proper coordination, the recreational opportunities offered by national forests, national wildlife refuges, and other Federal properties should be developed and utilized. Nor is there any important difference of opinion as to the responsibility and obligation of the municipality to provide needed parks, parkways, and play areas within its boundaries. It is within the vast field between these two extremes that we enter the realm of dispute and uncertainty. It may be indicated by such questions as these:

1.   Should the Federal Government assume, in addition to its task of development and administration of national parks and monuments, the cost of acquiring such areas.

2.   Should the Federal Government undertake to acquire, de-

velop, and administer those areas for which, because they are of recreational importance to a fairly large region comprising more than one State or parts of several States, the individual State cannot be expected to assume responsibility?

3. Should the State assume responsibility for acquisition, development, and administration of only those areas which may be said to possess Statewide significance, and, if so, how are the respective responsibilities of, for example, Texas and Rhode Island, to be determined?

4. Is an area of outstanding quality which happens to lie close to a large city and which is certain to be used mostly by those who live within that city, to be considered wholly the city's responsibility?

5. What is the place of the county and metropolitan park district in the recreation-land scheme of things?

6. To what extent, if any, should responsibility of acquisition, or development, or administration be shared by Federal or State government or by State and local government?

Let us seek to supply an answer to these questions, on which successful accomplishment of a national recreational area program so largely depends, realizing that because of the federal character of this country and the large degree of sovereignty still reserved to the States, there are certain to be always sharp differences in law and practice among State and local units of government, and considerable variations in their concepts of responsibility.

For a long period the practice of the Federal Government in connection with the acquisition of national parks and monuments from private ownership was to require that the necessary lands be furnished to it without cost—a practice based largely on the conviction that such a requirement would be effective in preventing the creation of second-rate parks. In addition to the doubtfulness of its logic, it may be claimed justly that this method by no means assures inclusion of genuinely worthy areas in the system—as witness the period of time which has passed since the Everglades National Park was authorized, without the State of Florida having taken any effective step either to acquire the land or to provide proper protection to its resources. It would seem that if the Nation, rather than just the State of Florida, has a concern for the preservation of this unique area, its concern may quite logically be indicated by provision of purchase funds, particularly if the State is to turn the lands it owns over to the Federal Government. Funds used to purchase Isle Royale National Park lands have come principally from the Federal Treasury, while the Great Smokies has reached the required acreage only because the Federal Government has supplied funds to supplement those furnished by the States of North Carolina and Tennessee and by private contributors. Standards of admission to the federal park system are safeguarded by better means than insistence on acquisition of lands by funds from other sources than the Federal Government.

For what might be called Federal or national recreational areas—a poor term since "recreational area" is generic and applies properly to any type of holding administered for recreation by the Federal Government—the answer is less easily given. It may be said, however, that for any area needed to meet the recreational requirements of the Nation as a whole and which, because of its regional importance, would place on the individual State a burden of initial cost greater than it could logically be expected to bear, there is justification for participation by the Federal Government in meeting that cost. The degree of participation would probably vary with respect to each individual area, perhaps on the basis of such calculations as might be made as to the relative benefits likely to be obtained. This type of calculation is in many ways analogous to that practiced with respect to local improvement district assessments or, specifically in the case of parks, by assessments levied on benefited communities for support of the Boston Metropolitan Park System. Normally it would be expected that, because of contingent benefits accruing from an influx of out-of-State travel, such as gasoline taxes, local purchases, and payment for use of special facilities in the park itself, the State would provide administration and such funds as were needed to support it, though Federal participation in the cost of adequate development might properly be provided on the same basis—and with the same justification—as participation in land purchase.

By analogy it would appear that the State's concern with provision of recreational areas and facilities extends beyond the State equivalents of national parks—the areas of genuinely State-wide significance; and every State system in existence has recognized that fact in actual practice. There seems to have developed a special State park type of undertaking which is based on such area qualities as topography or other natural features of considerably more than average interest, sufficient extent to provide satisfactory extensive recreation, and provision for those kinds of recreation—such as picnicking, camping, hiking, climbing, boating, fishing, and use of nature trails—which result in greatest enjoyment if carried on in a natural environment. Swimming facilities, valuable as they are both from the participants' standpoint and as an attraction to build up or retain attendance at a volume that makes the operation cost per visitor economically low, are not of the essence of it. Essentially it is the kind of place worth going a considerable distance to enjoy, if one is able to do it; but that fact does not preclude the location of parks of such character close to heavy populations. That is just so much the better; but the difficulty usually is to find qualified areas sufficient in extent to meet the heavy use such proximity to heavy population usually entails.

It is with respect to areas so located, which, in the natural course of events, are called upon to meet such day-by-day use as is usually provided by city park systems, that States would appear to be justified in expecting some degree of local participation in the costs of acquisition, development, and even of operation, since any such area provides a greater or less degree of relief for the pressure on the city to provide parks and playgrounds. The city is thus a special beneficiary and should pay something for the special benefit it receives.

Closely analogous, within the State-local field of action, to those regional areas previously discussed as a joint Federal-State responsibility, are those holdings, characterized by rather exceptional usefulness for physical recreation, serviceable to nearby populations of two or more counties and possibly several cities, for which no single local agency can be expected to assume full responsibility. They do not possess "such distinction as does, or is certain to, attract users from considerable distances in fair numbers," yet they are an essential part of the necessary all-around provision for recreational needs. Most States have included in their systems at least some properties such as these. Again, because of the heavy proportion of purely local use to which they are normally subjected, it is believed that some equitable contribution by local governmental agencies may properly be expected by the State, or even that the process may be reversed and the State contribute to the cost of acquisition, development, and maintenance in cases where a local government is disposed to take the initiative in these matters. While the practices are by no means general, at least one State, Mississippi, has approved legislation authorizing local units of government to contribute to the cost of developing and administering State parks, and several States have turned over to local governments the task of administering areas acquired and developed by them or with Federal assistance.

In certain situations, the county and the metropolitan park district have an important and legitimate place in the park and recreational picture. In the largely rural county, the county appears to be the logical agency to establish and manage systems of parks and recreational areas which will provide, reasonably near at hand, some of those means of recreation that are socially valuable but which otherwise the country dweller is likely to lack—areas that will appeal to or stimulate his feeling for beauty, that will preserve some typical sections of the native landscape, that will cater specifically to his gregarious instincts or provide him with the means of indulgence in group games and sports. That suggestion is made with full consciousness of the possibility that, because of the shockingly uneconomic multiplicity of counties, it may be impractical for many of them to undertake a park and recreational program and provide competent planning, development, and administration. This situation points to the vital necessity of State park and recreational agencies equipping themselves to supply State aid in the form of advisory service on both the technical and administrative phases of such county undertakings.

In complete contrast is the typical metropolitan park system, serving a city, or even two or more cities whose zones of influence overlap, and the nearby satellite communities with their varying degrees of urbanization. These are not infrequently established on county boundary bases, as in Milwaukee County, Wisconsin, and the Cook County Forest Preserve District in Illinois. They provide a reasonable day-by-day recreation to those whom they are designed to serve, and which would be unlikely to be adequately provided by the cities which form the core of the metropolitan community.

It will be seen that provision of adequate recreational facilities for the Nation as a whole is in very large degree dependent on joint agreement and action on the part of agencies at the several levels of government, and that this must be based upon understanding of and agreement on the logical degree of responsibility each bears toward situations in which there is a legitimate joint interest. As a means of bringing this about the application of the Federal aid principle to the relationship between the Federal Government and the States, and of the State aid principle to the relationship between State and local units of government is coming more and more to the fore as a possible solution. Its effectiveness is proved in a number of fields of governmental activity and, specifically, in other branches of conservation such as forestry and wildlife. In these it has served mightily to spur activity and accomplishment for purposes of genuine national or State significance, and it has tended to equalize the burden of the accomplishment on the public as a whole.

Few persons today are likely to discount seriously the value of a truly national highway system or to deny that, without Federal aid, there would be many States which would still be far back of the procession in supplying roads that are essential components of that national system. Every reason, of human need and of equalization of burden, which can be adduced in support of such aid for highways, education, forestry, wildlife, etc., applies also to the preservation of our scenic, historic, scientific, and outdoor recreational resources and their development for human use. It seems probable that such aid could be extended by such methods as would protect the national interest without involving any undesirable encroachment on the independence of action of State and local government. National, State and local interest and responsibility are inextricably intermingled in this as in almost every other field of human endeavor, but there appears to be no good reason why that community of interest and responsibility cannot be placed ultimately on a coordinated and sound basis.

Washington, D.C.: GPO, 1941, 122–132.

4

The Poverty Years

1942–1956

The bonanza of development, expansion, and reinterpretation that came to the national park system during the Depression years abruptly ended with Pearl Harbor. In a matter of weeks, federal work programs halted, funding dried up, and personnel began leaving for the military or wartime support industries. As in World War I, there were calls for use of park lands and resources to support the war effort. Pressure particularly focused on grazing, lumbering, and mining the great western parks. During the earlier conflict, grazing had been allowed, in part due to inadequate understanding of its ecological consequences but primarily because the fledgling Park Service felt politically incapable of blocking it.

By 1942, however, a mature national park system, beloved by the populace, handily rejected the requests for intrusion by consumptive users. In a major statement Director Newton Drury spelled out the reasons why, even in times of emergency, the park system should remain inviolate. Drawing upon letters from servicemen overseas he painted an image of the parks as icons for America to be preserved pure and unabused for their inspiration. This article represents one of the clearest statements of the NPS philosophy that preservation supersedes all extractive uses regardless of their monetary worth. The Park Service was able to resist such intrusions throughout World War II.

The war years gave the National Park Service time to reflect on the status and future of the parks, and those employees remaining in the system engaged in elaborate planning for the expected return to prosperity and development after the war. Many plans challenged the level of development already present in some parks. Reflecting on these questions in 1945, the National Parks Association released

165

a declaration of policy entitled "National Primeval Park Standards."
Its recommendations demonstrated the increasing division between
the Park Service's ideas of management and those of the conservation groups.

The expansion of the park system became a volatile issue in Congress
by 1950. Antagonism to parks creation and expansion focused especially in the West among the consumptive local users recently denied access during the war. One of the key fights of the period
culminated in the expansion of Grand Teton National Park in 1950.
In this enactment, Congress stipulated, under pressure from the
Wyoming delegation, that no further parks or monuments be established in that state without specific congressional authorization. This
is notable for its challenge to the Antiquities Act and demonstrative
of the backlash by western locals against the expansion of the park
system.

Three years later Congress acted again, this time by defining the
national park system and emphasizing that all units were to be managed under the Organic Act of 1916. This answered the criticism of
many that monuments, historic parks, etc., should not be as strin-
gently protected and denied to consumptive users.

By that same year, 1953, the park system Congress so defined
was in serious trouble. Years of wartime abandonment were followed
by neglect and underfunding as Congress grappled with ominous
international events. Annual maintenance of trails and facilities lagged,
and many features fell into serious, even dangerous, disrepair. Ag-
itation by the Park Service and conservation groups convinced a few
congressmen but it took an article in a major American periodical
by an eminent historian, Bernard DeVoto, to turn the trick. Adopting
the psychological ploy of shock, DeVoto suggested closing the national parks until they could be properly and safely operated. To a
public annually increasing its love affair with the parks, this was
indeed an appalling revelation.

Despite the now clear need for action it took one final hurdle for
the NPS to receive the funding it needed. Piecemeal project funding
was too easily pared when budget problems appeared. What the parks
needed was a massive, coordinated rescue plan with its own budget.
That rescue plan became known as Mission 66. Starting in 1956 and
projected to last a decade, Mission 66 would ultimately cost more
than one billion dollars and would substantially increase or renew
visitor facilities and access throughout the system. Director Conrad
Wirth's exhortation to Dwight Eisenhower's cabinet spelled out the
details and emphases of the plan which itself would come under fire
before the decade concluded.

THE NATIONAL PARKS IN WARTIME

By Newton B. Drury
1943

The impact of war upon the National Park Service and the areas it administers has of necessity altered its immediate program. At the same time it has served to highlight the primary function of the Service as trustee for that portion of the Federal estate in which are preserved examples of some of the greatest phases of America.

Questions that have been latent ever since the first "national park" was established in 1872 are now sharply brought to issue. Most important among these is whether we can justify, under wartime conditions, the concept that some reasonable percentage of the lands of the Nation should be held inviolate from commercial exploitation on the theory that, thus administered, they serve their highest purpose.

The test is whether this austere ideal can be maintained in war as in peace.

Today, the lands within the forty-eight States designated as the National Park System, which consists of 163 distinct areas, amount to approximately fifteen million acres, or about three-fourths of one percent of the Nation's total land area. The function of the National Park Service is to protect, administer, and interpret this nationwide system of significant and irreplaceable properties, which Congress, by the Act of August 25, 1916, establishing the National Park Service, indicated was to be so administered as "to conserve the scenery and the natural and historic objects, and the wildlife therein, and to provide for the enjoyment of the same in such manner as will leave them unimpaired for the enjoyment of future generations." The American people have every right to expect that so far as possible their national parks will be held intact as an important part of their cultural heritage.

Stress of war has compelled the National Park Service to take stock of its functions and responsibilities. Travel to the parks has declined to the point where it is only twenty-seven percent of the peak figure of 22,000,000 visitors in 1941, and it is obvious that increasing transportation restrictions will result in still further decreases. Appropriations and personnel have been curtailed and activities of many sorts have been suspended. Until after the war, the main task of the Service is one of protection and maintenance.

Travel to the parks, however, has not ceased, nor are they closed to visitors. A flexible system has been adopted which will adapt services and accommodations to the varying need as it develops under rapidly changing conditions. In war, no less than in peace, the national park areas serve as havens of refuge for those fortunate enough to be able to visit them. Affording an environment that gives relief from the tension of a warring world, the parks are, even now, being looked upon as a factor in the physical and mental rehabilitation that will be increasingly desirable as the war progresses. It has been necessary for the Secretary of the Interior to discourage civilian use of transportation resources involved in long-distance travel. Civilians not close by will find it exceedingly difficult to visit the parks. But even though the demands of war may sharply curtail civilian use for a time, Americans take pride and courage in the fact that the national parks and monuments are being protected and will be available for future enjoyment.

A new form of use has arisen. Since Pearl Harbor, approximately two million members of the armed forces have visited the areas administered by the National Park Service. There is significant meaning and definite justification of the national park concept in the fact that increasing thousands of members of the armed forces are being given opportunities they never had before and may never have again to see the inspiring beauty of this land of ours. A typical letter from the commanding officer of one battalion states: "Officers and enlisted men of this battalion join with me in expressing our appreciation to you and the members of your staff for the enjoyable time spent at Grand Canyon. I am sure the pleasant memories of this battalion's visit to Grand Canyon will long be remembered as a most worthwhile and educational trip."

Families of men on furlough have made great efforts to give their boys lasting impressions and experiences. One of the park rangers met an elderly gentleman on a Yosemite trail and asked him how he was enjoying himself. He said: "This trip is not for me. It is for my son who is a sailor in the United States Navy and has a furlough for a few days. I wanted to get him away from the war and give him something peaceful and beautiful to remember. One of the main reasons for this is that my other son was killed on the battleship *Arizona* at Pearl Harbor."

Another phase of the usefulness of the parks to the military organization is indicated in a letter the park superintendent of Hawaii National Park received from a major general at the time he was transferred: "When I arrived, you turned over to me your new park headquarters building. Without it and your very effective cooperation in assisting me to carry out my mission, the work of my headquarters would have been seriously hampered. It is my hope, that when peace has come again, I may have the pleasure of visiting this wonder of nature with the leisure to enjoy the many scenes and exhibits for which, up to now, there has not been time available."

In defining the purpose of this war, President Roosevelt has said that we are protecting a great past. Evidences of this past, both as to human history and natural history, are contained in areas of which the National Park Service is custodian. The march of freedom in the United States is realistically revealed at the scenes where our forefathers fought and won independence and national unity. The battlefields of Yorktown and Gettysburg, the Statue of Liberty, the sites of pioneer exploration and national expansion, are well-springs of patriotism from which thousands gain inspiration and renewed courage. Pride in America swells in the hearts of all who look upon the mile-deep chasm of Grand Canyon, the geysers and hot springs of Yellowstone, the thundering waterfalls of Yosemite, the towering Sequoias, and the sweep of mighty forests on the Olympic Peninsula. Consciously or unconsciously there is built up within all who have had such experiences an increased faith in our country. Can these experiences fail to strengthen the conviction that this is a nation worth fighting for?

Direct use of national park lands not contemplated in peace time has been arranged when required by the needs of war. Over 675 permits have been issued to the Army, Navy, and Coast Guard, and other war agencies for the utilization of lands, buildings, and facilities. These permits cover a wide range of uses, some of them confidential. In the nation's capital, for example, the National Capital Parks have turned over to war activities lands and buildings valued at $24,300,000, which represents a saving to the Government, since the alternative would have been to purchase others, probably at greater cost. Dollars cannot be used to evaluate all of what the National Park Service has done in connection with the war program. There is hardly an area in the National Park System that has not made some direct contribution to aid in winning the war.

Despite extensive military use of the parks, there has been little destruction of permanent values. The Service has applied the same criteria to such uses as it has to proposals for lumbering, mining, grazing, or other non-conforming activities within the national parks. Everyone is aware that the cost of victory in this war is going to be high and that the natural resources of the nation will be called upon more and more to meet the shortages in available strategic materials. Inconvenience to park administration and to park visitors or remediable damage to park property have not been considered sufficient reasons for denying the use of park resources for war purposes. Only where uses proposed would do irreparable damage and entail destruction or impairment of distinctive features and qualities in the parks has the Service felt justified in raising these questions: Have all reasonable alternatives been exhausted before invading the national parks areas? Is the demand based upon critical necessity?

A case very much in point at this time is the pressure being brought to bear for the logging of Sitka spruce in Olympic National Park. Sitka spruce furnishes the outstanding wood for aircraft construction and the easily available supply of this species suitable for aircraft material in the commercial forests of Washington and Oregon is rapidly decreasing. There are, however, further stands of limited extent that could be developed by construction of access roads. There is also an abundance of Sitka spruce in the national forests of Alaska, but the shortage of manpower and equipment and the difficulties of transportation in rafts to Puget Sound sawmills have prevented a sufficient output to overcome the reduced production in Washington and Oregon.

There surely is ample justification for the consensus among conservation leaders that the forests in the national parks should not be cut unless the trees are absolutely essential to the prosecution of the war, with no reasonable alternative. Critical necessity rather than convenience should be the governing reason for such a sacrifice of an important part of the federal estate. If Olympic National Park is opened to the logging of spruce to meet war needs for aircraft materials, there will exist great danger that pressure to widen this breach will be injected by local interests to maintain local industries after the war is over. That issue was given consideration by Congress and definitely decided on a basis of national good when the Olympic National Park was established.

If the war lasts several years, all of the readily available airplane spruce, not only in the park, but in Oregon and Washington, will be exhausted. In that event, the transition to the use of substitutes, or the fuller use of Alaska and British Columbia spruce, will have to be made.

It may fairly be asked whether, in view of the national importance of these last remnants of the once vast virgin forests of the Olympic Peninsula, the alternatives should not be exhausted before, rather than after, these forests are destroyed and an outstanding natural spectacle is lost to America forever.

From an economic appraisal of our natural resources, it should be remembered that approximately one-third of the land area of the continental United States, exclusive of Alaska, or 630 million acres is forested. Of this, some six and one-half million acres of forest are contained within the national parks and monuments, and 122 million acres within the national forests. Thus, the national park forests amount to approximately one percent of the total—certainly a very small fraction to be held inviolate in accordance with the national park pattern as established by Congress.

The greater part of the national park forests may be considered as "protection forests" which, because of the high elevations at which they grow, are of great value both scenically and for watershed protection, but are of little value for commercial use. The magnificent park forests at the lower elevations, however, though having potentially a high commercial value, are worthy of permanent preservation as outdoor museums for the benefit of this and future generations. If we should cut these forests, which include outstanding specimens hundreds of years of age, we not only would lose these forest giants forever, but also lose something far more fragile—the delicate ecological balance which exists among all the elements of the forest.

There are those who advocate the utilization of mature trees within the national parks on a selective logging basis. Such persons miss entirely the main point of the national park philosophy for they fail to realize that the removal of any portion of the forest under any system of logging, however restrictive, is contrary to the very principle upon which the national parks and monuments were established. *Once logging is introduced into any area, it no longer exists as a superlative virgin forest.* Future generations would be the poorer for the loss of the virgin forests in the parks, just as the present generations are poorer because they have no worthwhile illustrations of the magnificent white pine forests which were once a distinctive feature of the eastern portion of this nation.

The grazing interests also have sought entrance into the national parks and monuments in order to aid the war program. They are urged to increase production of food, hides and fiber, and, like the lumberman, are making sincere efforts to meet the nation's war requirements. In response to requests to open the national parks and monuments to grazing, the National Park Service recently undertook a careful analysis of all areas under its jurisdiction wherein grazing might be a possibility. The results of this exhaustive study proved that, within the great National Park System as a whole, the grazing potentiality is comparatively poor and is very limited. The study did reveal, however, that as a wartime emergency to meet the critical need for food and fiber, grazing could be increased on certain historic areas where livestock is not inimical to the preservation of the historical scene, and on recreational areas where it will not result in permanent physical injury.

A small amount of grazing still exists in ten national parks and a considerable amount in thirty-three of the national monuments and other areas. This situation was inherited at the time the areas were acquired, and is gradually being eliminated. Secretary Ickes, in appearing before the House Public Lands Committee on June 1, this year, said that the grazing rights of individuals in Jackson Hole National Monument would be guaranteed, and stated further that *"such action in no way nullifies the long-time policy, which I have recently reaffirmed, under which grazing in national parks and monuments will be gradually decreased and ultimately eliminated.* This policy is based upon long experience in the protection of scenery, wildlife, vegetation and other unique features in the national parks and monuments."

The damage to primeval areas resulting from grazing of cattle and sheep is extensive. Domestic animals in herds or flocks cause a definite change in the forests. Ecological balances are disrupted. Natural conditions are progressively altered and continued grazing will deteriorate the flowering meadows and grassy hillsides by changing the natural succession of plants. There is ample justification for the policy that grazing should never be allowed in truly superlative areas, and that it should be excluded as soon as possible from all national parks and from national monuments of the "wilderness" type.

Proposals have been made to mine certain critical minerals in national parks and monuments. The Service has taken the position that such invasion of a national park can be justified only when it would furnish strategic or critical minerals indispensable to the war and not obtainable in sufficient quantities elsewhere. In general, studies indicate that strategic minerals in national park areas are not of sufficient quantity or economic value to justify their extraction. There are exceptions, like one case in Yosemite National Park, California, where a valuable deposit of sheelite ore (tungsten) was discovered near the northwest boundary. After investigation by the U.S. Geological Survey, the Bureau of Mines and the National Park Service, in cooperation with the discoverer, it was decided that because of its scarcity this deposit should be mined. Through the Metals Reserve Corporation, a government agency, the ore is now being extracted under careful supervision with a minimum of damage to the primary park exhibit.

The National Park Service is not classed as a war agency, has made no attempt to be so classified, and has not looked upon the war emergency as an opportunity to expand its functions. Nevertheless, there are many important economic values and strategic facilities within the national parks and monuments. They are protected in harmony with the purposes for which these areas were established. Our program has to do with the interpretation of natural values for the public, with emphasis on their esthetic and cultural significance rather than on their industrial and financial worth. Even so, there exists a strong relationship between the national parks and the economy of the nation. The facts are that the vast mountain and plateau regions throughout the National Park System contribute to the sources of rivers which supply industrial and agricultural areas with water for domestic use, navigation, irrigation, power, and recreation. These watersheds must be protected. The economic stability of communities within a radius of several hundred miles of park areas depends upon water supplies originating in national park and monument areas. Many of these communities are the location of vital war industries. Should the forest cover of these watersheds be destroyed there are likely to occur flash floods, rapid erosion, and stream pollution which would threaten the future of many communities and industries dependent upon a constant supply of good water.

This is an integrated society. The national parks and monuments represent an institution that has its power and proportionate place in our national life. They are a segment of the federal estate that has been chosen for preservation so that this and future generations will see the untamed America that was, and understand the compelling influences that built and strengthened this nation. We cannot lightly abandon them, or the idea that gave them being, although we may have to sacrifice both in part if compelled to do so by the needs of war. Dr. John C. Merriam, in his recent book, "The Garment of God," points out that man cannot live in an isolated present separated from the past and future. The nation which forgets its past will have no future worth remembering.

The wisdom of the nation in preserving areas of the type represented by national parks and monuments is clearly evident on the American continent today as increased demands upon our natural resources are invading and forever changing the native landscape. As long as the basic law that created them endures, we are assured of at least those few places in the world, where forests continue to evolve normally, where animal life remains in harmonious relationship to its environment, and where the ways of nature and its works may still be studied in the original design.

Reprinted from *American Forests*, August 1943, with permission.

NATIONAL PRIMEVAL PARK STANDARDS
A DECLARATION OF POLICY

National Parks Association, 1945

I. DEFINITION

National primeval parks are spacious land and water areas essentially in their primeval condition and so outstandingly superior in quality and beauty to average examples of their several types as to make imperative their preservation intact and in their entirety for their enjoyment, education and inspiration of all the people for all time.

In the Convention on Nature Protection and Wildlife Preservation in the American Republics, the term "national parks" has been defined as denoting areas "established for the protection and preservation of superlative scenery, flora and fauna of national significance which the general public may enjoy and from which it may benefit when placed under public control."

It follows:

1. That primeval park areas must be of national importance to warrant their commitment to national care.

2. That the area of each primeval park must be a comprehensive unit embracing all territory required for effective administration and for continuing representation of its flora and fauna.

3. That each primeval park area shall be a sanctuary for the scientific study and preservation of all animal and plant life originally within its limits, to the end that all native species shall be preserved as nearly as possible in their aboriginal state.

4. That wilderness features within any primeval park shall be kept unmodified except insofar as the public shall be given reasonable access to outstanding spectacles.

5. That with respect to any unique geological formations or historic or prehistoric remains within its confines, each primeval park shall be regarded as an outdoor museum, the preservation of whose treasures is a sacred trust.

6. The educational and spiritual benefits to be derived from contact with pristine wilderness are of prime importance to all people, and call for the existence and vigilant maintenance of primeval park areas by responsible government agencies.

7. That primeval parks must be kept free from commercial use,

and that sanctuary, scientific and inspirational uses must always take precedence over non-conforming recreational uses.

II. RECOMMENDED POLICY

The areas to be included in the national primeval park group must conform to the standards for such parks herein set forth. Areas that may be added to this group must be units that will fully maintain or increase its supreme scenic magnificence, its scientific and educational superiority, and its character as a unique national institution.

It is desirable that, as a general principle, national primeval parks should differ as widely as possible from one another, and the National Primeval Park System should represent a wide range of typical areas of supreme quality.

To preserve the National Primeval Park System, it must be recognized: (1) that any infraction of standards in any primeval park constitutes an invasion of the system; (2) that the addition to the system, as a national primeval park, of any area below standard lowers the standard of the system. Every proposed use of any primeval park in defiance of national primeval park standards, and the admission to the system of any area falling short of the standards must be resisted. Areas primarily of local interest must not be admitted to the National Primeval Park System.

III. LEGISLATION

1.   Procedure: The first official act toward the creation of an national primeval park is usually the introduction of a bill in Congress. Since the beginning of the system in 1872, according to established precedent, the bill is referred to the Public Lands committees of Senate and House. These committees in turn refer it to the Secretary of the Interior for a report on the standards and availability of the proposed park. The Secretary of the Interior in due course refers the bill to the National Park Service for examination of the area and for a report to him. The Secretary embodies the recommendations of the National Park Service in his report to the Congress which is then in position to take action. Public hearings are often held by the appropriate committees prior to making their reports to the Congress.

2.   Recommendations: (1) The examination of an area to determine its suitability as a primeval park should be made at the expense of the federal government and not at the expense of the local community which would benefit by the park's creation. Committees to

consider boundary problems should be strictly advisory to the federal administration to which alone they should be empowered to report. (2) Exact metes and bounds based upon studies made by the National Park Service should be established by Congress in the organic act creating every new park. The federal government should purchase, as soon as practicable, alienated areas within the boundaries of an existing primeval park, and also areas necessary to round out such park. (3) No steps affecting an existing primeval park or concerned with the creation of a new primeval park should be taken without a prior study and approval of the National Park Service which alone possesses the requisite knowledge, tradition and experience united with responsibility to the people. No area offered for the creation of a new primeval park should be considered by Congress until a study has been made of the area by the National Park Service and its recommendations secured. On the recommendation of the National Park Service, park areas should be extended so as to include feeding grounds for the wildlife found therein. (4) Appropriations should be adequate to enable the National Park Service to protect existing parks and their forests against fire, vandalism and other agencies of destruction, and to maintain the system in accordance with national primeval park standards. (5) All existing national primeval parks now up to the standards set forth should remain as created, subject to modification only upon the favorable recommendation of the Secretary of the Interior and the Director of the National Park Service, based upon expert investigation.

IV. ADMINISTRATION

In administering national primeval parks it is recommended:

1.   That each park be administered with the primary objective of conserving its highest scientific and inspirational usefulness to the people of the nation.

2.   That no commercial use or activity such as logging, mining, grazing or damming of water courses should be permitted on primeval park lands, by exchange or otherwise.

3.   That scientific, educational and inspirational values dictate the major uses of primeval parks.

4.   That attracting crowds for the sake of records or profits, and the introduction of non-conforming recreational activities be regarded as violations of the national primeval park standards.

5.   That scientific administration be applied to all phases of park maintenance, and particularly to the preservation of wilderness, wildlife and geological features.

6.   That a suitable educational program be developed by the

National Park Service, using the natural features of the parks as instructional material. The National Park Service should inform the public concerning park purposes and functions, and emphasize the necessity of caring for and protecting irreplaceable objects of natural and scientific interest. No visitor to a primeval park area should leave without having been informed about the special significance of that particular area, as well as of the system as a whole.

7.   That roads be developed in each national primeval park only in order to bring the people in touch with its principal features and for the purpose of protecting the park. In every instance they should be constructed and placed so that they will cause the least possible impairment to natural features. Wilderness, sanctuary and research areas should be reached by trail only.

8.   That public airplane landing fields, as well as railroad stations, be located outside park boundaries. Flying across primeval parks, if permitted at all, should be closely regulated.

9.   That park buildings be as unobtrusive as possible, harmonizing with their surroundings. They should be erected only where necessary for the protection of the parks and for the comfort of visitors, and at locations where they will least interfere with natural conditions.

10.   That concessions be granted only for such business as is necessary for the care and comfort of visitors, and then in definitely localized areas. Such concessions should not interfere with the rights of individuals under park rules to provide for themselves while visiting the park.

11.   That the use of any primeval park interfere as little as possible with the rights of future generations to enjoy nature unmodified.

NATIONAL MONUMENTS

These standards should apply also to national monuments that are of similar character and purpose as the national primeval parks.

National Park Service Archives, Harpers Ferry, General Collection Box 5410. Reprinted with permission of the National Parks and Conservation Society.

AN ACT TO ESTABLISH A NEW GRAND TETON NATIONAL PARK IN THE STATE OF WYOMING, AND FOR OTHER PURPOSES,

Approved September 14, 1950 (64 Stat. 849)

Be it enacted by the State and House of Representatives of the United States of America in Congress assembled, That, for the purpose of including in one national park, for public benefit and enjoyment, the lands within the present Grand Teton National Park and a portion of the lands within the Jackson Hole National Monument, there is hereby established a new "Grand Teton National Park". The park shall comprise, subject to valid existing rights, all of the present Grand Teton National Park and all lands of the Jackson Hole National Monument that are not otherwise expressly provided for in this Act, and an order setting forth the boundaries of the park shall be prepared by the Secretary of the Interior and published in the Federal Register. The national park so established shall, so far as consistent with the provisions of this Act, be administered in accordance with the general statutes governing national parks, and shall supersede the present Grand Teton National Park and the Jackson Hole National Monument. The Act of February 26, 1929 (45 Stat. 1314), and any other provisions of law heretofore specifically applicable to such present park or monument, are hereby repealed: *Provided*, That no further extension or establishment of national parks or monuments in Wyoming may be undertaken except by express authorization of the Congress. (16 U.S.C. §§ 406d-1 and note, 431a, 451a.)

SEC. 2. The following described lands of the Jackson Hole National Monument are hereby made a part of the National Elk Refuge and shall be administered hereafter in accordance with the laws applicable to said refuge:

[description given]

Containing in all six thousand three hundred and seventy-six acres, more or less. (16 U.S.C. § 673b.)

SEC. 3. The following-described lands of the Jackson Hole National Monument are hereby made a part of the Teton National Forest and shall be administered hereafter in accordance with the laws applicable to said forest:

[description given]

containing in all two thousand eight hundred six and thirty-four one-hundredths acres, more or less. (16 U.S.C. § 482m.)

SEC. 4. With respect to those lands that are included by this Act within the Grand Teton National Park—

(a)   the Secretary of the Interior shall designate and open rights-of-way, including stock driveways, over and across Federal lands within the exterior boundary of the park for the movement of persons and property to or from State and private lands within the exterior boundary of the park and to or from national forest, State, and private lands adjacent to the park. The location and use of such rights-of-way shall be subject to such regulations as may be prescribed by the Secretary of the Interior;

(b)   all leases, permits, and licenses issued or authorized by any department, establishment, or agency of the United States with respect to the Federal lands within the exterior boundary of the park which are in effect on the date of approval of this Act shall continue in effect, subject to compliance with the terms and conditions therein set forth, until terminated in accordance with the provisions thereof;

(c)   where any Federal lands included within the park by this Act were legally occupied or utilized on the date of approval of this Act for residence or grazing purposes, or for other purposes not inconsistent with the Act of August 25, 1916 (39 Stat. 535), pursuant to a lease, permit, or license issued or authorized by any department, establishment, or agency of the United States, the person so occupying or utilizing such lands and the heirs, successors, or assigns of such person, shall, upon the termination of such lease, permit, or license, be entitled to have the privileges so possessed or enjoyed by him renewed from time to time, subject to such terms and conditions as the Secretary of the Interior shall prescribe, for a period of twenty-five years from the date of approval of this Act, and thereafter during the lifetime of such person and the lifetime of his heirs, successors, or assigns but only if they were members of his immediate family on such date, as determined by the Secretary of the Interior: *Provided,* That grazing privileges appurtenant to privately owned lands located within the Grand Teton National Park established by this Act shall not be withdrawn until title to lands to which such privileges are appurtenant shall have vested in the United States, except for failure to comply with the regulations applicable thereto after reasonable notice of default: *Provided further,* That nothing in this subsection shall apply to any lease, permit, or license for mining purposes or for public accommodations and services or to any occupancy or utilization of lands for purely temporary purposes. Nothing contained in this Act shall be construed as creating any vested right, title, interest, or estate in or to any Federal lands. (16 U.S.C. § 406d-2.)

SEC. 5. (a) In order to provide compensation for tax losses sustained as a result of any acquisition by the United States, subsequent to March 15, 1943, of privately owned lands, together with any improvements thereon, located within the exterior boundary of the Grand Teton National Park established by this Act, payments shall be made to the State of Wyoming for distribution to the county in which such lands are located in accordance with the following schedule of payments: For the fiscal year in which the land has been or may be acquired and nine years thereafter there shall be paid an amount equal to the full amount of annual taxes last assessed and levied on the land, together with any improvements thereon, by public taxing units in such county, less any amount, to be determined by the Secretary of the Interior, which may have been paid on account of taxes for any period falling within such fiscal year. For each succeeding fiscal year, until twenty years elapse, there shall be paid on account of such land an amount equal to the full amount of taxes referred to in the preceding sentence, less 5 per centum of such full amount for each fiscal year, including the year for which the payment is to be made: *Provided,* That the amount payable under the foregoing schedule for any fiscal year preceding the first full fiscal year following the approval of this Act shall not become payable until the end of such first full fiscal year.

(b) As soon as practicable after the end of each fiscal year, the amount then due for such fiscal year shall be computed and certified by the Secretary of the Interior, and shall be paid by the Secretary of the Treasury: *Provided,* That such amount shall not exceed 25 per centum of the fees collected during such fiscal year from visitors to the Grand Teton National Park established by this Act and the Yellowstone National Park. Payments made to the State of Wyoming under this section shall be distributed to the county where the lands acquired from private landowners are located and in such manner as the State of Wyoming may prescribe. (16 U.S.C. § 406d-3.)

SEC 6. (a) The Wyoming Game and Fish Commission and the National Park Service shall devise, from technical information and other pertinent data assembled or produced by necessary field studies or investigations conducted jointly by the technical and administrative personnel of the agencies involved, and recommend to the Secretary of the Interior and the Governor of Wyoming for their joint approval, a program to insure the permanent conservation of the elk within the Grand Teton National Park established by this Act. Such program shall include the controlled reduction of elk in such park, by hunters licensed by the State of Wyoming and deputized as rangers by the Secretary of the Interior, when it is found necessary for the purpose of proper management and protection of the elk.

(b)   At least once a year between February 1 and April 1, the Wyoming Game and Fish Commission and the National Park Service shall submit to the Secretary of the Interior and to the Governor of Wyoming, for their joint approval, their joint recommendations for the management, protection, and control of the elk for that year. The yearly plan recommended by the Wyoming Game and Fish Commission and the National Park Service shall become effective when approved by the Secretary of the Interior and the Governor of Wyoming, and thereupon the Wyoming Game and Fish Commission and the Secretary of the Interior shall issue separately, but simultaneously such appropriate orders and regulations as are necessary to carry out those portions of the approved plan that fall within their respective jurisdictions. Such orders and regulations, to be issued by the Secretary of the Interior and the Wyoming Game and Fish Commission, shall include provision for controlled and managed reduction by qualified and experienced hunters licensed by the State of Wyoming and deputized as rangers by the Secretary of the Interior, if and when a reduction in the number of elk by this method within the Grand Teton National Park established by this Act is required as a part of the approved plan for the year, provided that one elk only may be killed by each such licensed and deputized ranger. Such orders and regulations of the Secretary of the Interior for controlled reduction shall apply only to the lands within the park which lie east of the Snake River and those lands west of Jackson Lake and the Snake River which lie north of the present north boundaries of Grand Teton National Park, but shall not be applicable to lands within the Jackson Hole Wildlife Park. After the Wyoming Game and Fish Commission and the National Park Service shall have recommended to the Secretary of the Interior and the Governor of Wyoming in any specified year a plan, which has received the joint approval of the Secretary of the Interior and the Governor of Wyoming, calling for the controlled and managed reduction by the method prescribed herein of the number of elk within the Grand Teton National Park established by this Act, and after the Wyoming Game and Fish Commission shall have transmitted to the Secretary of the Interior a list of persons who have elk hunting licenses issued by the State of Wyoming and who are qualified and experienced hunters, on or before July 1 of that year the Secretary of the Interior, without charge, shall cause to be issued orders deputizing the persons whose names appear on such list, in the number specified by the plan, as rangers for the purpose of entering the park and assisting in the controlled reduction plan. Each such qualified hunter, deputized as a ranger, participating in the controlled reduction plan shall be permitted to remove from the park the carcass of the elk he has killed as a part of the plan. (16 U.S.C. § 673c.)

SEC. 7. The Secretary of the Interior is authorized to accept the donation of the following-described lands, which lands, upon acceptance by the United States, shall become a part of the national park:

SIXTH PRINCIPAL MERIDIAN

Township 41 north, range 116 west: Section 3, lots 1 and 2.

Containing seventy-eight and ninety-three one-hundredths acres, more or less. (16 U.S.C. § 406d-4.)

SEC. 8. All temporary withdrawals of public lands made by Executive order in aid of legislation pertaining to parks, monuments, or recreational areas, adjacent to the Grand Teton National Park as established by this Act are hereby revoked. (16 U.S.C. § 406d-1 note.)

SEC. 9. Nothing in this Act shall affect the use for reclamation purposes, in accordance with the Act of June 17, 1902 (32 Stat. 388), and Acts amendatory thereof or supplementary thereto, of the lands within the exterior boundary of the park as prescribed by this Act which have been withdrawn or acquired for reclamation purposes, or the operation, maintenance, rehabilitation, and improvement of the reservoir and other reclamation facilities located on such withdrawn or acquired lands. All provisions of law inconsistent with the provisions of this Act are hereby repealed to the extent of such inconsistency. The remaining unexpended balance of any funds appropriated for the present Grand Teton National Park and the Jackson Hole National Monument shall be available for expenditure in connection with the administration of the Grand Teton National Park established by this Act. (16 U.S.C. §§ 406d-5, 406d-1 note.)

LET'S CLOSE THE NATIONAL PARKS

By Bernard DeVoto
1953

The chief official of a national park is called the Superintendent. He is a dedicated man. He is also a patient, frustrated, and sorely harassed man. Sit in his office for an hour some morning and listen to what is said to him by the traveling public and by his administrative assistant, the Chief Ranger.

Some of his visitors are polite; some aren't; all have grievances. A middle-aged couple with a Cadillac make a formal protest: it is annoying that they must wait three-quarters of an hour to get a table at Lookout Point Lodge, but when it comes to queuing up in order to use the toilets at the Point—well, really! A woman in travel-stained denim is angry because Indian Creek Camp Ground is intolerably dusty. Clouds of dust hang over it, dust sifts into the sleeping bags at night, dust settles on the food and the children and the foliage, she has breathed dust throughout her two-weeks stay. Another woman reports that the toilet at Inspiration Cliff Camp Ground has been clogged since early last evening and that one of the tables there went to pieces at breakfast time. A man pounds the desk and shouts that he hit a chuck-hole on Rimrock Drive and broke a spring; the Drive, he says, is a car-killer and will soon be a man-killer. Another enraged tourist reports that a guardrail collapsed when his little girl leaned against it and that she nearly fell into the gorge. The representative of a nature society sums up his observations. He has hardly seen a ranger since he reached the park. (One reason is that most of the rangers are up in the high country fighting a forest fire.) Tourists have picked all the bear grass at Eyrie Overlook and the observer doubts if the species will come back there. Fifty-one names have been freshly carved in the vicinity of Cirque Falls, some of them actually on the famous Nine Centuries Tree itself. All but one of the camp grounds look like slums; in the observer's opinion, the reason why they look that way is that they are slums.

Such complaints must be distinguished from the irrational ones voiced to the Superintendent by tourists who are cantankerous, crackbrained, tired, or merely bewildered. They must be so distinguished because they are factual and true. (The Superintendent, not having a plumber, will send a ranger to clean out the toilet but replacing the guard rail will leave him too little money to buy lumber for a new table. He squeezed $1,200 from his budget to enlarge Indian Creek

Camp Ground and so reduce the dust there but Brawling River undercut fifty feet of main road and the emergency repairs cost $1350.) He answers all complaints courteously, as a representative of the National Park Service and the United States Government, but he has no effective answer. He is withheld from saying what would count, "Build a fire under your Congressman." He cannot go on and explain that the Service is suffering from financial anemia, that it is the impoverished stepchild of Congress, and that the lack of money has now brought our national park system to the verge of crisis. He cannot say this and neither can his superiors in the Washington office, but it is true.

Between visitors the Chief Ranger has been developing this theme. He got together a crew yesterday and put them to work on the decaying bridge inside the north entrance; it can be shored up for the rest of this season but next year it will be beyond help and the north entrance will have to be closed. He also went over Beaver Creek Trail again yesterday and he is scared; unless some work can be done on it at once it must be closed as unsafe. Costs on last week's rescue job are now in. Fourteen men worked three shifts a day for two days to bring that climber with a broken hip down from Deception Peak. A doctor had to be summoned from eighty miles away and an ambulance from a hundred and seventy-five miles. The episode cost just over a thousand dollars, which will have to come out of the budget, and this means one summer ranger less next year. (In 1936 the park had two more summer rangers than it has this year—and only one-twelfth as many visitors.) Furthermore, Ranger Doakes, an expert alpinist, has demanded overtime pay for that rescue—sacrilege in the Service, but the Chief Ranger cannot blame him. The recent increase in rents hit Ranger Doakes hard. He got only a 137 per cent increase, which was less than some others, but it brought his rent to 23.5 per cent of his annual salary.

Let's leave the Chief Ranger's remaining woes unprinted and look at this latest device for reducing pay by compelling personnel to subsidize the National Park Service budget. The most valuable asset the Service has ever had is the morale of its employees. I have said that the Superintendent is a dedicated man; all his permanent staff and all the temporary rangers and ranger-naturalists are dedicated men, too—they are all lovers and all fanatics or they would have quit long since. Ever since it was organized the Service has been able to do its difficult, complex, and highly expert job with great distinction because it could count on this ardor and devotion. The forty-hour week means nothing in a national park. Personnel have always worked sixteen hours a day and seven days a week whenever such labor was necessary. Superintendent, rangers, engineers, summer staff, fire lookouts—they all drop their specialties to join a garbage-disposal crew or a rescue party, to sweep up tourist litter, to clean a defouled spring, to do anything else that has to be done but can't be paid for. They are the most courteous and the most patient men in the United States and maybe once a week several of them get a full night's sleep. If you undermine their morale, you will destroy the Service. Well, the latest increase in rents has begun to undermine it.

By decree of the Bureau of the Budget the rents of government housing must be equalized with those of comparable housing in the same locality. In the end this amounts to some sleight of hand in the bookkeeping of the U.S. Treasury but it is probably sound in theory. Sound, that is, for a lot of government housing—but not for that which, to a varying degree, shields NPS employees from the weather. In the first place, the locality with which rents must be equalized is the nearest resort town outside the park, where rents are two or three times as high as in the nearest non-parasitical town. In the second place, there is practically no comparable housing. These are not the massive dwellings of a military installation, the imposing and luxurious ones that the Bureau of Reclamation erects, or the comfortable cabins of the Forest Service that were built by the CCC. Apart from a few such cabins by the CCC and a few new structures which the Service has been able to pay for from the pin-money that passes as its appropriations, they are either antiques or shacks. The best of them are usually inadequate—one-bedroom houses for couples with two or more small children, two-bedroom houses for couples with two or more adolescent children. Many of the rest of them belong in the Hoovervilles of 1931—CCC barracks built of tar-paper in 1934 and intended to last no more than five years, old warehouses and cook shacks built of slabs, curious structures hammered together from whatever salvaged lumber might be at hand. I have seen adobe huts in damp climate that were melting away from the rain, other quarters that were race-courses for rats, still others that would produce an egg shortage if you kept chickens in them.

Park Service employees are allowed an "isolation deduction" of from five to forty per cent, intended to compensate them for being forced to live at a galling and expensive distance from the service of civilization. Even so, the already high rents have been cruelly increased by the last directive from the Bureau of the Budget. On a list I have at hand of seventeen dwellings in Grand Teton National Park, the lowest increase *(after* the isolation deduction) is one hundred per cent, the highest two hundred per cent, the average one hundred and fifty-plus.

At this park there is an associated ingenuity. The park pays Teton

County, Wyoming, $26,000 a year in lieu of taxes; it produces God knows how much for the state in gasoline and sales taxes: the business brought in by its visitors is all that keeps the town of Jackson solvent or even alive. But a hangover from the controversy over Jackson Hole National Monument, a controversy created for profit by local politicians and the gamblers and land-speculators allied with them, has enabled the town of Jackson to pressure the state administration. By decree of the state Attorney General, park personnel are not residents of Wyoming, though any itinerant Okie who paused there would be, and must therefore pay for their children who attend public schools. They total $158 per pupil. It makes quite an item in the family finance of an underpaid public servant who has now had his rent increased, the rent of a leaky and rat-ridden crate which he cannot select but must take as assigned—and in which he gets no equity though he pays a fifth of his salary or more.

This last summer I visited some fifteen NPS areas. It was a commonplace to meet a park employee who had to bring a son or daughter back from college, as a result of the rent increase. It was even commoner to find one who had decided that the kids could not go to college when they finished high school. In many places, wives of park personnel are working for private firms licensed to operate businesses in the parks, and this is a highly undesirable practice. The chief clerk of one of the most important parks works weekends in a grocery store in order to stay fed while retaining the job he loves. I could add to these specimens indefinitely but let it go with the end-product: the most valuable asset of the National Park Service is beginning to erode away.

So are the parks and national monuments themselves. The deterioration of roads and plants that began with the war years, when proper maintenance was impossible, has been accelerated by the enormous increase in visitors, by the shrinkage of staffs, and by miserly appropriations that have prevented both repair and expansion of facilities. The Service is like a favorite figure of American legendry, the widow who scrapes and patches and ekes out, who by desperate expedients succeeds in bringing up her children to be a credit to our culture. (The boys work the graveyard shift in the mills; the girls' underwear is made of flour sacking.) Its general efficiency, the astonishingly good condition of its areas, its success at improvising and patching up is just short of miraculous. But it stops there, short of the necessary miracle. Congress did not provide money to rehabilitate the parks at the end of the war, it has not provided money to meet the enormously increased demand. So much of the priceless heritage which the Service must safeguard for the United States is beginning to go to hell.

Like a number of other small areas in the system, the Black Canyon of the Gunnison has *no* NPS personnel assigned to it. On one rim of this spectacular gorge there are a few inadequate guard rails, on the other and more precipitous rim there are none. When I visited it, one of the two registers for visitors and all the descriptive pamphlets had been stolen. The ranger force at Mesa Verde National Park is the same size it was in 1932; seven times as many people visited it in 1952; the figures for June 1953 were up 38 per cent from last year's. The park can man the entrance station for only one shift: automobiles which arrive in late afternoon cannot be charged the modest entrance fee. It cannot assign a ranger exclusively to fire-duty at headquarters, though it is in an arid region where destructive fire is a constant danger: the headquarters ranger must keep the fire-alert system operating while he attends to a dozen other jobs. All park facilities are strained to the utmost. Stretches of the main road keep sinking and must be repaired at excessive cost because there is not money enough to relocate them where the underlying strata are more stable. There is not even money enough to replace broken guard-rail posts along the edge of the canyon. Colorado and New Mexico are about to construct a new highway past the park to the famous Four Corners. On the day it is completed visitors to Mesa Verde will double in number and the park will be unable to take care of them. It will be paralyzed.

Last year Senator Hunt of Wyoming made a pleasure trip to Yellowstone Park, at least a trip that was intended to be pleasurable. He was so shocked by the condition of the roads that he wrote a letter of protest to President Truman. (It got buried under the election campaign.) And yet, considering the handicaps, Yellowstone has done magnificently with its roads; those of many other parks are in worse condition. (Of the *main* road system in the park 15 per cent is of pre-1920 standard, 42 per cent is pre-1930 standard, and only 27 percent of 1930-1940 standard. Exactly three miles of new road have been constructed since 1945 and those three complete a project that was begun before the war.) This is the oldest, most popular, and most important national park. In 1932, when 200,000 people visited it, its uniformed staff was large enough to permit just over 6,000 man-hours of work per week; last year, with one and one-third million visitors, the shrunken staff performed just over 4,000 man-hours per week. Like nearly every other popular park, it has reached the limit of performance and begun to slide downhill. There are not enough rangers to protect either the scenic areas from the depredations of tourists or the tourists from the consequences of their own carelessness—or to gather up the litter or to collect all the entrance fees that should be paid. Water and garbage and sewage systems are beginning to break down under the load put on them; already some sewage is being discharged in Yellowstone Lake. The park's high plateaus covered with lodgepole pine are natural fire-traps which some day will be burned out because the budget will not permit adequate fire-protection.

I have touched on only a few of Yellowstone's critical problems. What I have said is true also of all the most popular areas administered by the Service and in some degree of almost all the less accessible areas. There are true slum districts in Yellowstone, Rocky Mountain, Yosemite, Mesa Verde, various other parks. The National Park Service does a far better job on its starvation rations than it could reasonably be expected to do, but it falls increasingly short of what it must do. It is charged with the preservation, protection, maintenance, development, and administration of 28 national parks, 5 national historical parks, 85 national monuments, 56 acres of various other classifications, and 785 National Capital parks. Their importance to the American present and future is simply incalculable; they are inestimably valuable. But Congress made no proper provision for rehabilitating the areas at the end of the war or for preparing them for the enormous increase in use—more than thirty million people visited them last year. It could have provided for renovation and expansion at about a fourth or a fifth of what the job would cost now—but it didn't. It requires the Service to operate a big plant on a hot-dog-stand budget.

The crisis is now in sight. Homeopathic measures will no longer suffice; thirty cents here and a dollar-seventy-five there will no longer keep the national park system in operation. I estimate that an appropriation of two hundred and fifty million dollars, backed by another one to provide the enlarged staff of experts required to expend it properly in no more than five years, would restore the parks to what they were in 1940 and provide proper facilities and equipment to take care of the crowds and problems of 1953. After that we could take action on behalf of the expanding future—and save from destruction the most majestic scenery in the United States, and the most important field areas of archeology, history, and biological science.

No such sums will be appropriated. Therefore only one course seems possible. The national park system must be temporarily reduced to a size for which Congress is willing to pay. Let us, as a beginning, close Yellowstone, Yosemite, Rocky Mountain, and Grand Canyon National Parks—close and seal them, assign the Army to patrol them, and so hold them secure till they can be reopened. They have the largest staffs in the system but neither those staffs nor the budgets allotted them are large enough to maintain the areas at a proper level of safety, attractiveness, comfort, or efficiency. They are unable to do the job in full and so it had better not be attempted at all. If these staffs—and their respective budgets—were distributed among other areas, perhaps the Service could meet the demands now put on it. If not, additional areas could be temporarily closed and sealed, held in trust for a more enlightened future—say Zion, Big Bend, Great Smoky, Shenandoah, Everglades, and Gettysburg. Meanwhile letters from constituents unable to visit Old Faithful, Half Dome, the Great White Throne, and Bright Angel Trail would bring a nationally disgraceful situation to the really serious attention of the Congress which is responsible for it.

AN ACT TO FACILITATE THE MANAGEMENT OF THE NATIONAL PARK SYSTEM AND MISCELLANEOUS AREAS ADMINISTERED IN CONNECTION WITH THAT SYSTEM, AND FOR OTHER PURPOSES,

Approved August 8, 1953 (67 Stat. 495)

Be it enacted by the Senate and House of Representatives of the United States of America in Congress assembled, That, in order to facilitate the administration of the National Park System and miscellaneous areas administered in connection therewith, the Secretary of the Interior is hereby authorized to carry out the following activities, and he may use applicable appropriations for the aforesaid system and miscellaneous areas for the following purposes:

1.   Rendering of emergency rescue, fire fighting, and cooperative assistance to nearby law enforcement and fire prevention agencies and for related purposes outside of the National Park System and miscellaneous areas.

2.   The erection and maintenance of fire protection facilities, water lines, telephone lines, and other utility facilities adjacent to any area of the said National Park System and miscellaneous areas, where necessary, to provide service in such area.

3.   Transportation to and from work, outside of regular working hours, of employees of Carlsbad Caverns National Park, residing in or near the city of Carlsbad, New Mexico, such transportation to be between the park and the city, or intervening points, at reasonable rates to be determined by the Secretary of the Interior taking into consideration, among other factors, comparable rates charged by transportation companies in the locality for similar services, the amounts collected for such transportation to be credited to the appropriation current at the time payment is received: *Provided,* That if adequate transportation facilities are available, or shall be available by any common carrier, at reasonable rates, then and in that event the facilities contemplated by this paragraph shall not be offered.

4.   Furnishing, on a reimbursement of appropriation basis, all types of utility services to concessioners, contractors, permittees, or other users of such services, within the National Park System and miscellaneous areas: *Provided,* That reimbursements for cost of such utility services may be credited to the appropriation current at the time reimbursements are received.

5.   Furnishing, on a reimbursement of appropriation basis, supplies, and the rental of equipment to persons and agencies that in cooperation with, and subject to the approval of, the Secretary of the Interior, render services or perform functions that facilitate or supplement the activities of the Department of the Interior in the administration of the National Park System and miscellaneous areas: *Provided,* That reimbursements hereunder may be credited to the appropriation current at the time reimbursements are received.

6.   Contracting, under such terms and conditions as the said Secretary considers to be in the interest of the Federal Government, for the sale, operation, maintenance, repair, or relocation of Government-owned electric and telephone lines and other utility facilities used for the administration and protection of the National Park System and miscellaneous areas, regardless of whether such lines and facilities are located within or outside said system and areas.

7.   Acquiring such rights-of-way as may be necessary to construct, improve, and maintain roads within the authorized boundaries of any area of the said National Park System and miscellaneous areas, and the acquisition also of land and interests in land adjacent to such rights-of-way, when deemed necessary by the Secretary, to provide adequate protection of natural features or to avoid traffic and other hazards resulting from private road access connections, or when the acquisition of adjacent residual tracts, which otherwise would remain after acquiring such rights-of-way, would be in the public interest.

8.   The operation, repair, maintenance, and replacement of motor and other equipment on a reimbursable basis when such equipment is used on Federal projects of the said National Park System and miscellaneous areas, chargeable to other appropriations, or on work of other Federal agencies, when requested by such agencies. Reimbursement shall be made from appropriations applicable to the work on which the equipment is used at rental rates established by the Secretary, based on actual or estimated cost of operation, repair, maintenance, depreciation, and equipment management control, and credited to appropriations currently available at the time adjustment is effected, and the Secretary may also rent equipment for fire control purposes to State, county, private, or other non-Federal agencies that cooperate with the Secretary in the administration of the said National Park System and other areas in fire control, such rental to be under the terms of written cooperative agreements, the amount collected for such rentals to be credited to appropriations currently available at the time payment is received. (16 U.S.C. §1b.)

SEC. 2. (a) The term "National Park System" means all federally owned or controlled lands which are administered under the direction of the Secretary of the Interior in accordance with the provisions of the Act of August 25, 1916 (39 Stat. 535), as amended, and which are grouped into the following descriptive categories: (1) National parks, (2) national monuments, (3) national historical parks, (4) national memorials, (5) national parkways, and (6) national capital parks.

(b) The term "miscellaneous areas" includes lands under the administrative jurisdiction of another Federal agency, or lands in private ownership, and over which the National Park Service, under the direction of the Secretary of the Interior, pursuant to cooperative agreement, exercises supervision for recreational, historical, or other related purposes, and also any lands under the care and custody of the National Park Service other than those heretofore described in this section. (16 U.S.C. §1c.)

SEC. 3. Hereafter applicable appropriations of the National Park Service shall be available for the objects and purposes specified in the Act of August 7, 1946 (60 Stat. 885). (16 U.S.C. §1d.)

MISSION 66
SPECIAL PRESENTATION TO PRESIDENT
EISENHOWER AND THE CABINET BY
DIRECTOR CONRAD WIRTH

January 27, 1956

Mr. President and Members of the Cabinet: thank you for giving us this opportunity to report on the National Parks, and to discuss plans for their future.

As you know, these are the areas where the Nation preserves its irreplaceable treasures in lands, scenery—and its historic sites—to be used for the benefit and enjoyment of the people, and passed on unimpaired to future generations. The areas of the National Park System are among the most important vacation lands of the American people. Today, people flock to the parks in such numbers that it is increasingly difficult for them to get the benefits which parks ought to provide, or for us to preserve these benefits for Americans of tomorrow.

This is why the Department of the Interior and the National Park Service have surveyed the parks and their problems, and propose to embark upon MISSION 66—a program designed to place the national parks in condition to serve America and Americans, today and in the future.

National Parks are an American idea—Yellowstone became the first national park in the world when it was established during the administration of President U.S. Grant. Other parks, monuments, and historic sites have been added—each located for you on this map–until today there are 181 areas, widely scattered throughout the States and Territories. The national parks have become a real part of the American way of life, as attested by the phenomenal increase in their popularity.

But the intangible benefits of refreshment, understanding, and inspiration are not the only dividends. The national parks contribute substantially to the economic life of the Nation. With working hours going down and leisure time going up, vacation travel and vacation spending are, in fact, among the three biggest industries in the States pictured on this map. National park travel—as Estes Park Village— Gateway to Rocky Mountain National Park—suggests—benefits innumerable large and small enterprises throughout the Nation. The more the parks are used for their inherent, cultural, and recreational values, the more they contribute to the economy of the Nation. Here is one resource that earns its greatest human and economic profits the less it is used up.

The problem of today is simply that the parks are being loved to death. They are neither equipped nor staffed to protect their irreplaceable resources, nor to take care of their increasing millions of visitors.

Here is the attendance picture—358 thousand visitors in 1916—21 million in 1941—50 million last year—and by 1966, the parks will have at least 80 million visitors. Are all of these 50 million people finding, as this family group is, the unspoiled refreshment they seek and deserve?

[MOTION PICTURES OF SUMMER 1955]

What we have just seen is a plant operating at 200% capacity. How long would a business concern continue to operate its plant at a 100% overload? If it is to stay in business, it must plan and develop for the demand ahead. We must design to meet the needs of today and tomorrow. That is exactly what MISSION 66 will do. Fortunately, the problems of today can be solved without undue difficulty, without prolonged disruption of public use, and at surprising low cost, if handled on a long-range, one-package basis. By a one-package program I mean simply bringing each part of the program along together, so that when we build a road, the lodges, campgrounds, public use buildings, utility systems and the other things the road leads to will be ready for use at the same time, and, so that the parks will be staffed to meet their responsibilities to the visitor.

Now, specifically, what are the needs today, and how will MISSION 66 meet these needs? There is no point cataloguing every project, but let me give you enough examples to give you the broad picture of the things which MISSION 66 will accomplish: The most obvious is: modern accommodations so that the park visitor can be assured—as this couple was not—of a comfortable place to stay. Much of what we have is out-worn and out-moded, and will be replaced as we double total capacity. We will encourage private business to build more accommodations in the gateway communities near the parks.

Within the parks, we shall encourage greater participation by private enterprise, as illustrated by the new Jackson Lake Lodge in the Tetons. This encouragement will take two forms: Loan guarantees, such as now given to other enterprises through the Small Business Administration, should be extended to some of the smaller concessioners. But, the best encouragement the National Park Service can give is to go forward with its own part of the park development. For example, the Yellowstone Park Company is now ready to build 12 million dollars worth of new accommodations, but must wait until the Government installs utilities at a cost of 4-1/2 million dollars. We have to do our part before the park concessioners can go ahead.

Not all park visitors sleep under a roof. Americans heading for the wide-open spaces are justifiably frustrated when they find themselves intruded upon by the very crowds they intended to leave behind. Campground capacity will be doubled, so that this will be the typical camping picture. Picnic areas will be expanded.

But places to eat and to sleep comprise only a part of what it takes to run a park—many of the other necessary facilities are pitifully inadequate. Public use buildings, ranger stations, fire lookouts, and repair and storage buildings, and other facilities necessary for a complete park operation will be provided. Many of our employees live in shacks—eyesores that shame the Government, mar the scenery, and sap employee morale. Adequate employee housing will be supplied—its upkeep and replacement paid back through rentals. Narrow and dangerous roads must be modernized for safety, and to improve the flow of traffic; trails improved, and the presently authorized National Parkways—essentially completed.

MISSION 66 will give us an adequate staff, to protect the parks against forest fire and other dangers, to maintain the roads and buildings, operate the utility systems, and to serve the visiting public. This is a very important need, for while park attendance has increased from 21 million in 1941 to 50 million today, our field staff has actually decreased in term of man-years of employment.

Today some parks are checkerboarded with unsightly private inholdings—land parcels standing in the way of planned development and use. These should be acquired.

Right along with our own improvement, one of our major goals is the parallel development of County and State parks, Wildlife Refuges, Indian Reservations, National and State Forests, Public lands and Recreation Areas, so that each level of Government and private enterprise, will share its appropriate part of the expanding recreational use load.

These are the pressing needs today—the things MISSION 66 will accomplish—so that the American people can enjoy, in the best sense, their national parks. To put the National Parks in shape is an investment in the physical, mental, and spiritual well-being of Americans as individuals. It is a gainful investment contributing substantially to the national economy, as I have mentioned. It is an investment in good citizenship.

Where else do so many Americans under the most pleasant circumstances come face to face with their Government? Where else but on historic ground can they better renew the idealism that prompted the patriots to their deeds of valor? Where else but in the great out-of-doors as God made it can we better recapture the spirit and something of the qualities of the pioneers? Pride in their Government, love of the land, and faith in the American Tradition—these are the real products of our national parks.

[LIGHTS ON]

And here, gentlemen, are the costs outlined on this chart. As it now stands, the authorized budget for the National Park System for Fiscal Year 1957 totals 62 million, 888 thousand dollars. If we continue to operate at the present scale, the national parks will cost 628 million, 880 thousand dollars over the next ten year period—represented on this chart by the area in green. The yellow represents what MISSION 66 adds to present costs—154 million, 315 thousand, 6 hundred dollars in ten years—the entire one-package 10 year program.

In total new money, gentlemen, MISSION 66 represents a 10 year investment of less than 1/4 the cost of the Grand Coulee Dam. It seems a small amount to devote to the preservation of this national heritage for the use of the people. This is a program pared down to the essentials, and to the essentials only. If adequately financed, most of the MISSION 66—starting this year—can be carried out under present authority. As the program goes forward, additional legislation may be needed. This will be submitted for normal clearance through the Bureau of the Budget, as the need arises.

National Park Service Archives, Harpers Ferry, General Collection Box A8213, Folder entitled, "Cabinet Meeting, Mission 66 Presentation, January 27, 1956."

5

Questions of Resource Management

1957–1963

Mission 66 began in 1956 with great fanfare and optimism. However, within a few years this supposed panacea was mired in controversy, the result of changes in the conservation and science communities with which the National Park Service had not kept pace. The death of George Wright in 1936 had heralded the end of his Wildlife Division if not in name, then certainly in influence. Yet ecological science dramatically matured and diversified. Conservation groups began to take note by the early 1950s but the Park Service clung tenaciously to its concept of "atmosphere preservation" and to other ideals promoted decades earlier by Stephen Mather.

Over this short seven-year period, however, National Park Service management and philosophy would be challenged and ultimately overturned by a shocking series of reports and commentaries from scientists both within and outside the agency. At the same time, the search for additional recreation opportunities for the nation would be rekindled. Abandoned during World War II and the postwar funding crisis, the search for recreation areas, seashores, and state park sites led to a major government inquiry and new responsibilities for the Park Service.

It is often the case that one particular speech or article can catch an organization's attention when others have consistently failed. Throughout the decade of the 1950s pressure to reinvest the Wildlife Division with funds and influence and to manage the parks and monuments with attention to modern scientific data had increased. Yet it was a single speech by Stanley Cain at the Sixth Biennial Wilderness Conference that drove home the point. The eminent ecologist chastised the agency for "missing a bet" by not conducting a natural history research program and tailoring its management policies to reflect the findings. A transcript of the speech was widely circulated among NPS administrators with a note from an NPS scientist further exhorting his superiors.

The Cain speech gave impetus to a grudgingly developing research program which in turn led to two important in-house reports over the next several years. First prepared in 1960 and released servicewide three years later, *A Back Country Management Plan for Sequoia and Kings Canyon National Parks* was a document with far greater implications than its name might suggest. Always intended to be a blueprint for NPS backcountry planning, the Sequoia report began with a series of definitions and position statements about such issues as wilderness protection versus personal freedom, carrying capacities, population trends and wilderness use, and the ultimate aim of conservation.

In 1962, the agency produced a wildlife management report sometimes referred to as the Stagner report after its lead author. In the report, the authors propose a series of management principles which echo many of George Wright's from *Fauna of the National Parks* three decades earlier and anticipate those of the Leopold report a year later.

As important as these reports were to altering NPS policy, the year 1963 was the turning point. Two complementary reports by outside committees confirmed the findings of Cain, Stagner, and others and radically transformed policy priorities. The chief document in this transformation was officially known as the *Report of the Advisory Board on Wildlife Management in the National Parks*. Unofficially it was called the Leopold report after its committee chairman. Established by Secretary of the Interior Stewart Udall to comment on the problem of overgrazing by Yellowstone elk as well as other specific wildlife issues, the committee chose to go well beyond those tasks and define a basic management philosophy for the national parks. In a now familiar statement they suggested that the primary purpose of the parks was the maintenance or restoration of the biotic associations in each unit to the "condition that prevailed when the area was first visited by the white man." After further discussion and specific examples, the report called for a permanent staff of scientists in each park to oversee these management priorities. When published, a letter from Secretary Udall endorsing it accompanied the report.

Yet another blue-ribbon group took up the allied question of research in the parks. Known as the Robbins report after its chairman, this document added fuel to the fires of change and corroborated the Leopold committee, Cain, and the scientific and conservation communities. Together these two reports resulted directly in the revival of both extensive scientific research in the parks as well as new management for "ecosystem preservation."

While this management revolution transpired, the business of coping with the recreation needs of the nation also rekindled. A series of studies conducted in the late 1950s again identified park, parkway, and seashore opportunities for preservation. Congress, however, took a further step by creating the Outdoor Recreation Resources Review Commission to study the problem in a comprehensive way. Its report, issued in 1962, led to establishment of the Bureau of Outdoor Recreation (now absorbed into the National Park Service) as well as heightened responsibility for the National Park Service in planning a national recreation program and administering areas for that purpose.

The following year the Recreation Advisory Committee to the federal government further elaborated policy on national recreation areas. These units were to be primarily for outdoor recreation rather than natural or historic preservation. This clear mandate served notice on the Park Service of still greater diversity in its management responsibilities.

ECOLOGICAL ISLANDS AS
NATURAL LABORATORIES

By Stanley A. Cain, The University of Michigan
Presented at the Sixth Biennial Wilderness Conference
Fairmont Hotel, San Francisco, 20–21 March 1959

I do not know of another conference than this, the Sixth Biennial Wilderness Conference, that has addressed itself to the question of the meaning of wilderness to science. It seems to me that this is an appropriate and important theme, and I am complimented by your invitation to me to make some remarks on wilderness areas as islands for ecological field studies.

I have been caught, as many of you have, in what seems to some to be an untenable position when I have stated that the wilderness must be protected because of its value for scientific research, being unable in a given situation to point to any significant research that has already been done in the wilderness tract under question. One wins no adherents to the side of protection by only being able to say that someday someone will want to study something in the wilderness, that it might be important to do so, and that it would be a shameful loss were the wilderness no longer in existence. I believe this, but it does not provide a stout verbal cudgel. However, before saying more about wilderness as a natural ecological laboratory, I want to use my license as an invited speaker to make two or three preliminary points.

First, I would not in defense of the wilderness try to compete with the dollar values of the market place. The columns of marching pines that shoulder the mountainside in sunset silhouette do not add up like the columns of dollars for boardfeet of lumber. The leaping trout in the white-water stream can not compete with the sale price of kilowatts of hydro-power or the tons of sugar beets and bales of cotton that could be grown on lands irrigated by water impounded where the white water once flowed.

One believes in wilderness, or he doesn't. In your guts you know its value, or you don't. You can add up all of the expenditures of the people for out-door recreation focussed on wildlands—for arms and tackle, for gasoline and meals, for cameras, licenses, boats, and fancy clothing—and they amount to billions of dollars; yet the economic argument has never saved a single wilderness nor justified the saving of one. This is true for the simple reason that the economic argument cuts both ways. If your appraisal of the wilderness is made on an economic basis, all that has to be shown is that the dollar value of timber is greater, or that of minerals, or that of impounded water, and the wilderness value has been superseded. To put forward the argument of the economic value of the wilderness for appropriate kinds of recreation is to put the wilderness on the block, to be sold to the highest bidder.

I have tried to make it clear that I don't believe that it is wise or necessary, or that the public requires, that the wilderness be given economic justification. I am not saying that it is unimportant that perhaps one percent of the gross national product is attributable to expenditures related to wildland recreation. I am saying, rather, that the wilderness, broadly interpreted, would be valuable and its preservation justified even if its use contributed nothing to the gross national product.

Another point that sometimes embarrasses defenders of the wilderness is the fact that relatively few persons actually penetrate it on foot, by packtrain, or canoe—the vast majority seeing it only as it can from speeding automobiles or from points of tourist concentration.

I remember vividly at the hearings that were held by a Michigan legislative Committee and by the Conservation Commission last fall on a proposal to mine copper in the Porcupine Mountains State Park how some proponents of development spoke scathingly of the small number of users of the relatively vast wilderness parts of the Park. In connection with this sometimes bitter public debate it was said that scarcely anyone went into the undeveloped wilderness other than a few "odd-balls" and some outdoor writers for newspapers and magazines who were paid to do so, and that, if developed, a million tourists could be put in the woods and they wouldn't see one another. This would be approximately true if one tourist hid behind each tree.

Such an argument misses the point, in my opinion, just as truly as does the economic argument. Counting noses is a game we can't win at, and it is essentially irrelevant in any case. No one would expect attendance at the Corcoran Art Gallery or the Library of Congress to compete with that at the Rosebowl or with the Saturday night TV viewers of Gunsmoke.

Persons who defend wilderness areas from development and commercial exploitation are confronted with our pluralistic society which has many value systems. In addition to economic values, which have their proper and important place in any society, and to our inclination to evaluate many things according to size or numbers, I speak now of a quite different sort of value.

If we believe, as Robert Angell has said, that "serenity is a vanishing quality of life," and that "the world presses in upon us, insistent, confusing, often tragic," and if we believe, as many of us do, that a sojourn in the wilderness has restorative powers for our moral and social equanimity, then what other reason do we need to give for the establishment, scrupulous protection, and proper use of such wilderness as remains? If holding such a view is aristocratic snobbery, it is not that of money, family, or position, nor that of utility and progress, but is that of taste, of quality of sensitivity, and of reverence for nature. Could it not be that we weaken our position and do a disservice to such a value if we do not put it forward, stoutly and frankly in competition with the commoner uses of and benefits from the land?

I have made a value judgement concerning wilderness areas which is, concisely, that there are areas of landscape that are of higher usefulness to man when left alone than when they are exploited for the goods and services that have dollar value. So far I have not dealt with the wilderness as a scientist would, sticking to demonstrable fact, but have taken a value position as a humanist would. And yet this is not just my position, nor that only of us here for the Sixth Biennial Wilderness Conference, but it is also that of a large number of persons who have similar tacit standards as to what "ought" to be in wilderness preservation and use. These include the thousands who expressed opposition to development of state parks in the Porcupine Mountains case I referred to earlier, the tens of thousands who said "no" to an earlier proposal to flood Echo Canyon, and the hundreds of thousands of persons who would get fighting mad if Yellowstone or Yosemite were to be commercially exploited.

Rather than producing a lengthy polemic on wilderness preservation, the value of the wilderness for inducement of serenity, or to deny the validity of an indirect economic defense of the wilderness, I come at last to my specific subject: the value of the wilderness as an ecological laboratory. I am not going to claim that natural areas should be preserved because natural scientists need reservations where they can carry out their observations of and experiments with nature without disturbance. Nor am I going to claim that research in the wilderness is likely to produce results that will shatter the boundaries of ignorance and contribute to the revolution of natural history. I would not expect another Charles Darwin to emerge from the dripping Olympics or the prickly desert at Joshua Tree, nor a [James] Dana from Grand Canyon, or a Margaret Mead from Mesa Verde. My point is rather that the wilderness needs being studied, in breadth and depth, for the usefulness the resulting knowledge will have in enhancement of the enjoyment of the public which visits, loves, and cherishes the wilderness.

I believe that interest in and appreciation of nature lies at the heart of the general public's appraisal and valuation of the wilderness—each man to his own degree and with wonder at nature according to his own capacity. If this is true then some degree of interest in nature is the most prevalent motive of persons visiting national parks and other wildland areas. It suggests also that the most important function of agencies administering wild lands is that of interpretation.

The American public has amply demonstrated that it wants the wilderness left alone, even if it doesn't actually get into it in any numbers. The wilderness has real experience to offer, even at a distance. In any National Park, for one person who actually penetrates the wilderness there are hundreds who learn something about it on guided nature hikes, and thousands who have vicarious experiences at museums, information centers, and by listening to campfire talks. There are also the unknown numbers who get some, and often great, pleasure from magazine articles and books, from nature lectures and Disney films. This is the tremendously large public receptive to natural history information about the Nation's wilderness areas. These are the people who read, for example, <u>Arizona Highways</u>, attend Audubon lectures, and follow the outdoor and travel pages of newspapers and magazines.

I believe that there is a vast reservoir of opportunity for nature interpretation that is incompletely developed despite the fine work already done by Park Naturalists. I think that this is true because there is no adequate program of research designed to feed into the interpretation program the necessary and appropriate facts.

I am thinking especially of the National Park Service because of its responsibility for our most spectacular and precious remaining examples of wilderness, and because it is only the National Park Service that has legal responsibility for the protection of the entire environment and all that lives in it. Of course roads, camps, and other facilities are necessary for public use while the land and the people are protected, but the obligation of the Park Service can't stop there. Housekeeping is necessary, but it still is only housekeeping. I believe that the central function is that of interpretation. I believe this because I believe that interest in natural features of the parks is the principal attraction of them for the public.

I don't mean that we are a nation of ostensible nature lovers, and we are certainly much less open about it than, for example, the British with their Snail-Watching Society and the Friends of the Trees. But I do think that many a foolish seeming action of a tourist with a bear is not so much dumbness as it is fascination and perhaps naive curiosity about the animal. There is, for example, pretty good evidence that many a hunter is less interested in the chase and the kill than he is likely to admit casually, and is deep down more interested in his woodlore and his knowledge of the habits of animals and of terrain and weather, and his sense of relatedness with nature.

So I come again to the point that the National Park Service, and other federal and state agencies, are missing a bet in the lack of an adequate natural history research program that would regularly feed into their interpretation programs the basic information which they now do without or get only by happy chance.

As I have said, I will speak mainly of the National Park Service, although my point is more general. But I am speaking also to the National Park Service. After having accepted your invitation to speak here, and having thought some about what I would say (mostly a sort of incubation rather than a concentrated consideration), I discarded the idea of describing in detail some research that has been carried on in the wilderness areas, or that could be were circumstances propitious, and I decided to take this opportunity to promote the Service's activity in basic ecological and related research.

If such an effort at promotion were to have an effect (and I tried it some years ago without producing any effect), I needed to know the attitudes of those who make policy for the Service. The result was that I telephoned Mr. Ronald F. Lee, who is Chief of the Division of Interpretation, and found him so interested that a week ago today I flew to Washington to talk with him and his associates, including Eivind Scoyen, Associate Director of the Service. I needed to know how much interest, if any, existed in basic natural history research. I needed to know also whether outside pressure and cooperation would be welcomed. I am most happy to report that there is in the Service at the Washington level a real interest in basic research and a warm receptivity to suggestions, cooperation, and even pressure from the outside. I conclude, therefore, that it is possible for those of us who may be interested to be of real help to them.

I think the best summary of the scope of activities classified as research, as well as the limitations of the program, is to be found in the Annual Reports. I will, therefore, read the statement on research in the Annual Report for the fiscal year ended June 30, 1957.

"In the National Park Service, research is one of the important means which enable it better to meet its responsibilities. Historical research is done largely by its own historians. In other fields it is accomplished in various ways. Service employees contribute importantly to it, but it leans heavily also on the cooperation of other Federal agencies, State agencies, and publicly and privately supported institutions of higher learning.

"During calendar 1956, 160 research projects were performed by regular and seasonal members of the Service, collaborators and co-operating groups and individuals. Of these, 148 dealt directly with the areas administered by the National Park Service, four were in the category of investigations of proposed areas, and eight were general research projects. Graduate students and staff members of 23 colleges and universities, and four private individuals participated in these projects. Five Federal agencies, seven State agencies, and three professional societies also performed research or assisted field personnel in research projects.

"Natural science research during the year dealt with such matters as siltation at Mammoth Cave National Park; the ecology of Florida Bay, in Everglades; hydrothermal phenomena at Yellowstone; and the ecology of high mountain meadows and other fragile environments in several National Parks. The Coastal Studies Institute of Louisiana State University and the Office of Naval Research cooperated in detailed research on the geology, botany, archeology, and history of Cape Hatteras National Seashore; North Carolina State College is doing further research there in biology. Seven major geological research projects were in progress in Death Valley National Monument alone.

"Texas A. & M. College and the Texas Fish and Game Commission are participating in a 5-year study of Big Bend National Park ecology. Biological and geological studies are under way in Virgin Islands National Park, the University of Kansas and Princeton University cooperating.

"Historical research included studies of the appearance and use of Independence Hall at various periods; of details of colonial life and land ownership at Jamestown; of the locations of specific features at Fort McHenry, a necessary basis on which to prepare development plans; and a study of the historic structures remaining at Harpers Ferry.

"Archeological methods were employed at such historic places as Fort Frederica and Fort Union National Monuments, at Jamestown, at Independence and Cumberland Gap Historical Parks, and, on a minor scale, at a number of other areas. Important archeological research was continued by the University of Colorado at Mesa Verde, by the University of Southern California at Death Valley, by the University of California at Yosemite, and by the Bishop Museum of Honolulu at the proposed City of Refuge National Historical Park. Washington University, St. Louis, cooperated in the effort to find the original site of Arkansas Post. Service archeologists excavated at the site of Fort Clatsop, in Oregon.

"Under cooperative agreements, 16 universities and colleges and the Smithsonian Institution performed river-basin archeological salvage. In addition, the University of Utah, the School of American Research, and the Museum of Northern Arizona undertook such salvage in the Upper Colorado Basin. The Smithsonian, working at 11 reservoir sites with funds supplied by the Service, recorded more than 200 new archeological sites, excavated 19, and processed more than 179,000 excavated specimens."

The Annual Report for the fiscal year ended June 30, 1958, adds that historical research has been inaugurated at Harpers Ferry and Fort McHenry, and that the archeological salvage program for basins about to be inundated has been extended to the Glen Canyon and Navajo Indian Reservoir areas. A program of Alpine Wilderness research was initiated, a project on the biology of the United States Virgin Islands, and that a research conference in Everglades outlined needed studies. The Division of Interpretation was given specific responsibilities for developing the biological research program by stepping up Service-conducted investigations and encouraging cooperative research by qualified scientists and established research institutions.

At intervals in the past Service biologists have made significant basic studies. Of this type I will recall to you Adolf Murie's Ecology of the Coyote in the Yellowstone, 1940, and The Wolves of Mount McKinley, 1944, which were bulletins in the Service's now quiescent Fauna Series. The study by Coleman C. Newman, forester in the Branch of Park Forest and Wildlife Management, of the Roosevelt Elk of Olympic National Park, is an example of a recent Service-staff investigation that deals with basic biology. As an outstanding example of cooperative effort between the National Park Service and several other agencies, I would mention Carl B. Koford's Wildlife Monograph, Prairie Dogs, Whitefaces, and Blue Grama, 1958. The Service facilitated Koford's study by appointing him a Collaborator, giving him access to Service studies and maps, and by active assistance of Park personnel. Service personnel in the Division of Interpretation have in many cases contributed fundamental knowledge to the fauna and flora of the parks where they have been stationed, as in the case of Arthur Stupka, Chief Naturalist of the Great Smoky Mountains National Park, and they and their Park Superintendents have in many cases aided materially the research of independent scientists. They often have been graduate students gathering data for their Master's and Doctor's dissertations. A recent example here is that of Grant W. Sharpe whose PhD thesis was A taxonomical-ecological study of the vegetation by habitats in eight forest types of the Olympic rain forest, Olympic National Park, Washington.

Such a review of research activities carried out or facilitated by the National Park Service is, of course, not complete, yet I do not believe that it misrepresents the situation. I must conclude that the National Park Service does not have a program of basic ecological research. We come to the inevitable conjunction, what is done is fine as far as it goes, but it fails to approach at all closely the fundamental need of the Service itself.

The first difficulty, as I see, it, is that the research activities of the Service are directed toward immediate pressing problems. It is largely a matter of trouble-shooting. I will illustrate this point by reference to the program of <u>Ecological Research in the National Park Service on Alpine and other Wilderness Environments</u>. The following statement was used in justifying the appropriation of funds (described as modest) for the initiation of the ecological studies.

> "Increased visitor use of alpine and other wilderness areas in the parks present special problems of a related nature. The alpine meadows, which if once destroyed are irreplaceable, are particularly vulnerable to disturbances caused by man, and the unique wilderness values in the 'back country' of these parks generally require special attention. Studies on the ecology and the fundamental requirements for protecting these features are urgently needed."

Research contracts were executed with the Institute of Arctic and Alpine Research of the University of Colorado for studies at Rocky Mountains National Park; with the Jackson Hole Biological Research Station of the University of Wyoming for studies at Grand Teton National Park, and with Dr. Carl Sharsmith of San Jose State College for studies at Sequoia-Kings Canyon National Park. In addition to Dr. Sharsmith, project supervisors and investigators include doctors John W. Marr, L. Floyd Clarke, Charles C. Laing, and Beatrice Willard, all accomplished field ecologists. The purpose of this program, which may be extended to Yosemite, Mount Rainier, Olympic, and Glacier National Parks, is:

> "To determine the visitor impact on park vegetation by studying sections of the park which have been receiving extensive visitor use and by comparing these sections with portions of the same type which have received a minimum of visitor use. This would permit investigations of alpine meadows and other wilderness areas to determine the ecological status of fragile plant communities and to ascertain the basic requirements for their protection and perpetuation."

This is ecological research directed at trouble-shooting. It is programmed, and it will yield scientific information about alpine communities, but it is directed toward the needs of management, not toward the needs of the interpretation program.

I have no quarrel with such research. It is necessary and it follows an established park policy of management that was first enunciated in 1939 by Ira N. Gabrielson and approved by the Director of the National Park Service and the Secretary of the Interior, that "No management measure or other interference with biotic relationships shall be undertaken prior to a properly conducted investigation." I now turn to a suggestion of what, in my opinion, would be the ingredients of a National Park Service program of ecological research.

It would be presumptuous of me to state in detail or with finality what a basic ecological research program should include. I will do no more than discuss briefly some of the matters which could be given attention and for which research would yield information directly usable in the interpretation programs of the parks. That such investigations might have general interest and constitute contributions to basic knowledge or the advancement of scientific understanding or theory would be no more than a fortunate by-product from the point of view I am stressing.

I believe that every National Park should have a continuing program in faunistics and floristics, with suitable museum and herbarium vouchers for the species claimed to be present in the area. Such knowledge would be built up over time and would run the complete systematic gamut, including, for example, small mammals as well as large, molluscs as well [as] fish, mosses and lichens as well as trees and conspicuous flowers. Such knowledge is basic to everything else.

Inasmuch as the thousands of species of plants and animals do not occur in an area as a sort of chance agglomeration, one of the most satisfactory ways of organizing biotic information is by the communities which are formed. The research program should include investigation of the natural plant-animal communities that occur in each area, starting with the major cover types and progressing to the lesser and often very interesting communities of special habitats such as hot springs and serpentine rock outcrops. Using modern techniques for the study of vegetation, information should be gathered as to the composition and structure of all communities. How are the component species organized in layers and other synusiae? What are the dominants? What are the indicator species which are of high fidelity to a given community and ecological situation? Having gained information on the morphology of communities, investigation should then be directed to their ecology. What is their pattern with respect to altitude, slope and exposure, to microclimate, and to the conditions of the substratum? What are the important actions, reactions, and coactions in each community?

It seems to me that people are interested in biogeography. There may even be some excitement in knowing that a certain plant or animal is endemic and lives naturally no where else in the world, or that where it is found in the given area it is disjunct, perhaps hundreds of miles from its main range. Most species will be found to be intraneous and well within their full range, but many species are extraneous at the place under consideration and characteristically range northward or southward, eastward or westward, toward the coast or the mountains, etc. These irregularities of occurrence bring up matters of historical biogeography. Where are the ancestors of the endemics, and where are their relatives now? How did the disjuncts get to where we now find them? What are the relations of fossils found in the area to the kinds of plants and animals now occupying it? What are the forces of migration? Of evolution? Of extinction?

Life histories should be studied, not only of species causing management problems but of those which are interesting for any reason. How is the life cycle carried out? Through what stages, and with what relation to the environment? What are the diurnal and annual aspects of the life activity?

These matters, so briefly suggested, are basic. They ask the question: What is there? They ask what the conspicuous community arrangements are, and they go to the point of especially interesting life cycles and biogeographic features. Every interpretation program should be backed up by an abundance of information about such natural history matters. Such studies may not be on the frontier of scientific advancement, but they are eminently respectable and very practical in the National Park situation.

I'll speak now of several biological studies that can be made in any wildland area and which can produce information having appeal to the public that probably goes beyond that of names for species and knowledge of communities. In each case the examples should be local, that is, they should deal with plants and animals in the area where the interpretation effort is being made. There are such matters as the following.

What about rare or vanishing species? Why are they rare? Why are they in danger of extinction? What actions are necessary to protect them and to give them a chance to recover reasonable numbers? Why worry about extinction any way?

What are the regional prey-predator relations? Food chains and pyramids of numbers? And related to this in some ways, what is the nature of territories, ranges, and home ranges? Why must some wilderness areas be as large as they are? What is the concept of carrying capacity of land for given species? And what is the relationship to population size and density? And to habitat deterioration? What are population cycles? What causes them? How do they relate to prey-predator relations, and to producer and consumer relations? What are epidemics and infestations? What are some of the problems in a National Park in disease and pest control, and in the control of other populations when the balance of nature has been upset by reduction of predators or there are influential changes occurring outside the Park and beyond the Service's management?

What is the nature of community dynamics? What is succession? What is the nature of equilibrium in nature? What is a climax, and what are the climaxes of a given region?

Why do species of a genus have the patterns of occurrence that they do? Why are the most closely related species of a genus allopatric in their areas and only less closely related ones sympatric? What is the nature of some of the barriers to hybridization? What are clines? What is the subspecies as interpreted by mammalogists and ornithologists?

And so we might continue to ask questions the answers to which would be sought by National Park Service scientists were there a formal, continuing, and sufficiently massive program of ecological and systematic research.

Many persons will sit patiently and look at a few dozen Kodachrome slides of spring wildflowers or of nesting birds or of scenery, and have a good time, too, if the photography is good, but wouldn't they have a more exciting time if the intellectual content of the presentation were not static and, as is sometimes the case, at the "what-is-it" or "where-is-it" level? I suppose the point can be summed up succinctly by my saying:

Emphasize ideas, not things.

National Park Service Archives, Harpers Ferry, Box N16, Management Biology 1948–1962. Reprinted with permission of Mrs. Stanley Cain.

A BACK COUNTRY MANAGEMENT PLAN FOR SEQUOIA AND KINGS CANYON NATIONAL PARKS

Written 1960, General Release 1963

PART I–GENERAL PRINCIPLES AND GUIDELINES

WILDERNESS DEFINED

To avoid misunderstanding, the definition of wilderness as used in this report has been taken from The National Park Wilderness (Stagner, 1957). It can be summed up thus: A wilderness is a large, undeveloped, wild area extending beyond roads and developments for permanent occupancy, in which one can experience solitude, quiet, beauty, a sense of adventure, and feelings of remoteness from modern civilization, including mechanized transportation—and in which the drama of natural forces is permitted to unfold without interference except for such management practices as may be required to counteract major destructive influences.

THE NEEDS OF SCIENCE

The value of wilderness for scientific research used to be largely ignored, or at least not mentioned in reports such as this one. But the dates and titles of hundreds of research publications show that for many years the national parks have been recognized by scientists from all over the world as areas uniquely qualified for basic research, by virtue of their unmodified wilderness characteristics. A landmark along the route toward the growing appreciation of wilderness as a scientific resource was reached with the publication, in 1960, of "The Meaning of Wilderness to Science," a book on which the National Park Service collaborated with distinguished representatives of other conservation and research organizations and institutions.

Since that time, a growing awareness of the unique contributions that unmodified park environments can make to basic research has been reflected in increasingly adequate budgets in support of research programs in the parks. This trend is adding a new dimension to the concept of wilderness protection, and the responsibilities of the National Park Service in that connection.

THE ULTIMATE AIM OF CONSERVATION

The ultimate aim of conservation is to leave our earth as rich and productive as we found it. William Dean Howells said, "A nation is great not because it mines coals, cuts timber or builds railways, but rather because it has learned how to produce, build and grow without destroying the bases of its future existence" (The Conservation Foundation, 1958).

The National Park System wilderness is a culmination of processes and events that have been unfolding since the beginning of time. It presents the story of the evolution of the American land, and of the development of life upon it. It is a never-ending story that will continue to unfold, for the inspiration and general welfare of mankind and the advancement of science, so long as the integrity of these "islands of nature" is maintained.

In 1956 an international symposium of scientists emphasized how profoundly man has changed the face of the earth. Man's influence in this respect was compared in magnitude to a major geological-ecological force (W.L. Thomas, Jr., 1956). In 1961 a conference in Switzerland of the International Union for the Conservation of Nature and Natural Resources, bringing together ecologists from 12 countries, concluded that the national parks of the world offer the principal future hope of preserving some scattered fragments of primeval nature for fundamental scientific research. The large primeval parks of the National Park System of the United States were declared to be preeminent in this respect, and of international significance and value. This idea has been reiterated at other international forums.

As our society continues to increase in complexity and in population, such an understanding of natural, balanced environments will become increasingly essential for developing ways of living harmoniously, rather than destructively, on our Nation's lands, which are not increasing in area.

Science needs these environments as a point of reference and as a yardstick with which to measure man's success or failure in the countless land-management programs that he carries out in the rest of his environment.

Gaining worldwide recognition and acceptance is the belief that National Parks, preserved as natural ecological entities, can supply man with a more complete understanding of the natural laws that may govern his future—and possibly his ultimate survival—in lands and environments everywhere.

WE ARE THE CUSTODIANS

"The Nation's most treasured wilderness lands are set apart and dedicated as National Parks and Monuments." By these words are we reminded, in The National Park Wilderness, of our responsibilities.

The custodians of a library are obligated to preserve the qualities and atmosphere of a library. In like degree, are we as one of the acknowledged national custodians of wilderness, obligated to preserve and manage the back country—as back country—for the enjoyment, education, physical and spiritual refreshment of our people, for basic scientific research, for an understanding of the natural laws that govern man's existence. Recognition of this obligation is the basis for the present report.

POPULATION TRENDS AND WILDERNESS USE

Every day there are 7,000 more people in the United States than there were the day before; every morning, in California alone, there are 1,600 more people having breakfast (Life, October 19, 1962, p. 69). Recreation and land use planners are faced with the fact that a U.S. population of 100 million in 1920 grew to 180 million in 1960, is expected to reach 230 million in 1975, and 350 million by the year 2000.

Along with this anticipated quadrupling of the population in a period of about a century, there has already taken place an average increase in travel per person from 1500 miles in 1920 to 4700 miles in 1960. During this 40-year period, leisure time increased by nearly 40 percent, while use of state and national parks showed a more than sixfold increase, jumping from 50 million to over 331 million visits per year (Department of the Interior, "Fact Sheet," 1961). The rate of leisure-time increase can be expected to accelerate under the influence of automation.

Visitor use of the wilderness areas of Sequoia and Kings Canyon National Parks has nearly doubled since 1950. The current and anticipated population growth in the State of California over the next 20 years precludes any thought that there will be a slackening off in the numbers of persons who visit the National Park wilderness of the Sierra.

THE BACK COUNTRY IS BECOMING
THE FRONT COUNTRY

The gradual deterioration of Sierra meadows through overgrazing has been of serious concern as far back as 1940, when the first report was written concerning this problem in Sequoia and Kings Canyon National Parks.

Increasing visitor use has made serious inroads on the fish population of many of the high lakes and streams. Most of the fish habitat in the parks is limited by various physical factors. In earlier times of light fishing pressure it was a fisherman's paradise, but its accessibility threatens to bring its own doom. In several places this has already occurred. As county and state roads push ever closer to the exterior boundaries of the parks, mass invasions of visitors will extend the heavy fishing pressure almost the entire length of the fertile zone of fish habitat.

In 1958, over three tons of litter were removed from just one of the more popular back-country areas. In 1960, a full-scale Service-sponsored program was launched to begin initial cleanup work; and at season's end over 20 tons had been hauled out (Briggle, 1960).

The litter problem will always necessitate a program for back-country cleanup in much the same manner as litter is removed from front-country campgrounds, though it is hoped that wilderness users will show a greater sense of social responsibility in this regard than do the more predominantly urban users of park highways and other heavily developed facilities for motorists.

A review of these trends serves as a potent reminder that important areas of the parks that in an earlier day were wilderness or back country, now that they are easily accessible and serve heavy concentrations of visitors, are considered front country. The loss of the primeval wilderness qualities that appealed to an earlier generation of users has been gradual; some pockets of unchanged back country remain today, but they will suffer the same fate unless the trend is controlled soon. This report suggests ways to accomplish such control.

CARRYING CAPACITY DETERMINATIONS

Many deteriorating wilderness situations can be checked by appropriate management measures. But no management plan can be effective if it ignores the practical limitations of the natural environment. Ecologists point out that a basic requirement for intelligent management of any environment used by man, or influenced by him, is a determination of its "carrying capacity" (Fosberg, 1961).

This concept has long been applied with precision to the management of domestic livestock. Likewise, since the thirties, it has been standard procedure to base wildlife management programs on careful studies of the wildlife ranges to determine, and operate within, their carrying capacities. As human use of wilderness ranges begins to approach a saturation point, management has the responsibility of identifying basic factors that limit the carrying capacities of each area, and of tailoring the respective management programs to conform to these natural limitations.

When the present Committee determined from observation the minimum distance required for wilderness-type privacy between high country campsites, it was, in effect, determining the camper-carrying capacities of these areas. Similar carrying-capacity determinations and judgments were called for in preparing other phases of the management plan. Precedents and techniques for measuring wilderness carrying capacities are few; but present use trends clearly indicate the need for further application and refinement of known techniques, and the development of new ones.

WILDERNESS PROTECTION
VS PERSONAL FREEDOM

Oldtime use of wilderness was completely free of restrictions. Wilderness explorers could hunt and fish without limit, cut down trees at will, camp, make fires and graze their stock anywhere. The tradition of personal freedom in wilderness dies hard, and one of the foremost endeavors of the National Park Service is to respect and preserve the personal freedom of the wilderness user to the fullest extent possible within the framework of each area's carrying capacity.

But when human populations expand they become subject to the biological limitations that govern other dense populations: The greater the number of individuals the greater the loss of individual freedom of action. An illustration of this in the daily lives of all of us is afforded by the congestion, delays, and complicated regulations on today's crowded highways.

Wilderness users may with justice complain that they seek wilderness to escape the regimentation of daily life. But the time is past, for example, when a Boy Scout can use his axe to cut fresh pine branches for his bed in the old tradition. Today we have no choice but to agree with Snyder (1961) when he suggests that, ". . . when we speak loosely of an 'untouched' wilderness we must actually be reconciled and receptive to an area managed in a degree relative to the number of people who enter it."

The Committee recognizes a major responsibility to preserve all possible freedom in wilderness, but feels that the answer to complaints over present day restrictions is not "bureaucracy" but "Born too late."

REFERENCES CITED

Briggle, William J. 1960. "Back Country Cleanup." 10pp. Manuscript.

Conservation Foundation. 1958. Concepts of Conservation, A Guide to Discussion. 48pp.

Department of the Interior. 1961. White House Regional Conferences. Outdoor Recreation Opportunities for Urban Dwellers. Dept. of the Interior. 6 pp. plus "Fact Sheet" Appendix.

Fosberg, F. R. 1961. The Island Ecosystem. UNESCO Symposium on Man's Place in the Island Ecosystem. 11 pp. Mimeographed.

Life. 1962. The Call of California. Time Inc. Vol. 53, Oct. 19, p. 69.

Snyder, Arnold P. 1961. How Wild the Wilderness. American Forests. Vol. 67, May, 34-35, 62-63, illus.

Stagner, Howard R. 1957. The National Park Wilderness. National Park Service, Department of the Interior, Washington, D.C. 37 pp., 18 illus.

Thomas, W.L., Jr. (editor). 1956. Man's Role in Changing the Face of the Earth. Chicago: University of Chicago Press. xxxvii + 1193 pp.

National Park Service, Washington, D.C.—Report available in the office of the Chief of Resources Management, Sequoia and Kings Canyon National Parks, 3–9, 104–106.

WILDLIFE MANAGEMENT IN NATIONAL PARKS

(Often called the Stagner Report)
1962

INTRODUCTION

Many significant factors have affected wildlife populations during the past several decades. Western areas of the United States have experienced vast and often times drastic land-use changes that have dramatically altered the ecology of most wildlife species. These changes have proved beneficial to the existence of some faunal populations and detrimental to others.

In addition, the introduction of exotics has had varying effects upon native plant and animal species. Entire wildlife populations have sometimes been affected. Climatic conditions during the past decade that have generally been of a more mild nature have not only had decided effects upon the various vegetative processes, seasonal movements of migratory species, reproduction and decimating factors, but many other items that have definite effects upon wildland populations. The present population numbers of wildlife contained in State and Federal lands has also changed due to these and many other affecting forces of nature and actions by man.

Former preservation measures applied to the management of wildlife have in many instances become obsolete. The balance between land carrying capacity for wild populations and population numbers has reached and even surpassed the point where decided and definite action should be taken for a proper realignment. Only through attempts to realign these two items will other park resources such as soil and vegetation be conserved. The Service's obligation to conserve in an "unimpaired" manner is explicit. Waste or complete destruction is to be avoided.

This summary has been compiled to present field area data concerning wildlife management programs that were in effect during the July 1, 1961 to June 30, 1962 period. Certain special items and programs dealing with this management field are also included. An attempt to summarize activities and programs of each area in the National Park System has not been made. Only those areas which have significance, inherent or which possibly will be subjected to more extensive wildlife management problems are being considered. Limitations and omissions that should not have occurred are unavoidable due to this office's dependency upon submission of significant data from both field and regional offices.

The primary concern of this summary is to present management activities that developed with management of ungulates during early 1961. The controversy arising from some management programs continues to be the subject of considerable discussion and action among interested groups and individuals. The importance of direct reduction programs in National Parks for the conservation of several renewable and nonrenewable resources and maintenance of suitable biological relationships resulted in the appointment of a special wildlife advisory board by the Secretary of the Interior. Additional information on this group of eminent conservationists and their work as an advisory committee is included in the section dealing with Special Wildlife Programs.

For a number of years, annual summaries or reports were published on the status and condition of wildlife in areas of the National Park System. They also included accomplishments in the various fields of biological research. Resumption of this annual accomplishment summary, on a recurring basis, is anticipated for future years.

In attempts to comply with the act of August 25, 1916, areas comprising the National Park System are administered in such manner as ". . . to conserve the scenery and the natural and historic objects and the wild life therein and to provide for the enjoyment of the same in such manner and by such means as will leave them unimpaired for the enjoyment of future generations. . . ." The conservation principle of wise use and its application to certain wildlife populations has been determined to be necessary in order that other resources of America's priceless heritages, as found in the National Park System, may be enjoyed by future visitors.

During this reporting period new additions, such as Haleakala National Park, City of Refuge National Historical Park and Buck Island Reef National Monument, were added to the 189 existing Service administered parks and related areas. These additions all contain varying quantities and diversified compositions of wildlife that will add to visitor enjoyment and experiences. In addition, 82,276,000 visits and 91,758,000 days of visitor-use were experienced in the approximately 25,957,901 acres contained in scientific scenic, historic and prehistoric reservations of national significance. Recreational use of fishery resources, which originated from lands administered by this Service and as submitted by annual reports of thirty-five areas, indicated an angler use day figure of approximately 1,750,000.

POLICIES AND GUIDELINES

Adopted wildlife management policies as found in Volume VI,

Part 2, Chapter 5, of the <u>Administrative Manual</u>, have been issued as guidelines to all park activities or programs in this field. A review of this material by field offices will not only be of assistance in public presentations of current park activities, but should also assist in the formulation of an active management program.

Particular reference is made to the various regional office memoranda issued in conjunction with the Director's April 11, 1962 memorandum relating to Information on Bear Management Activities. Informational instructions contained therein should be sufficiently adequate and explicit for proper bear management in all applicable areas.

WILDLIFE

The animals indigenous to the parks shall be protected, restored, if practicable, and their welfare in the natural wild state perpetuated. Their management shall consist only of measures conforming with the basic laws and which are essential to the maintenance of populations and their natural environments in a healthy condition.

<u>Hunting</u>. Hunting in areas of the National Park System is incompatible with their preservation in the manner contemplated by the authorizations for their establishment and will not be permitted, except as specifically provided by law.

Wildlife problems, especially those in relation to overpopulation, are to be solved effectively, but use of public hunting as a method of wildlife management aimed at readjusting animal populations to approximate natural biotic conditions is definitely not to be a solution.

<u>Predatory Animals</u>. No native predator shall be destroyed because of its normal utilization of any other park animal or plant, unless such animal or plant is in immediate danger of extermination, and then only if the predator is not itself a vanishing form. When control is necessary, it shall be accomplished by transplanting, or if necessary, by killing offending individuals and not by campaigns to reduce the general population of a species.

Species predatory upon fish shall be allowed to continue in normal numbers and to share normally in the benefits of fish culture.

<u>Exotics</u>. Nonnative forms shall not be introduced into parks. Any exotic species which has already become established in a park shall be either eliminated or held to a minimum provided complete eradication is not feasible, and the possible invasion of the parks by other exotics shall be anticipated and steps taken to guard against the same.

<u>Native Forms</u>. Every native species in the areas of the National Park System shall be left to carry on its struggle for existence unaided as being to its greatest ultimate good, unless there is real cause to believe that it will perish if unassisted.

Where artificial feeding, control of natural enemies, or other protective measures are necessary to save a native species that is unable to cope with civilization's influences, every effort shall be made to place that species on a self-sustaining basis once more. The artificial aids, which themselves have unfortunate consequences, will then no longer be used.

Reintroduction. Any native species or subspecies which has been exterminated from a park shall be brought back if this can be done, but if a species has become extinct, no related species shall be considered a candidate for reintroduction in its place. If a subspecific variant of a species has become extinct, substitution of a closely related subspecies may be considered.

Adverse Biological Forces. Plants and animals which are inimical to the public health or welfare or which are destructive to historic, archeological or scientific structures, sites, features or records of primary importance shall be subject to neutralization or control.

Hoofed Animals. The numbers of native hoofed animals occupying a deteriorated range shall not be permitted to exceed its reduced capacity and, preferably shall be kept below the carrying capacity at every step until the range can be brought back to its original productiveness.

Artificial Feeding. No animal shall be encouraged to become dependent wholly or in part upon man for its support.

Captive Animals. Artificiality shall be avoided in the presentation of the animal life of the parks to the public. The preferred presentation shall be through wholly natural situations.

Management. Management measures or other interference with plant and animal relationships should be undertaken after properly conducted investigation. Approval of programs for the destruction and disposition of wild animals which are damaging the land, or its vegetative cover and of permits to collect rare or endangered species has not been delegated.

Endangered and Vanishing Species. The issuance of a scientific collector's permit must be based upon the abundance of the species in the park which the permit applies. Every request must be considered carefully, and the collection of endangered or vanishing species is restricted or prohibited.

FISHING

Recreational fishing within National Parks and Monuments shall be permitted under management programs directed toward the perpetuation, restoration and protection of native species and wild populations of fishes and the protection of the natural aquatic environments and the ecological relationships of the associated fauna and flora. This activity shall be directed so as to not decrease the wildlife, scenic, scientific or historic values of the park.

Where Fishing is Excluded. Fishing may be excluded from specific waters when necessary to preserve aquatic or terrestrial species or habitats which are limited in distribution or when such activity materially decreases the enjoyment of the areas by the general public.

Native Species. The perpetuation, protection and restoration of native species in safe numbers in waters where they originally were found shall be given primary consideration of any management plan whenever possible.

Native Nonsport Fishes. All species of fishes are fully protected, except those designated for recreational angling.

Native nonsport fishes shall not be reduced or eliminated except as may be unavoidable and incidental to the primary objective of extirpating an exotic unwanted population of fishes.

In any restoration plan, native nonsport fishes should be reintroduced as well as the sport fishes.

Hybrid Trout. Hybrid trout shall not be stocked in waters of National Parks and Monuments.

Stocking. Artificial replenishment of stocking may be employed:

1.   To reintroduce native species into waters where they have become eliminated or seriously depleted by natural or man-made causes.

2.   To maintain fish populations in selected and approved lakes which are capable of supporting fish life, but which lack sufficient natural spawning facilities to maintain an adequate fish population to meet the need of recreational angling.

Size of Fish to Stock.

1.   Fingerling trout may be planted in lakes where competent study had determined a need for supplementary stocking.

2.   The stocking of eyed-eggs, fry or fingerlings in streams shall not be practiced except to restore a depleted population of native trout. (Numerous qualified studies on streams of varying sizes throughout the country have demonstrated that where conditions are suitable for trout, natural populations are maintained at maximum carrying capacity by natural reproduction. Planting of eyed-eggs, fry or fingerling trout in streams to supplement this natural reproduction has proven to be of negligible or no benefit.)

3.   Stocking of catchable size trout to provide "put and take fishing" is not compatible with the fundamental concept of the National Park Service, therefore, the planting of fish for immediate recovery by the angler shall not be made in waters of national parks and monuments.

4. Adult wild trout may be transplanted to re-establish native species or depleted populations.

Stocking National Parkways. Recreational fishing within National Parkways is permitted under management programs and stocking procedures normally practiced by the State or States in which the Parkways are located. This activity shall be regulated by the National Park Service.

Each Parkway Superintendent shall designate Parkway fishing waters. When the impact of fishing pressure would create damage to Parkway features and facilities, would produce hazardous traffic congestion or would result in unusual enforcement problems, individual waters may be closed to fishing and to stocking.

Stocking Exotic Species. Exotic species of fishes or other exotic animals, or any exotic species of aquatic plants may not be introduced or stocked in waters of the National Parks and Monuments except:

1. In waters where exotic fishes are established and the restoration of native species is impracticable.

2. Where adequate investigations have demonstrated that additional planting is desirable and necessary to supplement limited or nonexisting natural reproduction.

Management of Exotic Sport Fishes. In waters where exotic sport species of fishes are established, and they are valuable for angling and are ecologically compatible with the existing environment, and their replacement by native species is impracticable, the fishery for the exotic species will be managed in a manner similar to that for native forms.

When replacement of the exotic by the native species is practicable, the latter shall be encouraged to take over its former place.

Removal of Exotic Species—Eradication or Control. Where exotic species have become dominantly established to the detriment of the native species, restoration of the original fish composition may be brought about by the removal of the undesirable exotics. Standard eradication methods; such as, chemical treatment or electric shocking may be employed. Also, these methods may be employed to control exotic species where complete elimination is not feasible.

The need for and techniques to be used for an eradication or control program shall be based upon adequate investigations by aquatic biologists.

Egg Taking. The taking of eggs from fishes for the purpose of artificial propagation within waters in national parks and monuments is rarely justified and should not be permitted until a thorough review has been made.

Protection of Virgin Waters. Lakes and streams which are barren of fish life shall remain in this virgin condition and shall not be stocked.

Artificial Improvement of Lakes and Streams. All forms of artificial improvement of streams or lakes for fishery management purposes which change the natural habitat and the surrounding landscape are prohibited, except that, when the aquatic environment has been so altered by man that restoration by natural means is improbable, measures may be taken to return the streams and lakes to a more natural condition.

Management by Regulations. To preserve the populations of native species and yet allow angling, sport fishing shall be controlled by regulations which provide for the conservation of native species of fishes and compatible management of introduced, established species. Limits shall be established so that the total catch will not exceed the natural productive capacity of the waters. Creel limits shall not be considered as "goals".

Fishery Investigations. The conservation and proper management of the fishery resources and angling as a recreational activity is dependent upon a complete knowledge of the status of the fish fauna and the angling pressures being exerted. Adequate and continuing investigations are vital to the successful preservation and management of this resource.

Commercial Fishing. Commercial fishing is generally noncompatible with National Park Service objective and shall be permitted only within national parks and monuments where this activity is specified by law. It will be conducted under restrictions which are designed to conserve and perpetuate the resource.

Publicity. Publicity regarding fishing within the areas of the National Park System shall be directed toward the recreational and aesthetic values, and the appreciation of the unspoiled environment as a whole rather than emphasis on the catch. Information regarding angling will be factual and realistic with respect to fishing conditions.

Promotional types of publicity are discouraged but this does not apply to release of information on subjects of conservation of aquatic resources, fish regulations, care of fish by anglers, or the place of angling in the national park experience.

National Park Service Archives, Harpers Ferry, Box N16, Management Biology 1946–1962, Introduction plus "Policies and Guidelines" section, 1–6.

OUTDOOR RECREATION FOR AMERICA

A Report to the President and to the Congress by the Outdoor Recreation Resources Review Commission Laurance S. Rockefeller, Chairman January 1962

This report is a study of outdoor recreation in America—its history, its place in current American life, and its future. It represents a detailed investigation of what the public does in the out-of-doors, what factors affect its choices, what resources are available for its use, what are the present and future needs, and what the problems are in making new resources available. The investigation involves the present and to some extent the past, but its principal concern is for the future—between now and the year 2000. It is a plan for coming generations, one that must be started now and carried forward so that the outdoors may be available to the Americans of the future as it has been to those of the past.

Americans have long been concerned with the values of the outdoors. From Thoreau, Olmsted, and Muir in the middle of the past century to the leaders of today, there has been a continuing tradition of love of the outdoors and action to conserve its values. Yet one of the main currents of modern life has been the movement away from the outdoors. It no longer lies at the back door or at the end of Main Street. More and more, most Americans must traverse miles of crowded highways to know the outdoors. The prospect for the future is that this quest will be even more difficult.

Decade by decade, the expanding population has achieved more leisure time, more money to spend, and better travel facilities; and it has sought more and better opportunities to enjoy the outdoors. But the public has also demanded more of other things. In the years following World War II, this process greatly accelerated as an eager Nation, released from wartime restrictions, needed millions of new acres for subdivisions, industrial sites, highways, schools, and airports. The resources for outdoor recreation—shoreline, green acres, open space, and unpolluted waters—diminished in the face of demands for more of everything else.

In Washington, this created legislative issues in the Congress and administrative problems within the agencies responsible for providing opportunities for outdoor recreation. Similar problems were faced in many State capitals across the country. In some cases, they stemmed from conflicts among different interests vying for use of the same resources. In others, it was the matter of responsibility—who should do the job, and who should pay the bill. Private landowners were faced with problems caused by the public seeking recreation on their land. The factors which brought about the increased need for outdoor recreation grew, and each year the problems intensified.

During the 1950's, the pressing nature of the problems of outdoor recreation had become a matter of deep concern for Members of Congress, State legislators, other public leaders, and many private citizens and organizations. Numerous problems, both foreign and domestic, were making demands upon the Nation's resources and energies. But it was felt that in making choices among these priorities, America must not neglect its heritage of the outdoors—for that heritage offers physical, spiritual, and educational benefits, which not only provide a better environment but help to achieve other national goals by adding to the health of the Nation.

By 1958, Congress had decided that an intensive nationwide study should be made of outdoor recreation, one involving all levels of government and the private contribution, and on June 28 of that year it established the Outdoor Recreation Resources Review Commission.

The authorizing act, Public Law 85-470, set forth the mission. It was essentially threefold:

To determine the outdoor recreation wants and needs of the American people now and what they will be in the years 1976 and 2000.

To determine the recreation resources of the Nation available to satisfy those needs now and in the years 1976 and 2000.

To determine what policies and programs should be recommended to ensure that the needs of the present and future are adequately and efficiently met.

The Commission that Congress established to carry out this task was composed of eight Congressional members, two representing each party from the Interior and Insular Affairs Committees of the Senate and of the House; and seven private citizens appointed by the President, one of whom was designated as Chairman.

In the fall of 1958, the Commission began recruiting a staff and in the following year launched its study program. The staff designed and coordinated the program and carried out some of the key studies, but many studies were assigned to outside contractors—Federal agencies, universities, and nonprofit research organizations—with particular skills, experience, or facilities. The reports resulting from these studies (listed in appendix C), with a full description of the techniques used in their conduct, are available in separate volumes because of their general public interest and potential value to officials at all levels of government and to others who may wish to pursue the subjects further. A few of the lines of investigation followed may be mentioned briefly.

To assess present resources for outdoor recreation, the Commission initiated an inventory of all the nonurban public designated recreation areas of the country. These numbered more than 24,000. Over a hundred items of information were analyzed in connection with 5,000 of the larger areas in order to evaluate present use and capacity and potential for development.

The Commission also carried out special studies to probe particular problems such as those connected with wilderness, water recreation, hunting and fishing, the densely populated Northeast, and sparsely populated Alaska.

To determine what the pressure is and will be on the resources, the Commission undertook a series of studies on the demand for outdoor recreation. At the base of these studies was a National Recreation Survey, conducted for the Commission by the Bureau of the Census. Some 16,000 persons were asked questions about their background, their economic status, what they presently do for outdoor recreation (if anything), what they would like to do more of, and why they do not do the things they want to do.

In further studies designed to complement and amplify the findings of the survey, the Commission investigated the effects on outdoor recreation of present and prospective changes—sectionally and nationally—in personal income, in population, in leisure time, and in travel facilities. To project future needs, the effects of such changes were applied to the present patterns as developed by the National Recreation Survey.

In order to have an effective method of working with the States, the Commission asked the Governor of each to appoint a State Contact Officer through whom it might channel all its requests. The Governors generally appointed the head of the State conservation, recreation, fish and game, or planning agency. These men and their associates made a major contribution in carrying out the inventory of State areas. This involved the laborious task of supplying detailed information on every area in the State. In other studies they provided financial, legal, and administrative data.

The Federal agencies in Washington and their field offices made available their valuable experience in the problems of outdoor recreation and provided specific data on their programs. In almost every study, the Commission began by consulting these agencies to determine what information was already available, and a great deal of valuable material was at hand.

The cooperation offered by their States and Federal agencies greatly expanded the reach of the Commission. Hundreds of people contributed significant time and effort and thus made it possible to do far more than otherwise could have been accomplished.

SOME FINDINGS OF THE STUDY

As results of the studies began flowing to the Commission, some old ideas were discarded, some were reinforced, and some new concepts evolved. The following are a few of the major conclusions.

The Simple Activities Are the
Most Popular.
Driving and walking for pleasure, swimming, and picnicking lead the list of the outdoor activities in which Americans participate, and driving for pleasure is most popular of all. This is generally true regardless of income, education, age, or occupation.

Outdoor Opportunities Are Most
Urgently Needed Near
Metropolitan Areas.
Three-quarters of the people will live in these areas by the turn of the century. They will have the greatest need for outdoor recreation, and their need will be the most difficult to satisfy as urban centers have the fewest facilities (per capita) and the sharpest competition for land use.

Across the Country, Considerable
Land Is Now Available for Outdoor
Recreation, But It Does Not
Effectively Meet the Need.
Over a quarter billion acres are public designated outdoor recreation areas. However, either the location of the land, or restrictive management policies, or both, greatly reduce the effectiveness of the land for recreation use by the bulk of the population. Much of the West and virtually all of Alaska are of little use to most Americans looking for a place in the sun for their families on a weekend, when the demand is overwhelming. At regional and State levels, most of the land is where people are not. Few places are near enough to metropolitan centers for a Sunday outing. The problem is not one of total acres but of *effective* acres.

Money Is Needed.
Most public agencies, particularly in the States, are faced with a lack of funds. Outdoor recreation opportunities can be created by acquiring new areas or by more intensive development of existing

resources, but either course requires money. Federal, State, and local governments are now spending about $1 billion annually for outdoor recreation. More will be needed to meet the demand.

Outdoor Recreation Is Often
Compatible With Other
Resource Uses.

Fortunately, recreation need not be the exclusive use of an area, particularly the larger ones. Recreation can be another use in a development primarily managed for a different purpose, and it therefore should be considered in many kinds of planning—urban renewal, highway construction, water resource development, forest and range management, to name only a few.

Water Is a Focal Point of
Outdoor Recreation.

Most people seeking outdoor recreation want water—to sit by, to swim and fish in, to ski across, to dive under, and to run their boats over. Swimming is now one of the most popular outdoor activities and is likely to be the most popular of all by the turn of the century. Boating and fishing are among the top 10 activities. Camping, picnicking, and hiking, also high on the list, are more attractive near water sites.

Outdoor Recreation Brings About
Economic Benefits.

Although the chief reason for providing outdoor recreation is the broad social and individual benefits it produces, it also brings about desirable economic effects. Its provision enhances community values by creating a better place to live and increasing land values. In some underdeveloped areas, it can be a mainstay of the local economy. And it is a basis for big business as the millions and millions of people seeking the outdoors generate an estimated $20 billion a year market for goods and services.

Outdoor Recreation Is a Major
Leisure Time Activity, and It Is
Growing in Importance.

About 90 percent of all Americans participated in some form of outdoor recreation in the summer of 1960. In total, they participated in one activity or another on 4.4 billion separate occasions. It is anticipated that by 1976 the total will be 6.9 billion, and by the year 2000 it will be 12.4 billion—a threefold increase by the turn of the century.

More Needs To Be
Known About the
Values of Outdoor Recreation.

As outdoor recreation increases in importance, it will need more land, but much of this land can be used, and will be demanded, for other purposes. Yet there is little research to provide basic information on its relative importance. More needs to be established factually about the values of outdoor recreation to our society, so that sounder decisions on allocation of resources for it can be made. More must be known about management techniques, so that the maximum social and economic benefit can be realized from these resources.

THE RECOMMENDATIONS

After 3 years of research, and an aggregate of some 50 days of discussion among the Commissioners, the Commission has developed specific recommendations for a recreation program. The 15 members brought differing political, social and resource-use opinions to the meeting table, and proposed recommendations were put through the test of this range of opinions. During the course of the study and discussion, views of individual members developed, and the collective opinion crystallized. The final recommendations are a consensus of the Commission.

In the process of evolving recommendations, the Commission's Advisory Council played an important role. It consisted of 25 individuals representative of mining, timber, grazing, business, and labor interests as well as of recreation and conservation groups. The Council also included top-level representatives of 15 Federal agencies which have a responsibility relating to the provision of outdoor recreation. In five 2-day meetings with the Commission, the Council reviewed tentative proposals and suggested alternative courses of action on several occasions. The advice of the Council had a marked effect on the final product.

State Contact Officers also contributed to the decision-making process. In a series of regional meetings, at which the Commission sought their advice on pressing issues, they put forward practical and urgent suggestions for action.

In many cases the recommendations are general; in others they are specific. For various reasons, the recommendations tend to be more detailed and more extensive regarding the Federal Government. The Commission wishes to emphasize, however, that the key elements in the total effort to make outdoor recreation opportunities available are private enterprise, the States, and local government. In relation to them, the role of the Federal agencies should be not one of domination but of cooperation and assistance in meeting their respective needs.

The recommendations of the Commission fall into five general categories—

A National Outdoor Recreation Policy.

Guidelines for the Management of Outdoor Recreation Resources.

Expansion, Modification, and Intensification of Present Programs to Meeting Increasing Needs.

Establishment of a Bureau of Outdoor Recreation in the Federal Government.

A Federal Grants-in-Aid Program to States.

The body of this report presents the reasoning and significance of these recommendations. To those who would like a quick over-all picture of the recommendations, the following digest will prove helpful.

A NATIONAL OUTDOOR RECREATION POLICY

It shall be the national policy, through the conservation and wise use of resources, to preserve, develop, and make accessible to all American people such quantity and quality of outdoor recreation as will be necessary and desirable for individual enjoyment and to assure the physical, cultural, and spiritual benefits of outdoor recreation.

Implementation of this policy will require the cooperative participation of all levels of government and private enterprise. In some aspects, the government responsibility is greater; in others, private initiative is better equipped to do the job.

The role of the Federal Government should be—

1. Preservation of scenic areas, natural wonders, primitive areas, and historic sites of national significance.
2. Management of Federal lands for the broadest possible recreation benefit consistent with other essential uses.
3. Cooperation with the States through technical and financial assistance.
4. Promotion of interstate arrangements, including Federal participation where necessary.
5. Assumption of vigorous, cooperative leadership in a nationwide recreation effort.

The States should play a pivotal role in making outdoor recreation opportunities available by—

1. Acquisition of land, development of sites, and provision and maintenance of facilities of State or regional significance.
2. Assistance to local governments.
3. Provision of leadership and planning.

Local governments should expand their efforts to provide outdoor recreation opportunities, with particular emphasis upon securing open space and developing recreation areas in and around metropolitan and other urban areas.

Individual initiative and private enterprise should continue to be the most important force in outdoor recreation, providing many and varied opportunities for a vast number of people, as well as the goods and services used by people in their recreation activities. Government should encourage the work of nonprofit groups wherever possible. It should also stimulate desirable commercial development, which can be particularly effective in providing facilities and services where demand is sufficient to return a profit.

GUIDELINES FOR MANAGEMENT

All agencies administering outdoor recreation resources—public and private—are urged to adopt a system of classifying recreation lands designed to make the best possible use of available resources in the light of the needs of people. Present jurisdictional boundaries of agencies need not be disturbed, but where necessary, use should be changed in accordance with the classification.

Implementation of this system would be a major step forward in a coordinated national recreation effort. It would provide a consistent and effective method of planning for all land-managing agencies and would promote logical adjustment of the entire range of recreation activities to the entire range of available areas. Under this approach of recreation zoning, the qualities of the respective classes of recreation environments are identified and therefore more readily enhanced and protected.

The following system of classifying outdoor recreation resources is proposed—

Class I—High-Density Recreation Areas
Areas intensively developed and managed for mass use.

Class II—General Outdoor Recreation Areas
Areas subject to substantial development for a wide variety of specific recreation uses.

Class III—Natural Environment Areas
Various types of areas that are suitable for recreation in a natural environment and usually in combination with other uses.

Class IV—Unique Natural Areas
Areas of outstanding scenic splendor, natural wonder, or scientific importance.

Class V—Primitive Areas
Undisturbed roadless areas characterized by natural, wild conditions, including "wilderness areas."

Class VI—Historic and Cultural Sites
Sites of major historic or cultural significance, either local, regional, or national.

EXPANSION, MODIFICATION, AND INTENSIFICATION OF PRESENT PROGRAMS

PLANNING, ACQUISITION, PROTECTION, AND ACCESS

1. Each State, through a central agency, should develop a long-range plan for outdoor recreation, to provide adequate opportunities for the public, to acquire additional areas where necessary, and to preserve outstanding natural sites.

2. Local governments should give greater emphasis to the needs of their citizens for outdoor recreation by considering it in all land-use planning, opening areas with recreation potential to use, and where necessary, acquiring new areas.

3. States should seek to work out interstate arrangements where the recreation-seeking public overflows political boundaries. The Federal Government should assist in meeting these interstate demand situations.

4. Systematic and continuing research, both fundamental and applied, should be promoted to provide the basis for sound planning and decisions.

5. Immediate action should be taken by Federal, State, and local governments to reserve or acquire additional water, beach, and shoreline areas, particularly near centers of population.

6. Full provision for acquiring shoreline lands for public access and use should be made in reservoir developments.

7. Surface rights to surplus Federal lands suitable for recreation should be transferred without cost to State or local governments with reversion clauses.

8. Open space programs for metropolitan areas should be continued.

9. Congress should enact legislation to provide for the establishment and preservation of certain primitive areas as "wilderness areas."

10. Certain rivers of unusual scientific, aesthetic, and recreation value should be allowed to remain in their free-flowing and natural setting without man-made alterations.

11. States should use their regulatory power to zone areas for maximum recreation benefit, maintain quality, and ensure public safety

in conflicts between recreation and other uses and in conflicts among recreation uses.

12. Recreation areas should be strongly defended against encroachments from nonconforming uses, both public and private. Where recreation land must be taken for another public use, it should be replaced with other land of similar quality and comparable location.

13. Public agencies should assure adequate access to water-based recreation opportunities by acquisition of access areas, easements across private lands, zoning of shorelines, consideration of water access in road design and construction, and opening of now restricted waters such as municipal reservoirs.

14. Interpretive and educational programs should be intensified and broadened to promote appreciation and understanding of natural, scientific, and historic values.

PROMOTING RECREATION VALUES IN RELATED FIELDS

15. Outdoor recreation should be emphasized in federally constructed or licensed multipurpose water developments and thus granted full consideration in the planning, design, and construction of such projects.

16. Recreation should be recognized as a motivating purpose in programs and projects for pollution control and as a necessary objective in the allocation of funds therefor.

17. Flood-plain zoning should be used wherever possible as a method to preserve attractive reaches of rivers and streams for public recreation in addition to the other benefits from such zoning.

18. The Federal Government and the States should recognize the potential recreation values in highway construction programs and assure that they are developed.

19. Activities under watershed and other agricultural conservation programs should be oriented toward greater recreation benefits for the public.

20. The States should encourage the public use of private lands by taking the lead in working out such arrangements as leases for hunting and fishing, scenic easements, and providing protection for landowners who allow the public to use their lands.

MEETING THE COSTS

21. All levels of government must provide continuing and adequate funds for outdoor recreation. In most cases, this will require a substantial increase over present levels.

22. State and local governments should consider the use of general obligation and revenue bonds to finance land acquisition and capital improvements for outdoor recreation.

23.   State and local governments should consider other financing devices such as season user fees, dedicated funds, and use of uncollected refunds of gasoline taxes paid by pleasure boat owners.

24.   States should take the lead in extending technical and financial assistance to local governments to meet outdoor recreation requirements.

25.   Public agencies should adopt a system of user fees designated to recapture at least a significant portion of the operation and maintenance costs of providing outdoor recreation activities that involve the exclusive use of a facility, or require special facilities.

26.   In addition to outright acquisition, local governments should consider the use of such devices as easements, zoning, cluster developments, and open-land tax policies to supplement the supply of outdoor recreation opportunities.

27.   Public agencies should stimulate desirable gifts of land and money from private individuals and groups for outdoor recreation purposes. The work of private, nonprofit organizations in providing and enhancing opportunities should be encouraged.

28.   Government should stimulate and encourage the provision of outdoor recreation opportunities by private enterprise.

29.   Where feasible, concessioners should be encouraged to provide facilities and visitor services on Federal lands under appropriate supervision. Where this is not feasible, the Federal Government should build facilities and lease them to private business for operation.

A BUREAU OF OUTDOOR RECREATION

A Bureau of Outdoor Recreation should be established in the Department of the Interior. This Bureau would have over-all responsibility for leadership of a nationwide effort by coordinating the various Federal programs and assisting other levels of government to meet the demands for outdoor recreation. It would not manage any land. This would continue to be the function of the existing managerial agencies.

Specifically, the new Bureau would—

1.   Coordinate the recreation activities of the more than 20 Federal agencies whose activities affect outdoor recreation.

2.   Assist State and local governments with technical aid in planning and administration, including the development of standards for personnel, procedures, and operations.

3.   Administer a grants-in-aid program to States for planning and for development and acquisition of needed areas.

4.   Act as a clearinghouse for information and guide, stimulate, and sponsor research as needed.

5. Encourage interstate and regional cooperation, including Federal participation where necessary.

To assure that recreation policy and planning receive attention at a high level and to promote interdepartmental coordination, there should be established a Recreation Advisory Council, consisting of the Secretaries of Interior, Agriculture, and Defense, with the Secretary of the Interior as Chairman. Other agencies would be invited to participate on an *ad hoc* basis when matters affecting their interests are under consideration by the Council.

The Recreation Advisory Council would provide broad policy guidance on all matters affecting outdoor recreation activities and programs carried out by the Bureau of Outdoor Recreation. The Secretary of the Interior should be required to seek such guidance in the administration of the Bureau.

Initially the new Bureau should be staffed where possible by transfer of experienced personnel from existing Federal agencies. It should have regional offices.

A Research Advisory Committee consisting of professional people from government, academic life, and private business should be established to advise the Bureau on its research activities.

It is urged that each State designate a focal point within its governmental structure to work with the Bureau. This focal point, perhaps one of the existing State agencies, could also serve to coordinate State recreation planning and activities and be responsible for a comprehensive State outdoor recreation plan.

A GRANTS-IN-AID PROGRAM

A Federal grants-in-aid program should be established to stimulate and assist the States in meeting the demand for outdoor recreation. This program, administered by the proposed Bureau of Outdoor Recreation, would promote State planning and acquisition and development of areas to meet the demands of the public. Projects would be approved in accordance with a statewide plan. They would be subject to review by the proposed Bureau of Outdoor Recreation to ensure conformance with Federal standards. This program would complement and would be closely coordinated with the open space aid provisions of recent legislation.

Initial grants of up to 75 percent of the total cost for planning would be made the first year and a reduced percentage thereafter. Grants for acquisition or development would be made up to 40 percent of the total cost. Federal participation could be raised to 50 percent where the State acquisition or development was part of an interstate plan.

Funds for the program would be allocated on a basis which would take into account State population, area, needs, and the amount of Federal land and Federal recreation programs in the State and region.

The grants-in-aid program should be supplemented by a program of loans to the States. This would assist in projects where the States did not have matching funds available but where the need for acquisition or development was particularly urgent, or where funds were needed beyond those available as grants-in-aid.

National Park Service Archives, Harpers Ferry, ORRRC Files, 1–10.

ADVISORY BOARD ON WILDLIFE MANAGEMENT APPOINTED BY SECRETARY OF THE INTERIOR UDALL

A. S. Leopold (Chairman), S. A. Cain,
C. M. Cottam, I. N. Gabrielson, T. L. Kimball
March 4, 1963

WILDLIFE MANAGEMENT IN THE NATIONAL PARKS

HISTORICAL

In the Congressional Act of 1916 which created the National Park Service, preservation of native animal life was clearly specified as one of the purposes of the parks. A frequently quoted passage of the Act states ". . . which purpose is to conserve the scenery and the natural and historic objects and the wild life therein and to provide for the enjoyment of the same in such manner and by such means as will leave them unimpaired for the enjoyment of future generations."

In implementing this Act, the newly formed Park Service developed a philosophy of wildlife *protection*, which in that era was indeed the most obvious and immediate need in wildlife conservation. Thus the parks were established as refuges, the animal populations were protected from wildfire. For a time predators were controlled to protect the "good" animals from the "bad" ones, but this endeavor mercifully ceased in the 1930's. On the whole, there was little major change in the Park Service practice of wildlife management during the first 40 years of its existence.

During the same era, the concept of wildlife management evolved rapidly among other agencies and groups concerned with the production of wildlife for recreational hunting. It is now an accepted truism that maintenance of suitable habitat is the key to sustaining animal populations, and that protection, though it is important, is not of itself a substitute for habitat. Moreover, habitat is not a fixed or stable entity that can be set aside and preserved behind a fence, like a cliff dwelling or a petrified tree. Biotic communities change through natural stages of succession. They can be changed deliberately through manipulation of plant and animal populations. In recent years the National Park Service has broadened its concept of wildlife conservation to provide for purposeful management of plant and animal communities as an essential step in preserving wildlife resources ". . . unimpaired for the enjoyment of future generations." In a few parks active manipulation of habitat is being tested, as for example in the Everglades where controlled burning is now used experimentally to maintain the open glades and piney woods with their interesting animal and plant life. Excess populations of grazing ungulates are being controlled in a number of parks to preserve the forage plants on which the animals depend. The question already has been posed—how far should the National Park Service go in utilizing the tools of management to maintain wildlife populations?

THE CONCEPT OF PARK MANAGEMENT

The present report proposes to discuss wildlife management in the national parks in terms of three questions which shift emphasis progressively from the general to the specific:

1) What should be the *goals* of wildlife management in the national parks?

2) What general *policies* of management are best adapted to achieve the pre-determined goals?

3) What are some of the *methods* suitable for on-the-ground implementation of policies?

It is acknowledged that this Advisory Board was requested by the Secretary of the Interior to consider particularly one of the methods of management, namely, the procedure of removing excess ungulates from some of the parks. We feel that this specific question can only be viewed objectively in the light of goals and operational policies, and our report is framed accordingly. In speaking of national parks we refer to the whole system of parks and monuments; national recreation areas are discussed briefly near the end of the report.

As a prelude to presenting our thoughts on the goals, policies, and methods of managing wildlife in the parks of the United States we wish to quote in full a brief report on "Management of National Parks and Equivalent Areas" which was formulated by a committee of the First World Conference on National Parks that convened in Seattle in July, 1962. The committee consisted of 15 members of the Conference, representing eight nations; the chairman was François Bourlière of France. In our judgment this report suggests a firm basis for park management. The statement of the committee follows:

"1. Management is defined as an activity directed toward achieving or maintaining a given condition in plant and/or animal populations and/or habitats in accordance with the conservation plan for the area. A prior definition of the purposes and objectives of each park is assumed.

Management may involve active manipulation of the plant and animal communities, or protection from modification or external influences.

"2. Few of the world's parks are large enough to be in fact self-

regulatory ecological units; rather, most are ecological islands subject to direct or indirect modification by activities and conditions in
the surrounding areas. These influences may involve such factors as
immigration and/or emigration of animal and plant life, changes in
the fire regime, and alterations in the surface or subsurface water.

"3.   There is no need for active modification to maintain large
examples of the relatively stable 'climax' communities which under
protection perpetuate themselves indefinitely. Examples of such
communities include large tracts of undisturbed rain-forests, tropi-
cal mountain paramos, and arctic tundra.

"4.   However, most biotic communities are in a constant state of
change due to natural or man-caused processes of ecological succession. In these 'successional' communities it is necessary to manage
the habitat to achieve or stabilize it at a desired stage. For example,
fire is an essential management tool to maintain East African open
savanna or American prairie.

"5.   Where animal populations get out of balance with their habitat
and threaten the continued existence of a desired environment,
population control becomes essential. This principle applies, for
example, in situations where ungulate populations have exceeded
the carrying capacity of their habitat through loss of predators,
immigration from surrounding areas, or compression of normal
migratory patterns. Specific examples include excess populations of
elephants in some African parks and of ungulates in some mountain
parks.

"6.   The need for management, the feasibility of management
methods, and evaluation of results must be based upon current and
continuing scientific research. Both the research and management
itself should be undertaken only by qualified personnel. Research,
management, planning, and execution must take into account, and if
necessary regulate, the human uses for which the park is intended.

"7.   Management based on scientific research is, therefore, not
only desirable but often essential to maintain some biotic communities in accordance with the conservation of a national park or
equivalent area."

THE GOAL OF PARK MANAGEMENT

Item 1 in the report just quoted specifies that "a prior definition
of the purposes and objectives of each park is assumed." In other
words, the goal must first be defined.

As a primary goal, we would recommend that the biotic associations within each park be maintained, or where necessary recreated, as nearly as possible in the condition that prevailed when the
area was first visited by the white man. A national park should represent a vignette of primitive America.

The implications of this seemingly simple aspiration are stupendous. Many of our national parks—in fact most of them—went through periods of indiscriminate logging, burning, livestock grazing, hunting and predator control. Then they entered the park system and shifted abruptly to a regime of equally unnatural protection from lightning fires, from insect outbreaks, absence of natural controls of ungulates, and in some areas elimination of normal fluctuations in water levels. Exotic vertebrates, insects, plants, and plant diseases have inadvertently been introduced. And of course lastly there is the factor of human use—of roads and trampling and camp grounds and pack stock. The resultant biotic associations in many of our parks are artifacts, pure and simple. They represent a complex ecologic history but they do not necessarily represent primitive America.

Restoring the primitive scene is not done easily nor can it be done completely. Some species are extinct. Given time, an eastern hardwood forest can be regrown to maturity but the chestnut will be missing and so will the roar of pigeon wings. The colorful drapanid finches are not to be heard again in the lowland forests of Hawaii, nor will the jack-hammer of the ivory-bill ring in southern swamps. The wolf and grizzly bear cannot readily be reintroduced into ranching communities, and the factor of human use of the parks is subject only to regulation, not elimination. Exotic plants, animals, and diseases are here to stay. All these limitations we fully realize. Yet, if the goal cannot be fully achieved it can be approached. A reasonable illusion of primitive America could be recreated, using the utmost in skill, judgment, and ecologic sensitivity. This in our opinion should be the objective of every national park and monument.

To illustrate the goal more specifically, let us cite some cases. A visitor entering Grand Teton National Park from the south drives across Antelope Flats. But there are no antelope. No one seems to be asking the question—why aren't there? If the mountain men who gathered here in rendezvous fed their squaws on antelope, a 20th century tourist at least should be able to see a band of these animals. Finding out what aspect of the range needs rectifying, and doing so, would appear to be a primary function of park management.

When the forty-niners poured over the Sierra Nevada into California, those that kept diaries spoke almost to a man of the wide-spaced columns of mature trees that grew on the lower western slope in gigantic magnificence. The ground was a grass parkland, in springtime carpeted with wildflowers. Deer and bears were abundant. Today much of the west slope is a dog-hair thicket of young pines, white fir, incense cedar, and mature brush—a direct function of overprotection from natural ground fires. Within the four national parks—Lassen, Yosemite, Sequoia, and Kings Canyon—the thickets are even more impenetrable than elsewhere. Not only is this accumulation of fuel dangerous to the giant sequoias and other mature trees but the animal life is meager, wildflowers are sparse, and to some at least the vegetative tangle is depressing, not uplifting. Is it possible that the primitive open forest could be restored, at least on a local scale? And if so, how? We cannot offer an answer. But we are posing a question to which there should be an answer of immense concern to the National Park Service.

The scarcity of bighorn sheep in the Sierra Nevada represents another type of management problem. Though they have been effectively protected for nearly half a century, there are fewer than 400 bighorns in the Sierra. Two-thirds of them are found in summer along the crest which lies within the eastern border of Sequoia and Kings Canyon National Parks. Obviously, there is some shortcoming of habitat that precludes further increase in the population. The high country is still recovering slowly from the devastation of early domestic sheep grazing so graphically described by John Muir. But the present limitation may not be in the high summer range at all but rather along the eastern slope of the Sierra where the bighorns winter on lands in the jurisdiction of the Forest Service. These areas are grazed in summer by domestic livestock and large numbers of mule deer, and it is possible that such competitive use is adversely affecting the bighorns. It would seem to us that the National Park Service might well take the lead in studying this problem and in formulating cooperative management plans with other agencies even though the management problem lies outside the park boundary. The goal, after all, is to restore the Sierra bighorn. If restoration is achieved in the Sequoia-Kings Canyon region, there might follow a program of reintroduction and restoration of bighorns in Yosemite and Lassen National Parks, and Lava Beds National Monument, within which areas this magnificent native animal is presently extinct.

We hope that these examples clarify what we mean by the goal of park management.

POLICIES OF PARK MANAGEMENT

The major policy change which we would recommend to the National Park Service is that it recognize the enormous complexity of ecologic communities and the diversity of management procedures required to preserve them. The traditional, simple formula of protection may be exactly what is needed to maintain such climax associations as arctic-alpine heath, the rain forests of Olympic peninsula, or the Joshua trees and saguaros of southwestern deserts. On the other hand, grasslands, savannas, aspen, and other successful shrub and tree associations may call for very different treatment. Reluctance to undertake biotic management can never lead to a realistic presentation of primitive America, much of which supported successional communities that were maintained by fires, floods, hurricanes, and other natural forces.

A second statement of policy that we would reiterate—and this one conforms with present Park Service standards—is that management be limited to native plants and animals. Exotics have intruded into nearly all the parks but they need not be encouraged, even those that have interest or ecologic values of their own. Restoration of antelope in Jackson Hole, for example, should be done by managing native forage plants, not by planting crested wheat grass or plots of irrigated alfalfa. Gambel quail in a desert wash should be observed in the shade of a mesquite, not a tamarisk. A visitor who climbs a volcano in Hawaii ought to see mamane trees and silver-swords, not goats.

Carrying this point further, observable artificiality in any form must be minimized and obscured in every possible way. Wildlife should not be displayed in fenced enclosures; this is the function of a zoo, not a national park. In the same category is artificial feeding of wildlife. Fed bears become bums, and dangerous. Fed elk deplete natural ranges. Forage relationships in wild animals should be natural. Management may at times call for the use of the tractor, chainsaw, rifle, or flame-thrower but the signs and sounds of such activity should be hidden from visitors insofar as possible. In this regard, perhaps the most dangerous tool of all is the roadgrader. Although the American public demands automotive access to the parks, road systems must be rigidly prescribed as to extent and design. Roadless wilderness areas should be permanently zoned. The goal, we repeat, is to maintain or create the mood of wild America. We are speaking here of restoring wildlife to enhance this mood, but the whole effect can be lost if the parks are overdeveloped for motorized travel. If too many tourists crowd the roadways, then we should ration the tourists rather than expand the roadways.

Additionally in this connection, it seems incongruous that there should exist in the national parks mass recreation facilities such as golf courses, ski lifts, motorboat marinas, and other extraneous developments which completely contradict the management goal. We urge the National Park Service to reverse its policy of permitting these non-conforming uses, and to liquidate them as expeditiously as possible (painful as this will be to concessionaires). Above all other policies, the maintenance of naturalness should prevail.

Another major policy matter concerns the research which must form the basis of all management programs. The agency best fitted to study park management problems is the National Park Service itself. Much help and guidance can be obtained from ecologic research conducted by other agencies, but the objectives of park management are so different from those of state fish and game departments, the Forest Service, etc., as to demand highly skilled studies of a very specialized nature. Management without knowledge would be a dangerous policy indeed. Most of the research now conducted by the National Park Service is oriented largely to interpretive functions rather than to management. We urge the expansion of the research activity in the Service to prepare for future management and restoration programs. As models of the type of investigation that should be greatly accelerated we cite some of the recent studies of elk in Yellowstone and of bighorn sheep in Death Valley. Additionally, however, there are needed equally critical appraisals of ecologic relationships in various plant associations and of many lesser organisms such as azaleas, lupines, chipmunks, towhees, and other non-economic species.

In consonance with the above policy statements, it follows logically that every phase of management itself be under the full jurisdiction of biologically trained personnel of the Park Service. This applies not only to habitat manipulation but to all facets of regulating animal populations. Reducing the numbers of elk in Yellowstone or of goats in Haleakala Crater is part of an overall scheme to preserve or restore a natural biotic scene. The purpose is single-minded. We cannot endorse the view that responsibility for removing excess game animals be shared with state fish and game departments whose primary interest would be to capitalize on the recreational value of the public hunting that could thus be supplied. Such a proposal imputes a multiple use concept of park management which was never intended, which is not legally permitted, nor for which we can find any impelling justification today.

Purely from the standpoint of how best to achieve the goal of park management, as here defined, unilateral administration directed to a single objective is obviously superior to divided responsibility in which secondary goals, such as recreational hunting, are introduced. Additionally, uncontrolled public hunting might well operate in opposition to the goal, by removing roadside animals and frightening the survivors, to the end that public viewing of wildlife would be materially impaired. In one national park, namely Grand Teton, public hunting was specified by Congress as the method to be used in controlling elk. Extended trial suggests this to be an awkward administrative tool at best.

Since this whole matter is of particular current interest is will be elaborated in a subsequent section on methods.

METHODS OF HABITAT MANAGEMENT

It is obviously impossible to mention in this brief report all the possible techniques that might be used by the National Park Service in manipulating plant and animal populations. We can, however, single out a few examples. In so doing, it should be kept in mind that the total area of any one park, or of the parks collectively, that may be managed intensively is a very modest part indeed. This is so for two reasons. First, critical areas which may determine animal abundance are often a small fraction of total range. One deer study on the west slope of the Sierra Nevada, for example, showed that the important winter range, which could be manipulated to support the deer, constituted less than two per cent of the year-long herd range. Roadside areas that might be managed to display a more varied and natural flora and fauna can be rather narrow strips. Intensive management, in short, need not be extensive to be effective. Secondly, manipulation of vegetation is often exorbitantly expensive. Especially will this be true when the objective is to manage "invisibly"—that is, to conceal the signs of management. Controlled burning is the only method that may have extensive application.

The first step in park management is historical research, to ascertain as accurately as possible what plants and animals and biotic associations existed originally in each locality. Much of this has been done already.

A second step should be ecologic research on plant-animal relationships leading to formulation of a management hypothesis.

Next should come small scale experimentation to test the hypothesis in practice. Experimental plots can be situated out of sight of roads and visitor centers.

Lastly, application of tested management methods can be undertaken on critical areas.

By this process of study and pre-testing, mistakes can be minimized. Likewise, public groups vitally interested in park management can be shown the results of research and testing before application, thereby eliminating possible misunderstandings and friction.

Some management methods now in use by the National Park Service seem to us potentially dangerous. For example, we wish to raise a serious question about the mass application of insecticides in the control of forest insects. Such application may (or may not) be justified in commercial timber stands, but in a national park the ecologic impact can have unanticipated effects on the biotic community that might defeat the overall management objective. It would seem wise to curtail this activity, at least until research and small scale testing have been conducted.

Of the various methods of manipulating vegetation, the controlled use of fire is the most "natural" and much the cheapest and easiest to apply. Unfortunately, however, forest and chaparral areas that have been completely protected from fire for long periods may require careful advance treatment before even the first experimental blaze is set. Trees and mature brush may have to be cut, piled, and burned before a creeping ground fire can be risked. Once fuel is reduced, periodic burning can be conducted safely and at low expense. On the other hand, some situations may call for a hot burn. On Isle Royale, moose range is created by periodic holocausts that open the forest canopy. Maintenance of the moose population is surely one goal of management on Isle Royale.

Other situations may call for the use of the bulldozer, the disc harrow, or the spring-tooth harrow to initiate desirable changes in plant succession. Buffalo wallows on the American prairie were the propagation sites of a host of native flowers and forbs that fed the antelope and the prairie chicken. In the absence of the great herds, wallows can be simulated.

Artificial reintroduction of rare native plants is often feasible. Overgrazing in years past led to local extermination of many delicate perennials such as some of the orchids. Where these are not reappearing naturally they can be transplanted or cultured in a nursery. A native plant, however small and inconspicuous, is as much a part of the biota as a redwood tree or a forage species for elk.

In essence, we are calling for a set of ecologic skills unknown in this country today. Americans have shown a great capacity for degrading and fragmenting native biotas. So far we have not exercised much imagination or ingenuity in rebuilding damaged biotas. It will not be done by passive protection alone.

CONTROL OF ANIMAL POPULATIONS

Good park management requires that ungulate populations be reduced to the level that the range will carry in good health and without impairment to the soil, the vegetation, or to habitats of other animals. This problem is world-wide in scope, and includes non-park as well as park lands. Balance may be achieved in several ways.

(a) *Natural predation*—Insofar as possible, control through natural predation should be encouraged. Predators are now protected in the parks of the United States, although unfortunately they were not in the early years and the wolf, grizzly bear, and mountain lion became extinct in many of the national parks. Even today populations of large predators, where they still occur in the parks, are kept below optimal level by programs of predator control applied outside the park boundaries. Although the National Park Service has attempted to negotiate with control agencies of Federal and local governments for the maintenance of buffer zones around the parks where predators are not subject to systematic control, these negotiations have been only partially successful. The effort to protect large predators in and around the parks should be greatly intensified. At the same time, it must be recognized that predation alone can seldom be relied upon to control ungulate numbers, particularly the larger species such as bison, moose, elk, and deer; additional artificial controls frequently are called for.

(b) *Trapping and transplanting*—Traditionally in the past the National Park Service has attempted to dispose of excess ungulates by trappings and transplanting. Since 1892, for example, Yellowstone National Park alone has supplied 10,478 elk for restocking purposes. Many of the elk ranges in the western United States have been restocked from this source. Thousands of deer and lesser numbers of antelope, bighorns, mountain goats, and bison also have been moved from the parks. This program is fully justified so long as breeding stocks are needed. However, most big game ranges of the United States are essentially filled to carrying capacity, and the cost of a continuing program of trapping and transplanting cannot be sustained solely on the basis of controlling populations within the parks. Trappings and handling of a big game animal usually costs from $50 to $150 and in some situations much more. Since annual surpluses will be produced indefinitely into the future, it is patently impossible to look upon trapping as a practical plan of disposal.

(c) *Shooting excess animals that migrate outside the parks*—Many park herds are migratory and can be controlled by public hunting outside the park boundaries. Especially is this true in mountain parks which usually consist largely of summer game range with relatively little winter range. Effective application of this form of control frequently calls for special regulations, since migration usually occurs after normal hunting dates. Most of the western states have cooperated with the National Park Service in scheduling late hunts for the specific purpose of reducing park game herds, and in fact most excess game produced in the parks is so utilized. This is by far the best and the most widely applied method of controlling park populations of ungulates. The only danger is that migratory habits may be eliminated from a herd by differential removal, which would favor survival of non-migratory individuals. With care to preserve, not eliminate, migratory traditions, this plan of control will continue to be the major form of herd regulation in national parks.

(d) *Control by shooting within the parks*—Where other methods of control are inapplicable or impractical, excess park ungulates must be removed by killing. As stated above in the discussion of park policy, it is the unanimous recommendation of this Board that such shooting be conducted by competent personnel, under the sole juris-diction of the National Park Service, and for the sole purpose of animal removal, not recreational hunting. If the magnitude of a given removal program requires the services of additional shooters beyond regular Park Service personnel, the selection, employment, training, deputization, and supervision of such additional personnel should be entirely the responsibility of the National Park Service. Only in this manner can the primary goal of wildlife management in the parks be realized. A limited number of expert riflemen, properly equipped and working under centralized direction, can selectively cull a herd with a minimum of disturbance to the surviving animals or to the environment. General public hunting by comparison is often non-selective and grossly disturbing.

Moreover, the numbers of game animals that must be removed annually from the parks by shooting is so small in relation to nor-mally hunted populations outside the parks as to constitute a minor contribution to the public bag, even if it were so utilized. All of these points can be illustrated in the example of the North Yellow-stone elk population which has been a focal point of argument about possible public hunting in national parks.

(e)   *The case of Yellowstone*—Elk summer in all parts of Yellow-stone Park and migrate out in nearly all directions, where they are subject to hunting on adjoining public and private lands. One herd, the so-called Northern Elk Herd, moves only to the vicinity of the park border where it may winter largely inside or outside the park, depending on the severity of the winter. This herd was estimated to number 35,000 animals in 1914 which was far in excess of the car-rying capacity of the range. Following a massive die-off in 1919-20 the herd has steadily decreased. Over a period of 27 years, the National Park Service removed 8,825 animals by shooting and 5,765 by live-trapping; concurrently, hunters took 40,745 elk from this herd out-side the park. Yet the range continues to deteriorate. In the winter of 1961-62 there were approximately 10,000 elk in the herd and carrying capacity of the winter range was estimated at 5,000. So the National Park Service at last undertook a definite reduction program, killing 4,283 elk by shooting, which along with 850 animals removed in other ways (hunting outside the park, trapping, winter kill) brought the herd down to 5,725 as censused from helicopter. The carcasses of the elk were carefully processed and distributed to Indian com-munities throughout Montana and Wyoming; so they were well used. The point at issue is whether this same reduction could or should have been accomplished by public hunting.

In autumn during normal hunting season the elk are widely scat-tered through rough inaccessible mountains in the park. Comparable areas, well stocked with elk, are heavily hunted in adjoining national forests. Applying the kill statistics from the forests to the park, a kill of 200-400 elk might be achieved if most of the available pack stock in the area were used to transport hunters within the park. Autumn hunting could not have accomplished the necessary reduction.

In mid-winter when deep snow and bitter cold forced the elk into the lower country along the north border of the park, the National Park Service undertook its reduction program. With snow vehicles, trucks, and helicopters they accomplished the unpleasant job in temperatures that went as low as -40 degrees F. Public hunting was out of the question. Thus, in the case most bitterly argued in the press and in legislative halls, reduction of the herd by recreational hunting would have been a practical impossibility, even if it had been in full conformance with park management objectives.

From now on, the annual removal from this herd may be in the neighborhood of 1,000 to 1,800 head. By January 31, 1963, removals had totalled 1,300 (300 shot outside the park by hunters, 600 trapped and shipped, and 400 killed by park rangers). Continued special hunts in Montana and other forms of removal will yield the desired reduction by spring. The required yearly maintenance kill is not a large operation when one considers that approximately 100,000 head of big game are taken annually by hunters in Wyoming and Montana.

(f) *Game control in other parks*—In 1961-62, excluding Yellowstone elk, there were approximately 870 native animals transplanted and 827 killed on 18 national parks and monuments. Additionally, about 2,500 feral goats, pigs and burros were removed from three areas. Animal control in the park system as a whole is still a small operation. It should be emphasized, however, that removal programs have not in the past been adequate to control ungulates in many of the parks. Future removals will have to be larger and in many cases repeated annually. Better management of wildlife habitat will naturally produce larger annual surpluses. But the scope of this phase of park operation will never be such as to constitute a large facet of management. On the whole, reductions will be small in relation to game harvests outside the parks. For example, from 50 to 200 deer a year are removed from a problem area in Sequoia National Park; the deer kill in California is 75,000 and should be much larger. In Rocky Mountain National Park 59 elk were removed in 1961-62 and the trim should perhaps be 100 per year in the future; Colorado kills over 10,000 elk per year on open hunting ranges. In part, this relates to the small area of the national park system which constitutes only 3.9 per cent of the public domain; hunting ranges under the jurisdiction of the Forest Service and Bureau of Land Management make up approximately 70 per cent.

In summary, control of animal populations in the national parks would appear to us to be an integral part of park management, best handled by the National Park Service itself. In this manner excess ungulates have been controlled in the national parks of Canada since 1943, and the same principle is being applied in the parks of many African countries. Selection of personnel to do the shooting likewise is a function of the Park Service. In most small operations this would logically mean skilled rangers. In larger removal programs, there might be included additional personnel, selected from the general public, hired and deputized by the Service or otherwise engaged, but with a view to accomplishing a task, under strict supervision and solely for the protection of park values. Examples of some potentially large removal programs where expanded crews may be needed are mule deer populations on plateaus fringing Dinosaur National Monument and Zion National Park (west side), and white-tailed deer in Acadia National Park.

WILDLIFE MANAGEMENT ON NATIONAL RECREATION AREAS

By precedent and logic, the management of wildlife resources on the national recreation areas can be viewed in a very different light than in the park system proper. National recreation areas are by definition multiple use in character as regards allowable types of recreation. Wildlife management can be incorporated into the operational plans of these areas with public hunting as one objective. Obviously, hunting must be regulated in time and place to minimize conflict with other uses, but it would be a mistake for the National Park Service to be unduly restrictive of legitimate hunting in these areas. Most of the existing national recreation areas are Federal holdings surrounding large water impoundments; there is little potentiality for hunting. Three national seashore recreational areas on the East Coast (Hatteras, Cape Cod, and Padre Island) offer limited waterfowl shooting. But some of the new areas being acquired or proposed for acquisition will offer substantial hunting opportunity for a variety of game species. This opportunity should be developed with skill, imagination, and (we would hopefully suggest) with enthusiasm.

On these areas as elsewhere, the key to wildlife abundance is a favorable habitat. The skills and techniques of habitat manipulation applicable to parks are equally applicable on the recreation areas. The regulation of hunting, on such areas as are deemed appropriate to open for such use, should be in accord with prevailing state regulations.

NEW NATIONAL PARKS

A number of new national parks are under [consideration]. One of the critical issues in the establishment of new parks will be the manner in which the wildlife resources are to be handled. It is our recommendation that the basic objectives and operating procedures of new parks be identical with those of established parks. It would seem awkward indeed to operate a national park system under two sets of ground rules. On the other hand, portions of several proposed parks are so firmly established as traditional hunting grounds that impending closure of hunting may preclude public acceptance of park status. In such cases it may be necessary to designate core areas as national parks in every sense of the word, establishing protective buffer zones in the form of national recreation areas where hunting is permitted. Perhaps only through compromises of this sort will the park system be rounded out.

SUMMARY

The goal of managing the national parks and monuments should be to preserve, or where necessary to recreate, the ecologic scene as viewed by the first European visitors. As part of this scene, native species of wild animals should be present in maximum variety and reasonable abundance. Protection alone, which has been the core of Park Service wildlife policy, is not adequate to achieve this goal. Habitat manipulation is helpful and often essential to restore or maintain animal numbers. Likewise, populations of the animals themselves must sometimes be regulated to prevent habitat damage; this is especially true of ungulates.

Active management aimed at restoration of natural communities of plants and animals demands skills and knowledge not now in existence. A greatly expanded research program, oriented to management needs, must be developed within the National Park Service itself. Both research and the application of management methods should be in the hands of skilled park personnel.

Insofar as possible, animal populations should be regulated by predation and other natural means. However, predation cannot be relied upon to control the populations of larger ungulates, which sometimes must be reduced artificially.

Most ungulate populations within the parks migrate seasonally outside the park boundaries where excess numbers can be removed by public hunting. In such circumstances the National Park Service should work closely with state fish and game departments and other interested agencies in conducting the research required for management and in devising cooperative management programs.

Excess game that does not leave a park must be removed. Trapping and transplanting has not proven to be a practical method of control, though it is an appropriate source of breeding stock as needed elsewhere.

Direct removal by killing is the most economical and effective way of regulating ungulates within a park. Game removal by shooting should be conducted under the complete jurisdiction of qualified park personnel and solely for the purpose of reducing animals to preserve park values. Recreational hunting is an inappropriate and non-conforming use of the national parks and monuments.

Most game reduction programs can best be accomplished by regular park employees. But as removal programs increase in size and scope, as well may happen under better wildlife management, the National Park Service may find it advantageous to employ or otherwise engage additional shooters from the general public. No objection to this procedure is foreseen so long as the selection, training, and supervision of shooting crews is under rigid control of the Service and the culling operation is made to conform to primary park goals.

Recreational hunting is a valid and potentially important use of national recreation areas, which are also under jurisdiction of the National Park Service. Full development of hunting opportunities on these areas should be provided by the Service.

UNITED STATES DEPARTMENT OF THE INTERIOR
Office of the Secretary
Washington, D.C. 20240

May 2, 1963

Memorandum

To: Director, National Park Service

From: Secretary of the Interior

Subject: Report of the Advisory Board on Wildlife Management

The report of the Advisory Board on Wildlife Management of the National Parks, dated March 4, 1963, has been reviewed. It emphasizes clearly the ecological principles involved, defines the esthetic, historical and scientific values of the parks, and sets forth the philosophy of management thus called for.

You should, accordingly, take such steps as appropriate to incorporate the philosophy and the basic findings into the administration of the National Park System.

STEWART L. UDALL, *Secretary of the Interior*

National Park Service Handbook of Administrative Policies for Natural Areas, 1968, 88–103.

A REPORT BY THE ADVISORY COMMITTEE
TO THE NATIONAL PARK SERVICE ON RESEARCH

National Academy of Sciences–National Research Council
Edward A. Ackerman, Marston Bates, Stanley A. Cain,
F. Fraser Darling, John M. Fogg, Jr., Tom Gill,
Joseph M. Gillson, E. Raymond Hall, Carl L. Hubbs,
William J. Robbins, Chairman and
C.J.S. Durham, Executive Secretary
1963

ABSTRACT

The report submitted to the Secretary describes how the Committee conducted its study and surveys the development of the national parks idea, which originated in the United States and has reached its fullest expression there. It calls attention to the responsibilities and obligations which stem from the worldwide recognition and appreciation of the leadership of the United States in this area.

It discusses some of the historical aspects of the establishment of national parks, the first of which was Yellowstone National Park in 1872, and highlights the characteristics of some of the 31 parks now in existence. The report asserts that the national parks of the United States are among the most valuable heritages of this country; that in setting these lands aside the people and the government of the United States demonstrated particular wisdom; and that the role of national parks in the lives of our citizens is dramatically enlarging.

The objectives or purposes of the National Park Service are discussed in the light of the origin of the national parks and the various Acts of Congress which deal with them. The conclusion is reached that the Service should strive first to preserve and conserve the national parks with due consideration for the enjoyment by their owners, the people of the United States, of the aesthetic, spiritual, inspirational, educational, and scientific values which are inherent in natural wonders and nature's creatures. The Service should be concerned with the preservation of nature in the national parks, the maintenance of natural conditions, and the avoidance of artificiality, with such provisions for the accommodation of visitors as will neither destroy nor deteriorate the natural features, which should be preserved for the enjoyment of future visitors who may come to the parks.

Each park should be regarded as a system of interrelated plants, animals, and habitat (an ecosystem) in which evolutionary processes will occur under such control and guidance as seems necessary to preserve its unique features. Naturalness, the avoidance of artificiality, should be the rule.

Each park should be dealt with individually, and the National Park Service in consultation with appropriate advisers should define their objectives and purposes for each park. These will vary from park to park and in general should be those for which the park was originally established, with special consideration for the specific natural phenomena (biological, geological, archeological) which instigated its establishment.

The report points out that the National Park Service has the responsibility of administering the national parks in accordance with the purposes for which they are or may be set aside by specific Acts of Congress and emphasizes that knowledge about the parks and their problems is needed to discharge this responsibility. Such knowledge comes from research, especially research in natural history.

An examination of natural history research in the National Park Service shows that it has been only incipient, consisting of many reports, numerous recommendations, vacillations in policy, and little action.

Research by the National Park Service has lacked continuity, coordination, and depth. It has been marked by expediency rather than by long-term considerations. It has in general lacked direction, has been fragmented between divisions and branches, has been applied piecemeal, has suffered because of a failure to recognize the distinctions between research and administrative decision-making, and has failed to insure the implementation of the results of research in operational management.

In fact, the Committee is not convinced that the policies of the National Park Service have been such that the potential contribution of research and a research staff to the solution of the problems of the national parks is recognized and appreciated. Reports and recommendations on this subject will remain futile unless and until the National Park Service itself becomes research-minded and is prepared to support research and to apply its findings.

It is inconceivable that property so unique and valuable as the national parks, used by such a large number of people, and regarded internationally as one of the finest examples of our national spirit should not be provided adequately with competent research scientists in natural history as elementary insurance for the preservation and best use of the parks.

It is pointed out, however, that the results of research can neither be predicted or prejudged. The results may not always be pleasant. They may indicate that a facility should not have been built, that a road should have been routed another way, that visitors into a particular region should not be encouraged in large numbers and without control. It may even indicate that a particular park has deteriorated so far that it can never be returned to its former state. It is the very integrity of these conclusions, however, that make it essential that they be brought to bear upon the management problems of the national parks.

The report presents the pressing need for research in the national parks by citing specific examples in which degradation or deterioration has occurred because research on which proper management operations should have been based was not carried out in time; because the results of research known to operational management were not implemented; or because the research staff was not consulted before action was taken. In still other situations problems are recognized for the solution of which research is needed, but where none has been undertaken or planned or, if planned, has not been financed.

Attention is called to the meager dollar support given to research and development in the natural sciences in the national parks. In the National Park Service as a whole less than one per cent of the appropriation in 1960, 1961, and 1962 was devoted to research and development while the proportion for comparable government agencies was in the neighborhood of 10 per cent. In fact, unless drastic steps are immediately taken there is a good possibility that within this generation several, if not all, the national parks will be degraded to a state totally different from that for which they were preserved and in which they were to be enjoyed.

Particular attention is called to the precarious conditions of the Everglades National Park and the big trees in California.

As a result of the study made by the Committee a series of twenty recommendations are made.

RECOMMENDATIONS

This Committee has stated that in its opinion the National Park Service must manage to some degree the lands which fall within the National Park System. The Committee has stated further that the management of any enterprise cannot be effective unless the objectives of the enterprise are clearly defined and well understood, and plans are devised to accomplish the objectives.

Plans must be based on information of the resources (inventory) of the activity, on its problems, and on its relation with other similar activities; and they must be implemented by adequate and competent personnel, properly organized, motivated, and supported financially.

Research is an essential part of the program outlined above and its use a necessity in each of the steps. These elementary principles apply to the national parks as well as to a business or any other organized activity.

The Committee has based its recommendations on these considerations, as well as on its acquaintance with the parks and their problems and begs leave to submit the following:

1. The objectives or purposes of each national park should be defined.

COMMENT: Each national park was established because of the potential esthetic, educational, scientific and cultural values of its natural history and/or its human history. The features of a park which make the values possible of attainment should be carefully defined to serve as the basis for operational management. They should be preserved and restored, where necessary, and provisions made for their proper enjoyment and use by the people. The objectives should exclude the use of the national parks for amusement or such mass recreation as requires elaborate facilities or extensive and/or artificial modification of the natural features of a park. The Committee endorses, in this respect, the conclusion of the report: "Wildlife Management in the National Parks." Zoning of a national park into, for example, natural and undisturbed areas, naturalistic areas, public use areas and Park Service facility areas is suggested.

2. Inventory and mapping of the natural history resources of each park should be made.

COMMENT: Such an inventory should cover the past as well as the present, and include information on topography, geology, climate, water regime, soil types, flora and fauna and natural communities. Mapping, including aerial maps, should cover species distributions, natural communities, land use, archeology and such other mappable features as may be of importance in the park.

An inventory serves as a basis for judging changes, good or bad, in the condition of a park, supplies the information necessary for interpreting the area to the public, and is essential for proper operational management, as well as for further research.

3. A distinction should be made between administration, operational management, and research management.

COMMENT: Research is essential to solve problems of operational management whether the latter concerns preservation, restoration, interpretation or the use of the parks by the public. Administration, the management of research and the management of operations require somewhat different though well recognized ad-

ministrative procedures. In most situations, the following steps are involved:
   1) Identification and definition of the problem or situation;
   2) Research, or fact finding, based on observation and/or experimentation;
   3) Administrative action which involves decision on a course of action, grounded on the findings and recommendations of research and such other considerations as may be involved; and
   4) Operational management, which means the implementation of the decisions by the appropriate operational division.

4.   A permanent, independent, and identifiable research unit should be established within the National Park Service to conduct and supervise research in natural history in the national parks, and to serve as consultant on natural history problems for the entire National Park System.

COMMENT: In order to maintain objectivity, the principal research organization should be independent of operational management. It should provide knowledge which would allow predictions of the consequences of alternate lines of action or inaction. Close liaison should be maintained between the research unit and the administrative and operating divisions in order that the results of research may be adequately applied. All branches of the service should participate fully in identifying problems and in preparing programs and budgets for research. The research staff should have complete freedom in the execution of an approved research program, in evaluating the results, in reporting the findings and in making recommendations based on the findings. There should be free communication on research ideas and research accomplishment from anywhere in the National Park Service to and from the top research staff. Provision should be made to enable the research staff to maintain close association with other scientists.

5.   The research unit in natural history in the National Park Service should be organized as a line arrangement with an "Assistant Director for Research in the Natural Sciences" reporting to the Director of the National Park Service.

COMMENT: A nucleus of highly competent scientists headed by a Chief Scientist should be assembled in the headquarters of the National Park Service. This nucleus should comprise at least 10 individuals—including the present staff. The scientific group in Washington should be supported by an appropriate staff of natural history specialists available for field assignments and other research. The committee emphasizes that quality is more important than numbers and that a selective and flexible approach to research problems is likely to be most profitable in the long term. Field research personnel should report directly to the Washington staff, and should be administered by personnel management policies compatible with their responsibilities.

6. Most of the research by the National Park Service should be mission-oriented.

COMMENT: The National Park Service should direct its in-service research mainly toward the problems involved in the preservation and/or restoration of the national parks for the esthetic, educational and scientific values and toward the adequate interpretation of these values. The solution of some of the problems may extend beyond the conventional bounds of natural history and involve, at least temporarily, contributions by, for example, economists, social scientists, and engineers. The problem should be emphasized and assistance for its solution sought wherever competence may be found. When appropriate, mission-oriented research should be carried out on a contract basis with universities or private research organizations.

7. The National Park Service should itself plan and administer its own mission-oriented research program directed toward the preservation, restoration, and interpretation of the national parks.

COMMENT: The mission of the Service in the preservation of the total environment is a unique responsibility. The research program necessary to support this objective is of a scope and character different from that of any other institution or land management agency. The Service must therefore accept the responsibility for the planning, administration and conduct of its own research program. While it may, and is encouraged to utilize the specialized services of other agencies and institutions, it cannot abrogate its responsibilities for the direction and execution of its own mission-oriented research program.

8. Research should be designed to anticipate and prevent problems in operational management as well as to meet those which have already developed.

COMMENT: A limited staff which has inadequate support can deal only with immediate "brush fire" problems; that is to say, it can deal only with situations which have already become critical and perhaps irreparable. A research staff adequate in competence and numbers can conduct research from long-term considerations, detect problems before they become critical and offer alternate choices of action for their solution.

9.   A research program should be prepared for each park.

COMMENT: A basic goal of management should be to perpetuate and where necessary restore the values which justified the parks' creation and maintenance. A program of research studies needed to provide management with the information required to reach this goal should be established and implemented with the requisite funds and personnel.

10.   Consultation with the research unit in natural history of the National Park Service should precede all decisions on management operations involving preservation, restoration, development, protection and interpretation, and the public use of a park.

COMMENT: The Committee discovered or had its attention called to numerous instances in which consultation with qualified scientists would have prevented or modified a development or operation which had harmful effects on a park or required expensive changes to prevent or correct such effects. Operational management is sensible of this need, as judged by frequent unsolicited comments to the Committee, but is handicapped by limited research staff available for consultation or by failures in communication.

11.   Research on aquatic life, as well as on that existing on and above the land, should be pursued to assist in determining general policies for the maintenance of natural conditions for their scientific, educational, and cultural values.

COMMENT: The Committee recognizes that serious management problems for the preservation and restoration of aquatic life in the parks exist and that research is needed to arrive at rational decisions on these problems. They arise in part from the use of rotenone or other poisons as a fish management tool, the effects on aquatic life of motorboat traffic, sport fishing, the introduction of exotic forms and their effects on native aquatic life. The so-called "barren" lakes and streams are devoid of game fish but are of considerable scientific interest because of that fact. Each of these raise questions which can be properly settled only through the result of research.

12.   Research should include specific attention to significant changes in land use, in other natural resource use, or in other economic activities on areas adjacent to national parks, and likely to affect the parks.

COMMENT: The problems of operating a park to meet objectives given the National Park Service by legislation are closely related to events in areas surrounding each of the parks. Effective, economical administration of each park could be materially aided by timely research of a modest extent on resource use in such surrounding areas. This research could be carried on jointly with the other agencies directly concerned.

13. Research laboratories or centers should be established for a national park when justified by the nature of the park and the importance of the research.

COMMENT: Such research laboratories or centers should not only serve the staff of the National Park Service but also scientists from universities and independent research organizations. Control of such centers should remain with the National Park Service. The location of such centers, and access to them, should be such as will not destroy other values of a park nor interfere with the proper use and enjoyment of a park by the public. Consideration should be given to establishing research centers, whenever possible, outside the limits of a park in some instances supported, administered and used jointly with other agencies or organizations.

14. The results of research undertaken by the National Park Service should be publishable and should be published.

COMMENT: Research in natural history carried out by the National Park Service should be of such quality that the results are worthy of publication and should be published. Although the research conducted by the National Park Service should be directed primarily toward park problems, it is in the public interest that the results be made available through publication, either in established journals or in a series sponsored by the National Park Service. It is recognized that on occasion research may be undertaken the results of which are not of general interest and do not require publication. Such investigations should be exceptions and not the rule.

15. Additional substantial financial support should be furnished the National Park Service for research in the national parks.

COMMENT: The Committee could not in the time available and from the data at hand, estimate the total cost of research, based upon the needs of each park. The Committee noted, however, that on the average, approximately 10 per cent of the annual budget was devoted in 1962 to research and development by those government agencies comparable to the National Park Service. The Committee considers this to be a reasonable basis for establishing a research budget and recommends that research in the National Park Service be supported at a level consistent with that of comparable agencies.

The Committee strongly urges that in future research appropriations and allotments within the National Park Service natural history research be given support commensurate with the key position of natural history in the preservation, restoration and interpretation of the parks. The number, variety and extent of the national parks, the unique character and international significance, as well as the complexity of their problems suggest that the allotment of money to research be of the order recommended above.

16.   Cooperative planning as a result of research should be fostered with other agencies which administer public and private lands devoted to conservation and to recreation.

COMMENT: Various agencies in the federal government, the states, municipalities, universities, and other private or public organizations administer lands devoted to conservation and to recreation of one type or another. The National Park Service should be fully cognizant of the resources, objectives, and activities of these areas, and cooperate fully with those responsible for their administration, especially as related to natural history research.

17.   Universities, private research institutions, and qualified independent investigators should be encouraged to use the national parks in teaching and research.

COMMENT: The national parks are a national and international scientific resource. In some respects, their natural history is unique or nearly so. They are outdoor laboratories of great scientific value and should be made available to independent investigators when the research work does not threaten deterioration of the park or interfere with its appropriate use by the public and when it can be effectively facilitated by the staff of the National Park Service.

18.   Consideration should be given to including in the budget of the National Park Service an item for aid to advanced students who wish to conduct research in the national parks.

COMMENT: A program of this character should be considered in part a training program and a practical source of future personnel. Support for field work by advanced students is frequently inadequate, especially in natural history. It is recognized that the supervision of students places responsibilities on park personnel, and that provision for adequate supervision should be a part of any plan of the nature recommended. An expansion of those aspects of the Student Conservation Program concerned with the support of advanced students as Assistant Ranger Naturalists should be considered.

19.   A Scientific Advisory Committee for the National Park Service should be established, and Scientific Advisory Committees for individual parks are desirable.

COMMENT: Such Advisory Committees should be working committees concerned with park problems. It should be clearly understood, however, that advisory committees are advisory, not decision-making bodies. The practice of engaging the assistance of ad hoc committees for special park problems should be continued.

20.   Action in implementing the recommendations of this Committee's report should be taken promptly.

COMMENT: Time is an essential factor in dealing with forces that threaten the existence of certain indigenous animal and plant species and threaten or otherwise degrade park values, in some instances beyond the possibility of restoration. Among these factors are excessive human use, overgrazing, the invasion of park areas by aggressive exotic flora and fauna and interference with water supply. Studies are urgently needed to provide the basis for prompt action.

Provided by the Natural Resources Division, National Park Service, Washington, D.C., ix–xii, 64–75.

FEDERAL EXECUTIVE BRANCH POLICY GOVERNING THE SELECTION, ESTABLISHMENT, AND ADMINISTRATION OF NATIONAL RECREATION AREAS BY THE RECREATION ADVISORY COUNCIL

Circular No. 1, March 26, 1963

PREAMBLE

The Recreation Advisory Council believes that:

1.   Greater efforts must be made by Federal, State, local governmental, and private interests to fulfill adequately the steeply mounting outdoor recreation demands of the American people;

2.   The Federal Government should provide leadership and stimulus to this effort, but does not have sole or primary responsibility for providing recreation opportunities;

3.   Present Federal programs should be augmented by a system of National Recreation Areas made up of a limited number of areas where the recreation demand is not being met through other programs.

The system of National Recreation Areas should:

1.   Provide for Federal investment in outdoor recreation that is more clearly responsive to recreation demand than other investments that are based primarily upon considerations of preserving unique natural or historical resources, the need to develop and conserve public lands and forests, or the requirements of major water resource development undertakings;

2.   Be areas which have natural endowments that are well above the ordinary in quality and recreation appeal, being of lesser significance than the unique scenic and historic elements of the National Park System, but affording a quality of recreation experience which transcends that normally associated with areas provided by State and local governments;

3.   Be consistent with Federal programs relating to national parks, national forests, public lands, fish and wildlife, water resource development, grants for urban open space, recreation programs on private agricultural lands, and programs for financial assistance to States in providing recreation opportunity.

In order to provide a rational basis for planning and evaluating proposed projects where outdoor recreation use is the dominant or primary purpose, the Recreation Advisory Council hereby sets forth the guidelines it believes should govern the selection, establishment, and administration of National Recreation Areas.

Under authority bestowed upon the Council by Executive Order 11017, of April 27, 1962, the Council commends this policy to all concerned Federal agencies, and by mutual agreement makes it binding upon the member agencies of the Recreation Advisory Council. It shall be applied to the existing backlog of National Recreation Area proposals, as well as to all future proposals.

TERMINOLOGY AND DEFINITION OF SCOPE

Many names have been used in the past in describing areas to be acquired and developed, or to be administratively designated, predominantly for recreation use. Some of these are National Seashore, National Lakeshore, National Waterway, National Riverway, National Recreation Demonstration Areas, and similar names which embody either the physical resource base or the functional purpose to be served. This policy statement includes such areas.

The following criteria are not intended to apply to (a) the classical elements of the National Park System; (b) the standard recreation use areas designated under National Forest practices; (c) the normal scale of recreation development associated with Federal multiple-purpose impoundments; (d) the National Wildlife Refuges, Game Ranges, and National Fish Hatcheries; (e) military and national defense installations; and (f) sites within the zone of metropolitan responsibility, such as provided through the Open Space program of the Housing and Home Finance Agency, or which primarily serve massive day use requirements that properly should be met by local and State agencies of government. On the other hand, it is conceivable that National Recreation Areas may include within their boundaries portions of any existing Federal real property.

PRIMARY CRITERIA FOR SELECTION
OF NATIONAL RECREATION AREAS

These criteria represent an essential test. National Recreation Areas are conceived of as consisting of a limited number of areas. Therefore, the Council recognizes that a high degree of judgment will have to be exercised in the choice of priorities among qualifying areas.

Application of the following seven primary criteria shall be mandatory for all proposals:

1.   National Recreation Areas should be spacious areas, including within their perimeter an aggregate gross area of not less than

20,000 acres of land and water surface, except for riverways, narrow coastal strips, or areas where total population within a 250-mile radius is in excess of 30 million people.

2.   National Recreation Areas should be located and designed to achieve a comparatively high recreation carrying capacity, in relation to type of recreation primarily to be served.

3.   National Recreation Areas should provide recreation opportunities significant enough to assure interstate patronage within the region of service, and to a limited extent should attract patronage from outside of the normal service region.

4.   The scale of investment, development, and operational responsibility should be sufficiently high to require either direct Federal involvement, or substantial Federal participation to assure optimum public benefit.

5.   Although nonurban in character, National Recreation Areas should nevertheless be strategically located within easy driving distance, i.e., not more than 250 miles from urban population centers which are to be served. Such areas should be readily accessible at all times, for all-purpose recreational use.

6.   Within National Recreation Areas, outdoor recreation shall be recognized as the dominant or primary resource management purpose. If additional natural resource utilization is carried on, such additional use shall be compatible with fulfilling the recreation mission, and none will be carried on that is significantly detrimental to it.

7.   National Recreation Areas should be established in only those areas where other programs (Federal and non-Federal) will not fulfill high priority recreation needs in the foreseeable future.

SECONDARY CRITERIA FOR SELECTION OF NATIONAL RECREATION AREAS

Application of the following six secondary criteria will be given weight in situations where they bear a meaningful relationship to a specific proposal:

1.   Preference should be given to proposed National Recreation Areas that:

    a.   Are within or closely proximate to those official U.S. Census Divisions having the highest population densities;

    b.   Are in areas which have a serious deficiency in supply of both private and public outdoor recreation areas and facilities as determined by the National Recreation Plan;

    c.   Are in areas which have a comparatively low amount of federally provided recreation carrying capacity;

    d.   Show an optimum ratio of carrying capacity to estimated cost.

2. National Recreation Areas may be based upon existing or proposed Federal water impoundments where it can be shown that significant increases in the scale of recreation development are required, beyond the level normally justified under standard multiple-purpose project development, in order to assure that full recreational potential is provided for projected needs.

3. National Recreation Areas may include within their boundaries scenic, historic, scientific, scarce or disappearing resources, provided the objectives of their preservation and enjoyment can be achieved on a basis compatible with the recreation mission.

4. National Recreation Areas should be in conformity with the National Recreation Plan prepared by the Bureau of Outdoor Recreation, and shall take into consideration State, regional, and local comprehensive plans.

5. Whenever possible, National Recreation Areas should be selected, developed, and managed to provide maximum compatibility with the recreation potential of adjacent rural areas in private ownership.

6. Preference should be given to areas within or proximate to a Redevelopment Area as officially designated by the Department of Commerce and deemed significant in the economic improvement of such a Redevelopment Area.

ESTABLISHMENT OF NATIONAL
RECREATION AREAS

National Recreation Areas shall be established by an Act of Congress. Legislation to establish National Recreation Areas will be processed in accordance with established procedures for handling legislation. Upon request of the Executive Office of the President, the Recreation Advisory Council will review specific National Recreation Area proposals, based upon studies made or prescribed by the Bureau of Outdoor Recreation. For those proposals referred to it, the Council will recommend appropriate action regarding authorization, modification; priority of establishment; and the responsible management agency or agencies.

ADMINISTRATION OF NATIONAL
RECREATION AREAS

National Recreation Area proposals should include recommendations as to the agency or agencies responsible for their management. In making this recommendation, sponsoring organizations should take into account the proximity of the proposed area to other publicly administered areas, along with such other factors as will assure effective and economical administration of the new area. Where deemed feasible and desirable, a joint Federal-State management arrangement may be recommended.

National Park Service Handbook of Administrative Policies for Recreation Areas, 1968, 69–72.

6

The Ecological
Revolution

1964–1969

Adoption of the Leopold report's recommendations as well as continued pressure to diversify the system to include recreation as well as preservation strained a National Park Service already undergoing change and growth from Mission 66. In the next six years these two issues would demand continual adjustment and reinterpretation. Secretary of the Interior Stewart Udall signaled the new tone with his 1964 letter on national park management. In it the secretary reaffirmed the Leopold report as a guideline and differentiated the management prescriptions for natural, historic, and recreational areas. The latter was a tacit admission of the growing complexity of the agency's mission.

That complexity would increase over the next several years. A web of legislation further diversified the contents and duties of the system as well as the strictures it would be required to follow. First came the Wilderness Act of 1964 over the resistance of the Park Service. The agency had long argued that its application to parks was redundant because they were already managed for roadless preservation. In fact, the Wilderness Act, where implemented, added a legal buffer against the possibility of further projects like Tioga Road in Yosemite National Park.

The following year saw the addition of two more laws of profound importance. First came the landmark Land and Water Conservation Fund Act which established a fund for acquisition of new recreation lands either within or adjacent to existing park units or as new parks themselves. A portion of the money to be provided to the fund would come from fees charged at existing parks. Congress also chose that year to further define the always delicate policy of private enterprise in the parks with the Concession Policies Act of

1965. Although Mather's ideas of monopoly concessions with first rights of contract renewal were maintained, the new law reiterated the stipulation that concession operations be of minimum area and extent necessary and specified that the government could force a movement of such operations to another location upon payment of compensatory funds.

In 1966 the complicated duties of the National Park Service with regard to historical sites and structures would be magnified by passage of the National Historic Preservation Act. This authorized the Secretary of the Interior to create and maintain a national register of historic districts, sites, and structures and to establish programs of matching grants to states and to the National Trust for Historic Preservation. The National Park Service became the coordinating agency for these activities and its director the executive director of the Advisory Council on Historic Preservation.

In 1967 the Clean Air Act (summarized in the Appendix) laid another layer of protection on the parks but also demanded management compliance. Identified as areas of desired maximum air purity, the parks' airsheds would presumably be more tightly constrained in production of pollutants. The parks themselves would fall under this legislation however, necessitating additional planning in the incipient fire program.

The following year, 1968, saw a flurry of important legislation and policy statements continuing the rapid evolution of the park system and its management program. Two laws, the Wild and Scenic Rivers Act and the National Trails System Act, further diversified the units managed by the Park Service. The rivers act in particular has added extensively to the system and created long, sinuous areas with complicated management problems.

Secretary Udall's 1964 prescription for separate administrative policies for areas of natural, historic, and recreation protection became reality in 1968. Three handbooks specifying general policy purposes and specific prescriptions for common problems, and containing appendixes of related legislation and agency orders, sought to cope with the confusing myriad of duties and responsibilities the agency faced.

While this flurry of legislative and policy maneuvering occupied Washington, D.C., the parks continued to adjust to the Leopold report and its attendant infusion of scientists and natural resource programs. At Sequoia National Park decades of vigorous protection of giant sequoias from fire had halted reproduction of the species. Acting on the advice of one of Stanley Cain's students, the park administration experimented with controlled burns in 1964. The success of those trial burns led in 1968 to a policy for their continuation in the forest ecosystems of the park. This adoption of systematic provision of fire in a forest marked a substantive turning point in resource protection in the parks.

The final law of this busy period was perhaps the farthest reaching of all. The National Environmental Policy Act of 1969 (NEPA) forms the basic national charter for environmental protection. It ordered federal agencies to carry out their duties in such a way as to avoid or minimize environmental degradation. It required those agencies to conduct planning with studies of potential environmental impact for all development projects. The planning procedure, further, was to be open for public input. This latter provision was to have extraordinary results as conservation organizations in particular became powerful players at the required hearings. NEPA rounded out a short period during which the duties and ground rules of the NPS evolved with dizzying speed especially for old-time employees, hired at a time when the parks were distant, serene enclaves of natural landscape architecture.

SECRETARY UDALL'S LETTER ON
NATIONAL PARK MANAGEMENT

July 10, 1964

Memorandum

To: Director, National Park Service
From: Secretary of the Interior
Subject: Management of the National Park System

As the golden anniversary of the National Park Service draws near, and we approach the final years of the MISSION 66 program, it is appropriate to take stock of the events of the past and to plan for the future. The accomplishments of the past are not only a source of pride—they are also a source of guidance for the future.

The accelerating rate of change in our society today poses a major challenge to the National Park Service and its evolving responsibilities for the management of the National Park System. The response to such changes calls for clarity of purpose, increasing knowledge, speedier action and adaptability to changing needs and demands upon our diverse resources.

In recognition of this need, a year ago I approved a comprehensive study of the long-range objectives, organization and management of the National Park Service. Moreover, I was pleased to have had the opportunity to participate in the CONFERENCE OF CHALLENGES at Yosemite National Park, at which this study was discussed by the personnel of the Service.

In looking back at the legislative enactments that have shaped the National Park System, it is clear that the Congress has included within the growing System three different categories of areas—natural, historical, and recreational.

Natural areas are the oldest category, reaching back to the establishment of Yellowstone National Park almost a century ago. A little later historical areas began to be authorized, culminating in the broad charter for historical preservation set forth in the Historic Sites Act of 1935. In recent decades, with exploding population and diminishing open space, the urgent need for national recreation areas is receiving new emphasis and attention.

The long-range study has brought into sharp focus the fact that a single, broad management concept encompassing these three categories of areas within the System is inadequate either for their proper preservation or for realization of their full potential for public use as embodied in the expressions of Congressional policy. Each of these categories requires a separate management concept and a separate set of management principles coordinated to form one organic management plan for the entire System.

Following the Act of August 25, 1916, establishing the National Park Service, the then Secretary of the Interior Franklin K. Lane, in a letter of May 13, 1918, to the first Director of the National Park Service, Stephen T. Mather, outlined the management principles which were to guide the Service in its management of the areas then included within the System. That letter, sometimes called the Magna Carta of the National Parks, is quoted, in part, as follows:

> For the information of the public an outline of the administrative policy to which the new Service will adhere may now be announced. This policy is based on three broad principles: First, that the national parks must be maintained in absolutely unimpaired form for the use of future generations as well as those of our own time; second, that they are set apart for the use, observation, health, and pleasure of the people; and third, that the national interest must dictate all decisions affecting public or private enterprise in the parks.

The principles enunciated in this letter have been fully supported over the years by my predecessors. They are still applicable for us today, and I reaffirm them.

Consistent with specific Congressional enactments, the following principles are approved for your guidance in the management of the three categories of areas now included within the system. Utilizing the results of the new broad program of resource studies, you should proceed promptly to develop such detailed guidelines as may be needed for the operation of each of these categories of areas.

NATURAL AREAS

Resource Management: The management and use of natural areas shall be guided by the 1918 directive of Secretary Lane. Additionally, management shall be directed toward maintaining, and where necessary re-establishing, indigenous plant and animal life, in keeping with the March 4, 1963, recommendations of the Advisory Board on Wildlife Management.

In those areas having significant historical resources, management shall be patterned after that of the historical areas category to the extent compatible with the primary purpose for which the area was established.

Resource Use: Provide for all appropriate use and enjoyment by the people, that can be accommodated without impairment of the natural values. Park management shall recognize and respect wilderness as a whole environment of living things whose use and enjoyment depend on their continuing interrelationship free of man's spoilation.

Physical Developments: They shall be limited to those that are necessary and appropriate, and provided only under carefully controlled safeguards against unregulated and indiscriminate use, so that the least damage to park values will be caused. Location, design, and material, to the highest practicable degree, shall be consistent with the preservation and conservation of the grandeur of the natural environment.

HISTORICAL AREAS

Resource Management: Management shall be directed toward maintaining and where necessary restoring the historical integrity of structures, sites and objects significant to the commemoration or illustration of the historical story.

Resource Use: Visitor uses shall be those which seek fulfillment in authentic presentations of historic structures, objects and sites, and the memorialization of historic individuals or events. Visitor use of significant natural resources should be encouraged when such use can be accommodated without detriment to historical values.

Physical Developments: Physical developments shall be those necessary for achieving the management and use objectives.

RECREATIONAL AREAS

Resource Management: Outdoor recreation shall be recognized as the dominant or primary resource management objective. Natural resources within the area may be utilized and managed for additional purposes where such additional uses are compatible with fulfilling the recreation mission of the area. Scenic, historical, scientific, scarce, or disappearing resources within recreational areas shall be managed compatible with the primary recreation mission of the area.

Resource Use: Primary emphasis shall be placed on active participation in outdoor recreation in a pleasing environment.

Physical Developments: Physical developments shall promote the realization of the management and use objectives. The scope and type of developments, as well as their design, materials, and construction, should enhance and promote the use and enjoyment of the recreational resources of the area.

LONG-RANGE OBJECTIVES

While the establishment of management principles to guide the operation of the three categories of areas within the System is vital, I believe it is of equal consequence that we now identify the long-range objectives of the National Park Service. The objectives developed by the Service have been recommended to me by my Advisory Board on National Parks, Historic Sites, Buildings and Monuments. I am approving these objectives, as follows:

1. To provide for the highest quality of use and enjoyment of the National Park System by increased millions of visitors in years to come.
2. To conserve and manage for their highest purpose the Natural, Historical and Recreational resources of the National Park System.
3. To develop the National Park System through inclusion of additional areas of scenic, scientific, historical and recreational value to the Nation.
4. To participate actively with organizations of this and other Nations in conserving, improving and renewing the total environment.
5. To communicate the cultural, inspirational, and recreational significance of the American Heritage as represented in the National Park System.
6. To increase the effectiveness of the National Park Service as a "people serving" organization dedicated to park conservation, historical preservation, and outdoor recreation.

You should develop such goals and procedures as may be necessary to implement these objectives.

In the development of these goals and procedures, I think it is important to emphasize that effective management of the National Park System will not be achieved by programs that look only within the parks without respect to the pressures, the influences, and the needs beyond park boundaries. The report of my Advisory Board on Wildlife Management emphasizes this observation.

The concern of the National Park Service is the wilderness, the wildlife, the history, the recreational opportunities, etc., within the areas of the System and the appropriate uses of these resources. The responsibilities of the Service, however, cannot be achieved solely within the boundaries of the areas it administers.

The Service has an equal obligation to stand as a vital, vigorous, effective force in the cause of preserving the total environment of our Nation. The concept of the total environment includes not only the land, but also the water and the air, the past as well as the present, the useful as well as the beautiful, the wonders of man as well as the wonders of nature, the urban environment as well as the natural landscape. I am pleased that among its contributions, the Service is identifying National Historic and Natural History Landmarks throughout the country and is cooperating in the Historic American Buildings Survey.

It is obvious that the staggering demand for outdoor recreation projected for this country will eventually inundate public park areas unless public and private agencies and individuals join in common effort. National park administrators must seek methods to achieve close cooperation with all land-managing agencies, considering broad regional needs, if lands for public outdoor recreation sufficient to the future needs of the Nation are to be provided.

The national parklands have a major role in providing superlative opportunities for outdoor recreation, but they have other "people serving" values. They can provide an experience in conservation education for the young people of the country; they can enrich our literary and artistic consciousness; they can help create social values; contribute to our civic consciousness; remind us of our debt to the land of our fathers.

Preserving the scenic and scientific grandeur of our Nation, presenting its history, providing healthful outdoor recreation for the enjoyment of our people, working with others to provide the best possible relationships of human beings to their total environment; this is the theme which binds together the management principles and objectives of the National Park Service—this, for the National Park Service, is the ROAD TO THE FUTURE.

Stewart L. Udall, Secretary of the Interior

National Park Service Handbook of Administrative Policies for Natural Areas, 1968, 76–80.

AN ACT TO ESTABLISH A NATIONAL WILDERNESS PRESERVATION SYSTEM FOR THE PERMANENT GOOD OF THE WHOLE PEOPLE, AND FOR OTHER PURPOSES,

1964 (78 Stat. 890)

Be it enacted by the Senate and House of Representatives of the United States of America in Congress assembled,

SHORT TITLE

SECTION 1. This Act may be cited as the "Wilderness Act".

WILDERNESS SYSTEM ESTABLISHED STATEMENT OF POLICY

SEC. 2. (a) In order to assure that an increasing population, accompanied by expanding settlement and growing mechanization, does not occupy and modify all areas within the United States and its possessions, leaving no lands designated for preservation and protection in their natural condition, it is hereby declared to be the policy of the Congress to secure for the American people of present and future generations the benefit of an enduring resource of wilderness. For this purpose there is hereby established a National Wilderness Preservation System to be composed of federally owned areas designated by Congress as "wilderness areas", and these shall be administered for the use and enjoyment of the American people in such manner as will leave them unimpaired for future use and enjoyment as wilderness, and so as to provide for the protection of these areas, the preservation of their wilderness character, and for the gathering and dissemination of information regarding their use and enjoyment as wilderness; and no Federal lands shall be designated as "wilderness areas" except as provided for in this Act or by a subsequent Act.

(b) The inclusion of an area in the National Wilderness Preservation System notwithstanding, the area shall continue to be managed by the Department and agency having jurisdiction thereover immediately before its inclusion in the National Wilderness Preservation System unless otherwise provided by Act of Congress. No appropriation shall be available for the payment of expenses or salaries for the administration of the National Wilderness Preservation System as a separate unit nor shall any appropriations be available for additional personnel stated as being required solely for the purpose of managing or administering areas solely because they are included within the National Wilderness Preservation System.

(c)   A wilderness, in contrast with those areas where man and his own works dominate the landscape, is hereby recognized as an area where the earth and its community of life are untrammeled by man, where man himself is a visitor who does not remain. An area of wilderness is further defined to mean in this Act an area of undeveloped Federal land retaining its primeval character and influence, without permanent improvements or human habitation, which is protected and managed so as to preserve its natural conditions and which (1) generally appears to have been affected primarily by the forces of nature, with the imprint of man's work substantially unnoticeable; (2) has outstanding opportunities for solitude or a primitive and unconfined type of recreation; (3) has at least five thousand acres of land or is of sufficient size as to make practicable its preservation and use in an unimpaired condition; and (4) may also contain ecological, geological, or other features of scientific, educational, scenic, or historical value.

NATIONAL WILDERNESS PRESERVATION
SYSTEM—EXTENT OF SYSTEM

SEC. 3. (a) All areas within the national forests classified at least 30 days before the effective date of this Act by the Secretary of Agriculture or the Chief of the Forest Service as "wilderness", "wild", or "canoe" are hereby designated as wilderness areas. The Secretary of Agriculture shall—

(1)   Within one year after the effective date of this Act, file a map and legal description of each wilderness area with the Interior and Insular Affairs Committees of the United States Senate and the House of Representatives, and such descriptions shall have the same force and effect as if included in this Act: *Provided, however,* That correction of clerical and typographical errors in such legal descriptions and maps may be made.

(2)   Maintain, available to the public, records pertaining to said wilderness areas, including maps and legal descriptions, copies of regulations governing them, copies of public notices of, and reports submitted to Congress regarding pending additions, eliminations, or modifications. Maps, legal descriptions, and regulations pertaining to wilderness areas within their respective jurisdictions also shall be available to the public in the offices of regional foresters, national forest supervisors, and forest rangers.

(b)   The Secretary of Agriculture shall, within ten years after the enactment of this Act, review, as to its suitability or nonsuitability for preservation as wilderness, each area in the national forests classified on the effective date of this Act by the Secretary of Agriculture or the Chief of the Forest Service as "primitive" and report his findings to the President. The President shall advise the United States Senate and House of Representatives of his recommendations with respect to the designation as "wilderness" or other reclassification of each area on which review has been completed, together with maps and a definition of boundaries. Such advice shall be given with respect to not less than one-third of all the areas now classified as "primitive" within three years after the enactment of this Act, not less than two-thirds within seven years after the enactment of this Act, and the remaining areas within ten years after the enactment of this Act. Each recommendation of the President for designation as "wilderness" shall become effective only if so provided by an Act of Congress. Areas classified as "primitive" on the effective date of this Act shall continue to be administered under the rules and regulations affecting such areas on the effective date of this Act until Congress has determined otherwise. Any such area may be increased in size by the President at the time he submits his recommendations to the Congress by not more than five thousand acres with no more than one thousand two hundred and eighty acres of such increase in any one compact unit: if it is proposed to increase the size of any such area by more than five thousand acres or by more than one thousand two hundred and eighty acres in any one compact unit the increase in size shall not become effective until acted upon by Congress. Nothing herein contained shall limit the President in proposing, as part of his recommendations to Congress, the alteration of existing boundaries of primitive areas or recommending the addition of any contiguous area of national forest lands predominantly of wilderness value. Notwithstanding any other provisions of this Act, the Secretary of Agriculture may complete his review and delete such area as may be necessary, but not to exceed seven thousand acres, from the southern tip of the Gore Range-Eagles Nest Primitive Area, Colorado, if the Secretary determines that such action is in the public interest.

(c)   Within ten years after the effective date of this Act the Secretary of the Interior shall review every roadless area of five thousand contiguous acres or more in the national parks, monuments and other units of the national park system and every such area of, and every roadless island within, the national wildlife refuges and game ranges, under his jurisdiction on the effective date of this Act and shall report to the President his recommendation as to the suitability or nonsuitability of each such area or island for preservation as wilderness. The President shall advise the President of the Senate and the Speaker of the House of Representatives of his recommendation with respect to the designation as wilderness of each such area or island on which review has been completed, together with a map thereof and a definition of its boundaries. Such advice shall be given with respect to not less than one-third of the areas and islands to be reviewed under this subsection within three years after enactment of this Act, not less than two-thirds within seven years of enactment of this Act, and the remainder within ten years of enactment of this Act. A recommendation of the President for designation as wilderness shall become effective only if so provided by an Act of Congress. Nothing contained herein shall, by implication or otherwise, be construed to lessen the present statutory authority of the Secretary of the Interior with respect to the maintenance of roadless areas within units of the national park system.

(d) (1)   The Secretary of Agriculture and the Secretary of the Interior shall, prior to submitting any recommendations to the President with respect to the suitability of any area for preservation as wilderness—

(A)   give such public notice of the proposed action as they deem appropriate, including publication in the Federal Register and in a newspaper having general circulation in the area or areas in the vicinity of the affected land;

(B)   hold a public hearing or hearings at a location or locations convenient to the area affected. The hearings shall be announced through such means as the respective Secretaries involved deem appropriate, including notices in the Federal Register and in newspapers of general circulation in the area: *Provided,* That if the lands involved are located in more than one State, at least one hearing shall be held in each State in which a portion of the land lies;

(C)   at least thirty days before the date of a hearing advise the Governor of each State and the governing board of each county, or in Alaska the borough, in which the lands are located, and Federal departments and agencies concerned, and invite such officials and Federal agencies to submit their views on the proposed action at the hearing or by no later than thirty days following the date of the hearing.

(2)   Any views submitted to the appropriate Secretary under the provisions of (1) of this subsection with respect to any area shall be included with any recommendations to the President and to Congress with respect to such area.

(e)   Any modification or adjustment of boundaries of any wilderness area shall be recommended by the appropriate Secretary after public notice of such proposal and public hearing or hearings as provided in subsection (d) of this section. The proposed modification or adjustment shall then be recommended with map and description thereof to the President. The President shall advise the United States Senate and the House of Representative of his recommendations with respect to such modification or adjustment and such recommendations shall become effective only in the same manner as provided for in subsections (b) and (c) of this section.

USE OF WILDERNESS AREAS

SEC. 4. (a) The purposes of this Act are hereby declared to be within and supplemental to the purposes for which national forests and units of the national park and national wildlife refuge systems are established and administered and—

(1)   Nothing in this Act shall be deemed to be in interference with the purpose for which national forests are established as set forth in the Act of June 4, 1897 (30 Stat. 11), and the Multiple-Use Sustained-Yield Act of June 12, 1960 (74 Stat. 215).

(2)   Nothing in this Act shall modify the restrictions and provisions of the Shipstead-Nolan Act (Public Law 539, Seventy-first Congress, July 10, 1930; 46 Stat. 1020), the Thye-Blatnik Act (Public Law 733, Eightieth Congress, June 22, 1948; 62 Stat. 568), and the Humphrey-Thye-Blatnik-Andresen Act (Public Law 607, Eighty-fourth Congress, June 22, 1956; 70 Stat. 326), as applying to the Superior National Forest or the regulations of the Secretary of Agriculture.

(3)   Nothing in this Act shall modify the statutory authority under which units of the national park system are created. Further, the designation of any area of any park, monument, or other unit of the national park system as a wilderness area pursuant to this Act shall in no manner lower the standards evolved for the use and preservation of such park, monument, or other unit of the national park system in accordance with the Act of August 25, 1916, the statutory authority under which the area was created, or any other Act of Congress which might pertain to or affect such area, including, but not limited to, the Act of June 8, 1906 (34 Stat. 225; 16 U.S.C. 432 et seq.); section 3(2) of the Federal Power Act (16 U.S.C. 796(2)); and the Act of August 21, 1935 (49 Stat. 666; 16 U.S.C. 461 et seq.).

(b)   Except as otherwise provided in this Act, each agency administering any area designated as wilderness shall be responsible for preserving the wilderness character of the area and shall so administer such area for such other purposes for which it may have been established as also to preserve its wilderness character. Except as otherwise provided in this Act, wilderness areas shall be devoted to the public purposes of recreational, scenic, scientific, educational, conservation, and historical use.

PROHIBITION OF CERTAIN USES

(c)   Except as specifically provided for in this Act, and subject to existing private rights, there shall be no commercial enterprise and no permanent road within any wilderness area designated by this Act and, except as necessary to meet minimum requirements for the administration of the area for the purpose of this Act (including measures required in emergencies involving the health and safety of persons within the area), there shall be no temporary road, no use of motor vehicles, motorized equipment or motorboats, no landing of aircraft, no other form of mechanical transport, and no structure or installation within any such area.

SPECIAL PROVISIONS

(d)   The following special provisions are hereby made:

(1)   Within wilderness areas designated by this Act the use of aircraft or motorboats, where these uses have already become established, may be permitted to continue subject to such restrictions as the Secretary of Agriculture deems desirable. In addition, such measures may be taken as may be necessary in the control of fire, insects, and diseases, subject to such conditions as the Secretary deems desirable.

(2)   Nothing in this Act shall prevent within national forest wilderness areas any activity, including prospecting, for the purpose of gathering information about mineral or other resources, if such activity is carried on in a manner compatible with the preservation of the wilderness environment. Furthermore, in accordance with such program as the Secretary of the Interior shall develop and conduct in consultation with the Secretary of Agriculture, such areas shall be surveyed on a planned, recurring basis consistent with the concept of wilderness preservation by the Geological Survey and the Bureau of Mines to determine the mineral values, if any, that may be present; and the results of such surveys shall be made available to the public and submitted to the President and Congress.

(3)   Notwithstanding any other provisions of this Act, until midnight December 31, 1983, the United States mining laws and all laws pertaining to mineral leasing shall, to the same extent as applicable

prior to the effective date of this Act, extend to those national forest lands designated by this Act as "wilderness areas"; subject, however, to such reasonable regulations governing ingress and egress as may be prescribed by the Secretary of Agriculture consistent with the use of the land for mineral location and development and exploration, drilling, and production, and use of land for transmission lines, waterlines, telephone lines, or facilities necessary in exploring, drilling, producing, mining, and processing operations, including where essential the use of mechanized ground or air equipment and restoration as near as practicable of the surface of the land disturbed in performing prospecting, location, and, in oil and gas leasing, discovery work, exploration, drilling, and production, as soon as they have served their purpose. Mining locations lying within the boundaries of said wilderness areas shall be held and used solely for mining or processing operations and uses reasonably incident thereto; and, hereafter, subject to valid existing rights, all patents issued under the mining laws of the United States affecting national forest lands designated by this Act as wilderness areas shall convey title to the mineral deposits within the claim, together with the right to cut and use so much of the mature timber therefrom as may be needed in the extraction, removal, and benefication of the mineral deposits, if needed timber is not otherwise reasonably available, and if the timber is cut under sound principles of forest management as defined by the national forest rules and regulations, but each such patent shall reserve to the United States all title in or to the surface of the lands and products thereof, and no use of the surface of the claim or the resources therefrom not reasonably required for carrying on mining or prospecting shall be allowed except as otherwise expressly provided in this Act: *Provided,*That, unless hereafter specifically authorized, no patent within wilderness areas designated by this Act shall issue after December 31, 1983, except for the valid claims existing on or before December 31, 1983. Mining claims located after the effective date of this Act within the boundaries of wilderness areas designated by this Act shall create no rights in excess of those rights which may be patented under the provisions of this subsection. Mineral leases, permits, and licenses covering lands within national forest wilderness areas designated by this Act shall contain such reasonable stipulations as may be prescribed by the Secretary of Agriculture for the protection of the wilderness character of the land consistent with the use of the land for the purposes for which they are leased, permitted, or licensed. Subject to valid rights then existing, effective January 1, 1984, the minerals in lands designated by this Act as wilderness areas are withdrawn from all forms of appropriation under the mining laws and from disposition under all laws pertaining to mineral leasing and all amendments thereto.

(4) Within wilderness areas in the national forests designated by this Act, (1) the President may, within a specific area and in accordance with such regulations as he may deem desirable, authorize prospecting for water resources, the establishment and maintenance of reservoirs, water-conservation works, power projects, transmission lines, and other facilities needed in the public interest, including the road construction and maintenance essential to development and use thereof, upon his determination that such use or uses in the specific area will better serve the interests of the United States and the people thereof than will its denial; and (2) the grazing of livestock, where established prior to the effective date of this Act, shall be permitted to continue subject to such reasonable regulations as are deemed necessary by the Secretary of Agriculture.

(5) Other provisions of this Act to the contrary notwithstanding, the management of the Boundary Waters Canoe Area, formerly designated as the Superior, Little Indian Sioux, and Caribou Roadless Areas, in the Superior National Forest, Minnesota, shall be in accordance with regulations established by the Secretary of Agriculture in accordance with the general purpose of maintaining, without unnecessary restrictions on other uses, including that of timber, the primitive character of the area, particularly in the vicinity of lakes, streams, and portages: *Provided*, That nothing in this Act shall preclude the continuance within the area of any already established use of motorboats.

(6) Commercial services may be performed within the wilderness areas designated by this Act to the extent necessary for activities which are proper for realizing the recreational or other wilderness purposes of the areas.

(7) Nothing in this Act shall constitute an express or implied claim or denial on the part of the Federal Government as to exemption from State water laws.

(8) Nothing in this Act shall be construed as affecting the jurisdiction or responsibilities of the several States with respect to wildlife and fish in the national forests.

STATE AND PRIVATE LANDS
WITHIN WILDERNESS AREAS

SEC. 5. (a) In any case where State-owned or privately owned land is completely surrounded by national forest lands within areas designated by this Act as wilderness, such State or private owner shall be given such rights as may be necessary to assure adequate access to such State-owned or privately owned land by such State or private owner and their successors in interest, or the State-owned land or privately owned land shall be exchanged for federally owned land in the same State of approximately equal value under authorities available to the Secretary of Agriculture: *Provided, however,* That the United States shall not transfer to a State or private owner any mineral interests unless the State or private owner relinquishes or causes to be relinquished to the United States the mineral interest in the surrounded land.

(b)   In any case where valid mining claims or other valid occupancies are wholly within a designated national forest wilderness area, the Secretary of Agriculture shall, by reasonable regulations consistent with the preservation of the area as wilderness, permit ingress and egress to such surrounded areas by means which have been or are being customarily enjoyed with respect to other such areas similarly situated.

(c)   Subject to the appropriation of funds by Congress, the Secretary of Agriculture is authorized to acquire privately owned land within the perimeter of any area designated by this Act as wilderness if (1) the owner concurs in such acquisition or (2) the acquisition is specifically authorized by Congress.

GIFTS, BEQUESTS, AND CONTRIBUTIONS

SEC. 6. (a) The Secretary of Agriculture may accept gifts or bequests of land within wilderness areas designated by this Act for preservation as wilderness. The Secretary of Agriculture may also accept gifts or bequests of land adjacent to wilderness areas designated by this Act for preservation as wilderness if he has given sixty days advance notice thereof to the President of the Senate and the Speaker of the House of Representatives. Land accepted by the Secretary of Agriculture under this section shall become part of the wilderness area involved. Regulations with regard to any such land may be in accordance with such agreements, consistent with the policy of this Act, as are made at the time of such gift, or such conditions, consistent with such policy, as may be included in, and accepted with, such bequest.

(b)   The Secretary of Agriculture or the Secretary of the Interior is authorized to accept private contributions and gifts to be used to further the purposes of this Act.

ANNUAL REPORTS

SEC. 7. At the opening of each session of Congress, the Secretaries of Agriculture and Interior shall jointly report to the President for transmission to Congress on the status of the wilderness system, including a list and descriptions of the areas in the system, regulations in effect and other pertinent information, together with any recommendations they may care to make.

Approved September 3, 1964.

16 U.S.C. 1131 et seq.

AN ACT TO ESTABLISH A LAND AND WATER CONSERVATION FUND TO ASSIST THE STATES AND FEDERAL AGENCIES IN MEETING PRESENT AND FUTURE OUTDOOR RECREATION DEMANDS AND NEEDS OF THE AMERICAN PEOPLE, AND FOR OTHER PURPOSES,

1964 (78 Stat. 897)

Be it enacted by the Senate and House of Representatives of the United States of America in Congress assembled,

TITLE I—LAND AND WATER CONSERVATION PROVISIONS

SHORT TITLE AND STATEMENT OF PURPOSES

SECTION 1. (a) CITATION: EFFECTIVE DATE.—This Act may be cited as the "Land and Water Conservation Fund Act of 1965" and shall become effective on January 1, 1965.

(b) PURPOSES.—The purposes of this Act are to assist in preserving, developing, and assuring accessibility to all citizens of the United States of America of present and future generations and visitors who are lawfully present within the boundaries of the United States of America such quality and quantity of outdoor recreation resources as may be available and are necessary and desirable for individual active participation in such recreation and to strengthen the health and vitality of the citizens of the United States by (1) providing funds for and authorizing Federal assistance to the States in planning, acquisition, and development of needed land and water areas and facilities and (2) providing funds for the Federal acquisition and development of certain lands and other areas.

CERTAIN REVENUES PLACED IN SEPARATE FUND

SEC. 2. SEPARATE FUND.—During the period ending June 30, 1989, and during such additional period as may be required to repay any advances made pursuant to section 4(b) of this Act, there shall be covered into the land and water conservation fund in the Treasury of the United States, which fund is hereby established and is hereinafter referred to as the "fund", the following revenues and collections:

(a)   ENTRANCE AND USER FEES: ESTABLISHMENT: REGULATIONS.—All proceeds from entrance, admission, and other recreation user fees or charges collected or received by the National Park Service, the Bureau of Land Management, the Bureau of Sport Fisheries and Wildlife, the Bureau of Reclamation, the Forest Service, the Corps of Engineers, the Tennessee Valley Authority, and the United States section of the International Boundary and Water Commission (United States and Mexico), notwithstanding any provision of law that such proceeds shall be credited to miscellaneous receipts of the Treasury; *Provided,* That nothing in this Act shall affect any rights or authority of the States with respect to fish and wildlife, nor shall this Act repeal any provision of law that permits States or political subdivisions to share in the revenues from Federal lands or any provision of law that provides that any fees or charges collected at particular Federal areas shall be used for or credited to specific purposes or special funds as authorized by the provision of law; but the proceeds from fees or charges established by the President pursuant to this subsection for entrance or admission generally to Federal areas shall be used solely for the purposes of this Act.

The President is authorized, to the extent and within the limits hereinafter set forth, to designate or provide for the designation of land or water areas administered by or under the authority of the Federal agencies listed in the preceding paragraph at which entrance admission, and other forms of recreation user fees shall be charged and to establish and revise or provide for the establishment and revision of such fees as follows:

(i)   An annual fee of not more than $7 payable by a person entering an area so designated by private noncommercial automobile which, if paid, shall excuse the person paying the same and anyone who accompanies him in such automobile from payment of any other fee for admission to that area and other areas administered by or under the authority of such agencies, except areas which are designated by the President as not being within the coverage of the fee, during the year for which the fee has been paid.

(ii)   Fees for a single visit or a series of visits during a specified period of less than a year to an area so designated payable by persons who choose not to pay an annual fee under clause (i) of this paragraph or who enter such an area by means other than private noncommercial automobile.

(iii)   Fees payable for admission to areas not within the coverage of a fee paid under clause (i) of this paragraph.

(iv)   Fees for the use within an area of sites, facilities, equipment, or services provided by the United States.

Entrance and admission fees may be charged at areas administered primarily for scenic, scientific, historical, cultural, or recreational purposes. No entrance or admission fee shall be charged except at such areas or portions thereof administered by a Federal agency where recreation facilities or services are provided at Federal expense. No fee of any kind shall be charged by a Federal agency under any provision of this Act for use of any waters. All fees established pursuant to this subsection shall be fair and equitable, taking into consideration direct and indirect cost to the Government, benefits to the recipient, public policy or interest served, and other pertinent factors. Nothing contained in this paragraph shall authorize Federal hunting or fishing licenses or fees or charges for commercial or other activities not related to recreation. No such fee shall be charged for travel by private noncommercial vehicle over any national parkway or any road or highway established as a part of the national Federal-aid system, as defined in section 101, title 23, United States Code, or any road within the National Forest system or a public land area, which, though it is part of a larger area, is commonly used by the public as a means of travel between two places either or both of which are outside the area. No such fee shall be charged any person for travel by private noncommercial vehicle over any road or highway to any land in which such person has any property right if such land is within any such designated area.

No fees established under clause (ii) or clause (iii) of the second paragraph of this subsection shall become effective with respect to any area which embraces lands more than half of which have heretofore been acquired by contribution from the government of the State in which the area is located until sixty days after the officer of the United States who is charged with responsibility for establishing such fees has advised the Governor of the affected State, or an agency of the State designated by the Governor for this purpose, of his intention so to do, and said officer shall, before finally establishing such fees, give consideration to any recommendation that the Governor or his designee may make with respect thereto within said sixty days and to all obligations, legal or otherwise, that the United States may owe to the State concerned and to its citizens with respect to the area in question. In the Smoky Mountains National Park, unless fees are charged for entrance into said park on main highways and thoroughfares, fees shall not be charged for entrance on other routes into said park or any part thereof.

There is hereby repealed the third paragraph from the end of the division entitled "National Park Service" of section 1 of the Act of March 7, 1928 (45 Stat. 238) and the second paragraph from the end of the division entitled "National Park Service" of section 1 of the

Act of March 4, 1929 (45 Stat. 1602; 16 U.S.C. 14). Section 4 of the Act entitled "An Act authorizing the construction of certain public works on rivers and harbors for flood control, and for other purposes", approved December 24, 1944 (16 U.S.C. 460d), as amended by the Flood Control Act of 1962 (76 Stat. 1195) is further amended by deleting "without charge," in the third sentence from the end thereof. All other provisions of law that prohibit the collection of entrance, admission, or other recreation user fees or charges authorized by this Act or that restrict the expenditure of funds if such fees or charges are collected are hereby also repealed: *Provided,* That no provision of any law or treaty which extends to any person or class of persons a right of free access to the shoreline of any reservoir or other body of water, or to hunting and fishing along or on such shoreline, shall be affected by this repealer.

The heads of departments and agencies are authorized to prescribe rules and regulations for the collection of any entrance, admission, and other recreation user fees or charges established pursuant to this subsection for areas under their administration: *Provided further,* That no free passes shall be issued to any Member of Congress or other government official. Clear notice that a fee or charge has been established shall be posted at each area to which it is applicable. Any violation of any rules or regulations promulgated under this title at an area so posted shall be punishable by a fine of not more than $100. Any person charged with the violation of such rules and regulations may be tried and sentenced by any United States commissioner specially designated for that purpose by the court by which he was appointed, in the same manner and subject to the same condition as provided for in title 18, United States Code, section 3401, subsections (b), (c), (d), and (e), as amended.

(b)   SURPLUS PROPERTY SALES.—All proceeds (except so much thereof as may be otherwise obligated, credited, or paid under authority of those provisions of law set forth in section 485(b)-(e), title 40, United States Code, or the Independent Offices Appropriations Act, 1963 (76 Stat. 725) or in any later appropriation Act) hereafter received from any disposal of surplus real property and related personal property under the Federal Property and Administrative Services Act of 1949, as amended, notwithstanding any provision of law that such proceeds shall be credited to miscellaneous receipts of the Treasury. Nothing in this Act shall affect existing laws or regulations concerning disposal of real or personal surplus property to schools, hospitals, and States and their political subdivisions.

(c)   MOTORBOAT FUELS TAX.—The amounts provided for in section 201 of this Act.

SEC. 3. APPROPRIATIONS.—Moneys covered into the fund shall be available for expenditure for the purposes of this Act only when appropriated therefor. Such appropriations may be made without fiscal-year limitation. Moneys covered into this fund not subsequently authorized by the Congress for expenditures within two fiscal years following the fiscal year in which such moneys had been credited to the fund, shall be transferred to miscellaneous receipts of the Treasury.

ALLOCATION OF LAND AND WATER CONSERVATION FUND FOR STATE AND FEDERAL PURPOSES: AUTHORIZATION FOR ADVANCE APPROPRIATIONS

SEC. 4. (a) ALLOCATIONS.—There shall be submitted with the annual budget of the United States a comprehensive statement of estimated requirements during the ensuing fiscal year for appropriations from the fund. In the absence of a provision to the contrary in the Act making an appropriation from the fund, (i) the appropriation therein made shall be available in the ratio of 60 per centum for State purposes and 40 per centum for Federal purposes, but (ii) the President may, during the first five years in which appropriations are made from the fund, vary said percentages by not more than 15 points either way to meet, as nearly as may be, the current relative needs of the States and the Federal Government.

(b) ADVANCE APPROPRIATIONS; REPAYMENT.—Beginning with the third full fiscal year in which the fund is in operation, and for a total of eight years, advance appropriations are hereby authorized to be made to the fund from any moneys in the Treasury not otherwise appropriated in such amounts as to average not more than $60,000,000 for each fiscal year. Such advance appropriations shall be available for Federal and State purposes in the same manner and proportions as other moneys appropriated from the fund. Such advance appropriations shall be repaid without interest, beginning at the end of the next fiscal year after the first ten full fiscal years in which the fund has been in operation, by transferring, annually until fully repaid, to the general fund of the Treasury 50 per centum of the revenues received by the land and water conservation fund each year under section 2 of this Act prior to July 1, 1989, and 100 per centum of any revenues thereafter received by the fund. Revenues received from the sources specified in section 2 of this Act after July 1, 1989, or after payment has been completed as provided by this subsection, whichever occurs later, shall be credited to miscellaneous receipts of the Treasury. The moneys in the fund that are not required for repayment purposes may continue to be appropriated and allocated in accordance with the procedures prescribed by this Act.

FINANCIAL ASSISTANCE TO STATES

SEC. 5. GENERAL AUTHORITY; PURPOSES.—(a) The Secretary of the Interior (hereinafter referred to as the "Secretary") is authorized to provide financial assistance to the States from moneys available for State purposes. Payments may be made to the States by the Secretary as hereafter provided, subject to such terms and conditions as he considers appropriate and in the public interest to carry out the purposes of this Act, for outdoor recreation: (1) planning, (2) acquisition of land, waters, or interests in land or waters, or (3) development.

(b) APPORTIONMENT AMONG STATES; NOTIFICATION.— Sums appropriated and available for State purposes for each fiscal year shall be apportioned among the several States by the Secretary, whose determination shall be final, in accordance with the following formula:

(1) two-fifths shall be apportioned equally among the several States; and

(2) three-fifths shall be apportioned on the basis of need to individual States by the Secretary in such amounts as in his judgment will best accomplish the purposes of this Act. The determination of need shall include among other things a consideration of the proportion which the population of each State bears to the total population of the United States and of the use of outdoor recreation resources of individual States by persons from outside the State as well as a consideration of the Federal resources and programs in the particular States.

The total allocation to an individual State under paragraphs (1) and (2) of this subsection shall not exceed 7 per centum of the total amount allocated to the several States in any one year.

The Secretary shall notify each State of its apportionments; and the amounts thereof shall be available thereafter for payment to such State for planning, acquisition, or development projects as hereafter prescribed. Any amount of any apportionment that has not been paid or obligated by the Secretary during the fiscal year in which such notification is given and for two fiscal years thereafter shall be reapportioned by the Secretary in accordance with paragraph (2) of this subsection.

The District of Columbia, Puerto Rico, the Virgin Islands, Guam, and American Samoa shall be treated as States for the purposes of this title, except for the purpose of paragraph (1) of this subsection.

Their population also shall be included as a part of the total population in computing the apportionment under paragraph (2) of this subsection.

(c) MATCHING REQUIREMENTS.—Payments to any State shall cover not more than 50 per centum of the cost of planning, acquisition, or development projects that are undertaken by the State. The remaining share of the cost shall be borne by the State in a manner and with such funds or services as shall be satisfactory to the Secretary. No payment may be made to any State for or on account of any cost or obligation incurred or any service rendered prior to the date of approval of this Act.

(d) COMPREHENSIVE STATE PLAN REQUIRED; PLANNING PROJECTS.—A comprehensive statewide outdoor recreation plan shall be required prior to the consideration by the Secretary of financial assistance for acquisition or development projects. The plan shall be adequate if, in the judgment of the Secretary, it encompasses and will promote the purposes of this Act. The plan shall contain—

(1) the name of the State agency that will have authority to represent and act for the State in dealing with the Secretary for purposes of this Act;

(2) an evaluation of the demand for and supply of outdoor recreation resources and facilities in the State;

(3) a program for the implementation of the plan; and

(4) other necessary information, as may be determined by the Secretary.

The plan shall take into account relevant Federal resources and programs and shall be correlated so far as practicable with other State, regional, and local plans. Where there exists or is in preparation for any particular State a comprehensive plan financed in part with funds supplied by the Housing and Home Finance Agency, any statewide outdoor recreation plan prepared for purposes of this Act shall be based upon the same population, growth, and other pertinent factors as are used in formulating the Housing and Home Finance Agency financed plans.

The Secretary may provide financial assistance to any State for projects for the preparation of a comprehensive statewide outdoor recreation plan when such plan is not otherwise available or for the maintenance of such plan.

(e) PROJECTS FOR LAND AND WATER ACQUISITION; DEVELOPMENT.—In addition to assistance for planning projects, the Secretary may provide financial assistance to any State for the following types of projects or combinations thereof if they are in accordance with the State comprehensive plan;

(1) ACQUISITION OF LAND AND WATERS.—For the acquisi-

tion of land, waters, or interests in land or waters (other than land, waters, or interests in land or waters acquired from the United States for less than fair market value), but not including incidental costs relating to acquisition.

(2) DEVELOPMENT.—For development, including but not limited to site planning and the development of Federal lands under lease to States for terms of twenty-five years or more.

(f) REQUIREMENTS FOR PROJECT APPROVAL; CONDITION.—Payments may be made to States by the Secretary only for those planning, acquisition, or development projects that are approved by him. No payment may be made by the Secretary for or on account of any project with respect to which financial assistance has been given or promised under any other Federal program or activity, and no financial assistance may be given under any other Federal program or activity for or on account of any project with respect to which such assistance has been given or promised under this Act. The Secretary may make payments from time to time in keeping with the rate of progress toward the satisfactory completion of individual projects: *Provided,* That the approval of all projects and all payments, or any commitments relating thereto, shall be withheld until the Secretary receives appropriate written assurance from the State that the State has the ability and intention to finance its share of the cost of the particular project, and to operate and maintain by acceptable standards, at State expense, the particular properties or facilities acquired or developed for public outdoor recreation use.

Payments for all projects shall be made by the Secretary to the Governor of the State or to a State official or agency designated by the Governor or by State law having authority and responsibility to accept and to administer funds paid hereunder for approved projects. If consistent with an approved project, funds may be transferred by the State to a political subdivision or other appropriate public agency.

No property acquired or developed with assistance under this section shall, without the approval of the Secretary, be converted to other than public outdoor recreation uses. The Secretary shall approve such conversion only if he finds it to be in accord with the then existing comprehensive statewide outdoor recreation plan and only upon such conditions as he deems necessary to assure the substitution of other recreation properties of at least equal fair market value and of reasonably equivalent usefulness and location.

No payment shall be made to any State until the State has agreed to (1) provide such reports to the Secretary, in such form and containing such information, as may be reasonably necessary to enable the Secretary to perform his duties under this Act, and (2) provide such fiscal control and fund accounting procedures as may be necessary to assure proper disbursement and accounting for Federal funds paid to the State under this Act.

Each recipient of assistance under this Act shall keep such records as the Secretary of the Interior shall prescribe, including records which fully disclose the amount and the disposition by such recipient of the proceeds of such assistance, the total cost of the project or undertaking in connection with which such assistance is given or used, and the amount and nature of that portion of the cost of the project or undertaking supplied by other sources, and such other records as will facilitate an effective audit.

The Secretary of the Interior, and the Comptroller General of the United States, or any of their duly authorized representatives, shall have access for the purpose of audit and examination to any books, documents, papers, and records of the recipient that are pertinent to assistance received under this Act.

(g) COORDINATION WITH FEDERAL AGENCIES.—In order to assure consistency in policies and actions under this Act, with other related Federal programs and activities (including those conducted pursuant to title VII of the Housing Act of 1961 and section 701 of the Housing Act of 1954) and to assure coordination of the planning, acquisition, and development assistance to States under this section with other related Federal programs and activities, the President may issue such regulations with respect thereto as he deems desirable and such assistance may be provided only in accordance with such regulations.

ALLOCATION OF MONEYS
FOR FEDERAL PURPOSES

SEC. 6. (a) Money appropriated from the fund for Federal purposes, shall, unless otherwise allotted in the appropriation Act making them available, be allotted by the President to the following purposes and subpurposes in substantially the same proportion as the number of visitor-days in areas and projects hereinafter described for which admission fees are charged under section 2 of this Act:

(1) For the acquisition of land, waters, or interests in land or waters as follows:

NATIONAL PARK SYSTEM; RECREATION AREAS.—Within the exterior boundaries of areas of the national park system now or hereafter authorized or established and of areas now or hereafter authorized to be administered by the Secretary of the Interior for outdoor recreation purposes.

NATIONAL FOREST SYSTEM.—Inholdings within (a) wilderness areas of the National Forest System, and (b) other areas of national forests as the boundaries of those forests exist on the effective date of this Act which other areas are primarily of value for outdoor recreation purposes: *Provided,* That lands outside of but adjacent to an existing national forest boundary, not to exceed five hundred acres in the case of any one forest, which would comprise an integral part of a forest recreational management area may also be acquired with moneys appropriated from this fund; *Provided further,* That not more than 15 per centum of the average added to the National Forest System pursuant to this section shall be west of the 100th meridian.

THREATENED SPECIES.—For any national area which may be authorized for the preservation of species of fish or wildlife that are threatened with extinction.

RECREATION AT REFUGES.—For the incidental recreation purposes of section 2 of the Act of September 28, 1962 (76 Stat. 653, 16 U.S.C. 460 k-1); and

(2)   For payment into miscellaneous receipts of the Treasury as a partial offset for those capital costs, if any, of Federal water development projects hereafter authorized to be constructed by or pursuant to an Act of Congress which are allotted to public recreation and the enhancement of fish and wildlife values and financed through appropriations to water resource agencies.

(b)   ACQUISITION RESTRICTION.—Appropriations from the fund pursuant to this section shall not be used for acquisition unless such acquisition is otherwise authorized by law.

FUNDS NOT TO BE USED FOR PUBLICITY

SEC. 7. Moneys derived from the sources listed in section 2 of this Act shall not be available for publicity purposes.

TITLE II—MOTORBOAT FUEL TAX PROVISIONS

TRANSFERS TO AND FROM LAND AND WATER CONSERVATION FUND

Sec. 201. (a) There shall be set aside in the land and water conservation fund in the Treasury of the United States provided for in title I of this Act the amounts specified in section 209 (f)(5) of the Highway Revenue Act of 1956 (relating to special motor fuels and gasoline used in motorboats).

(b)   There shall be paid from time to time from the land and water conservation fund into the general fund of the Treasury amounts estimated by the Secretary of the Treasury as equivalent to—

(1)   the amounts paid before July 1, 1973, under section 6421 of the Internal Revenue Code of 1954 (relating to amounts paid in respect of gasoline used for certain nonhighway purposes or by local transit systems) with respect to gasoline used after December 31, 1964, in motorboats, on the basis of claims filed for periods ending before October 1, 1972; and

(2)   80 percent of the floor stocks refunds made before July 1, 1973, under section 6412 (a)(2) of such Code with respect to gasoline to be used in motorboats.

AMENDMENTS TO HIGHWAY REVENUE ACT OF 1956

SEC. 202. (a) Section 209 (f) of the Highway Revenue Act of 1956 (relating to expenditures from highway trust fund) is amended by adding at the end thereof the following new paragraph:

"(5)   TRANSFERS FROM THE TRUST FUND FOR SPECIAL MOTOR FUELS AND GASOLINE USED IN MOTORBOATS.— The Secretary of the Treasury shall pay from time to time from the trust fund into the land and water conservation fund provided for in title I of the Land and Water Conservation Fund Act of 1965 amounts as determined by him in consultation with the Secretary of Commerce equivalent to the taxes received, on or after January 1, 1965, under section 4041 (b) of the Internal Revenue Code of 1954 with respect to special motor fuels used as fuel for the propulsion of motorboats and under section 4081 of such Code with respect to gasoline used as fuel in motorboats."

(b) Section 209 (f) of such Act is further amended—

(1)   by adding at the end of paragraph (3) the following new sentence: "This paragraph shall not apply to amounts estimated by the Secretary of the Treasury as paid under section 6421 of such Code with respect to gasoline used after December 31, 1964, in motorboats."; and

(2)   by inserting after "such Code" in paragraph (4) (C) the following: "(other than gasoline to be used in motorboats, as estimated by the Secretary of the Treasury)".

Approved September 3, 1964.

16 U.S.C. 460d et seq.

AN ACT RELATING TO THE ESTABLISHMENT OF CONCESSION POLICIES IN THE AREAS ADMINISTERED BY NATIONAL PARK SERVICE, AND FOR OTHER PURPOSES,

1965 (79 Stat. 969)

Be it enacted by the Senate and House of Representatives of the United States of America in Congress assembled, That in furtherance of the Act of August 25, 1916 (39 Stat. 535), as amended (16 U.S.C. 1), which directs the Secretary of the Interior to administer national park system areas in accordance with the fundamental purposes of conserving their scenery, wildlife, natural and historic objects, and providing for their enjoyment in a manner that will leave them unimpaired for the enjoyment of future generations, the Congress hereby finds that the preservation of park values requires that such public accommodations, facilities, and services as have to be provided within those areas should be provided only under carefully controlled safeguards against unregulated and indiscriminate use, so that the heavy visitation will not unduly impair these values and so that development of such facilities can best be limited to locations where the least damage to park values will be caused. It is the policy of the Congress that such development shall be limited to those that are necessary and appropriate for public use and enjoyment of the national park area in which they are located and that are consistent to the highest practicable degree with the preservation and conservation of the areas.

SEC. 2. Subject to the findings and policy stated in section 1 of this Act, the Secretary of the Interior shall take such action as may be appropriate to encourage and enable private persons and corporations (hereinafter referred to as "concessioners") to provide and operate facilities and services which he deems desirable for the accommodation of visitors in areas administered by the National Park Service.

SEC. 3. (a) Without limitation of the foregoing, the Secretary may include in contracts for the providing of facilities and services such terms and conditions as, in his judgment, are required to assure the concessioner of adequate protection against loss of investment in structures, fixtures, improvements, equipment, supplies, and other tangible property provided by him for the purposes of the contract (but not against loss of anticipated profits) resulting from discretionary acts, policies, or decisions of the Secretary occurring after the contract has become effective under which acts, policies, or decisions the concessioner's authority to conduct some or all of his authorized operations under the contract ceases or his structures, fixtures, and improvements, or any of them, are required to be transferred to another party or to be abandoned, removed, or demolished. Such terms and conditions may include an obligation of the United States to compensate the concessioner for the loss of investment, as aforesaid.

(b) The Secretary shall exercise his authority in a manner consistent with a reasonable opportunity for the concessioner to realize a profit on his operation as a whole commensurate with the capital invested and the obligations assumed.

(c) The reasonableness of a concessioner's rates and charges to the public shall, unless otherwise provided in the contract, be judged primarily by comparison with those current for facilities and services of comparable character under similar conditions, with due consideration for length of season, provision for peakloads, average percentage of occupancy, accessibility, availability and costs of labor and materials, type of patronage, and other factors deemed significant by the Secretary.

(d) Franchise fees, however stated, shall be determined upon consideration of the probable value to the concessioner of the privileges granted by the particular contract or permit involved. Such value is the opportunity for net profit in relation to both gross receipts and capital invested. Consideration of revenue to the United States shall be subordinate to the objectives of protecting and preserving the areas and of providing adequate and appropriate services for visitors at reasonable rates. Appropriate provisions shall be made for reconsideration of franchise fees at least every five years unless the contract is for a lesser period of time.

SEC. 4. The Secretary may authorize the operation of all accommodations, facilities, and services for visitors, or of all such accommodations, facilities, and services of generally similar character, in each area, or portion thereof, administered by the National Park Service by one responsible concessioner and may grant to such concessioner a preferential right to provide such new or additional accommodations, facilities, or services as the Secretary may consider necessary or desirable for the accommodation and convenience of the public. The Secretary may, in his discretion, grant extensions, renewals, or new contracts to present concessioners, other than the concessioner holding a preferential right, for operations substantially similar in character and extent to those authorized by their current contracts or permits.

SEC. 5. The Secretary shall encourage continuity of operation and facilities and services by giving preference in the renewal of contracts or permits and in the negotiation of new contracts or permits to the concessioners who have performed their obligations under prior contracts or permits to the satisfaction of the Secretary. To this end, the Secretary, at any time in his discretion, may extend or renew a contract or permit, or may grant a new contract or permit to the same concessioner upon the termination or surrender before expiration of a prior contract or permit. Before doing so, however, and before granting extensions, renewals or new contracts pursuant to the last sentence of section 4 of this Act, the Secretary shall give reasonable public notice of his intention so to do and shall consider and evaluate all proposals received as a result thereof.

SEC. 6. A concessioner who has heretofore acquired or constructed or who hereafter acquires or constructs, pursuant to a contract and with the approval of the Secretary, any structure, fixture, or improvement upon land owned by the United States within an area administered by the National Park Service shall have a possessory interest therein, which shall consist of all incidents of ownership except legal title, and except as hereinafter provided, which title shall be vested in the United States. Such possessory interest shall not be construed to include or imply any authority, privilege, or right to operate or engage in any business or other activity, and the use or enjoyment of any structure, fixture, or improvement in which the concessioner has a possessory interest shall be wholly subject to the applicable provisions of the contract and of laws and regulations relating to the area. The said possessory interest shall not be extinguished by the expiration or other termination of the contract and may not be taken for public use without just compensation. The said possessory interest may be assigned, transferred, encumbered, or relinquished. Unless otherwise provided by agreement of the parties, just compensation shall be an amount equal to the sound value of such structure, fixture, or improvement at the time of taking by the United States determined upon the basis of reconstruction cost less depreciation evidenced by its condition and prospective serviceability in comparison with a new unit of like kind, but not to exceed fair market value. The provisions of this section shall not apply to concessioners whose current contracts do not include recognition of a possessory interest, unless in a particular case the Secretary determines that equitable considerations warrant recognition of such interest.

SEC. 7. The provisions of section 321 of the Act of June 30, 1932 (47 Stat. 412; 40 U.S.C. 303(b)), relating to the leasing of buildings and properties of the United States, shall not apply to privileges, leases, permits, and contracts granted by the Secretary of the Interior for the use of lands and improvements thereon, in areas administered by the National Park Service, for the purpose of providing accommodations, facilities, and services for visitors thereto, pursuant to the Act of August 25, 1916 (39 Stat. 535), as amended, or the Act of August 21, 1935, chapter 593 (49 Stat. 666; 16 U.S.C. 461-467), as amended.

SEC. 8. Subsection (h) of section 2 of the Act of August 21, 1935, the Historical Sites, Buildings, and Antiquities Act (49 Stat. 666; 16 U.S.C. 462(h)), is amended by changing the proviso therein to read as follows: "*Provided,* That the Secretary may grant such concessions, leases, or permits and enter into contracts relating to the same with responsible persons, firms, or corporations without advertising and without securing competitive bids."

SEC. 9. Each concessioner shall keep such records as the Secretary may prescribe to enable the Secretary to determine that all terms of the concession contract have been and are being faithfully performed, and the Secretary and his duly authorized representatives shall, for the purpose of audit and examination, have access to said records and to other books, documents, and papers of the concessioner pertinent to the contract and all the terms and conditions thereof.

The Comptroller General of the United States or any of his duly authorized representatives shall, until the expiration of five (5) calendar years after the close of the business year of each concessioner or subconcessioner have access to and the right to examine any pertinent books, documents, papers, and records of the concessioner or subconcessioner related to the negotiated contract or contracts involved.

Approved October 9, 1965, 6:35 a.m.

16 U.S.C. 20–20g, 16 U.S.C. 462.

AN ACT TO ESTABLISH A PROGRAM
FOR THE PRESERVATION OF
ADDITIONAL HISTORIC PROPERTIES
THROUGHOUT THE NATION,
AND FOR OTHER PURPOSES,

1966 (80 Stat. 915)

Be it enacted by the Senate and House of Representatives of the United States of America in Congress assembled,
The Congress finds and declares—

(a) that the spirit and direction of the Nation are founded upon and reflected in its historic past;

(b) that the historical and cultural foundations of the Nation should be preserved as a living part of our community life and development in order to give a sense of orientation to the American people;

(c) that, in the face of ever-increasing extensions of urban centers, highways, and residential, commercial, and industrial developments, the present governmental and nongovernmental historic preservation programs and activities are inadequate to insure future generations a genuine opportunity to appreciate and enjoy the rich heritage of our Nation; and

(d) that, although the major burdens of historic preservation have been borne and major efforts initiated by private agencies and individuals, and both should continue to play a vital role, it is nevertheless necessary and appropriate for the Federal Government to accelerate its historic preservation programs and activities, to give maximum encouragement to agencies and individuals undertaking preservation by private means, and to assist State and local governments and the National Trust for Historic Preservation in the United States to expand and accelerate their historic preservation programs and activities.

TITLE I

SEC. 101. (a) The Secretary of the Interior is authorized—

(1) to expand and maintain a national register of districts, sites, buildings, structures, and objects significant in American history, architecture, archeology, and culture, hereinafter referred to as the National Register, and to grant funds to States for the purpose of preparing comprehensive statewide historic surveys and plans,

in accordance with criteria established by the Secretary, for the preservation, acquisition, and development of such properties;

(2) to establish a program of matching grants-in-aid to States for projects having as their purpose the preservation for public benefit of properties that are significant in American history, architecture, archeology, and culture; and

(3) to establish a program of matching grants-in-aid to the National Trust for Historic Preservation in the United States, chartered by act of Congress approved October 26, 1949 (63 Stat. 927), as amended, for the purpose of carrying out the responsibilities of the National Trust.

(b) As used in this Act—

(1) The term "State" includes, in addition to the several States of the Union, the District of Columbia, the Commonwealth of Puerto Rico, the Virgin Islands, Guam, and American Samoa.

(2) The term "project" means programs of State and local governments and other public bodies and private organizations and individuals for the acquisition of title or interests in, and for the development of, any district, site, building, structure, or object that is significant in American history, architecture, archeology, and culture, or property used in connection therewith, and for its development in order to assure the preservation for public benefit of any such historical properties.

(3) The term "historic preservation" includes the protection, rehabilitation, restoration, and reconstruction of districts, sites, buildings, structures, and objects significant in American history, architecture, archeology, or culture.

(4) The term "Secretary" means the Secretary of the Interior.

SEC. 102. (a) No grant may be made under this Act—

(1) unless application therefor is submitted to the Secretary in accordance with regulations and procedures prescribed by him;

(2) unless the application is in accordance with the comprehensive statewide historic preservation plan which has been approved by the Secretary after considering its relationship to the comprehensive statewide outdoor recreation plan prepared pursuant to the Land and Water Conservation Fund Act of 1965 (78 Stat. 897);

(3) for more than 50 per centum of the total cost involved, as determined by the Secretary and his determination shall be final;

(4) unless the grantee has agreed to make such reports, in such form and containing such information as the Secretary may from time to time require;

(5) unless the grantee has agreed to assume, after completion of the project, the total cost of the continued maintenance, repair,

and administration of the property in a manner satisfactory to the Secretary; and

(6) until the grantee has complied with such further terms and conditions as the Secretary may deem necessary or advisable.

(b) The Secretary may in his discretion waive the requirements of subsection (a), paragraphs (2) and (5) of this section for any grant under this Act to the National Trust for Historic Preservation in the United States, in which case a grant to the National Trust may include funds for the maintenance, repair, and administration of the property in a manner satisfactory to the Secretary.

(b) No State shall be permitted to utilize the value of real property obtained before the date of approval of this Act in meeting the remaining cost of a project for which a grant is made under this Act.

SEC. 103. (a) The amounts appropriated and made available for grants to the States for comprehensive statewide historic surveys and plans under this Act shall be apportioned among the States by the Secretary on the basic of needs as determined by him: *Provided, however,* That the amount granted to any one State shall not exceed 50 per centum of the total cost of the comprehensive statewide historic survey and plan for that State, as determined by the Secretary.

(b) The amounts appropriated and made available for grants to the States for projects under this Act for each fiscal year shall be apportioned among the States by the Secretary in accordance with needs as disclosed in approved statewide historic preservation plans.

The Secretary shall notify each State of its apportionment, and the amounts thereof shall be available thereafter for payment to such State for projects in accordance with the provisions of this Act. Any amount of any apportionment that has not been paid or obligated by the Secretary during the fiscal year in which such notification is given, and for two fiscal years thereafter, shall be reapportioned by the Secretary in accordance with this subsection.

SEC. 104. (a) No grant may be made by the Secretary for or on account of any survey or project under this Act with respect to which financial assistance has been given or promised under any other Federal program or activity, and no financial assistance may be given under any other Federal program or activity for or on account of any survey or project with respect to which assistance has been given or promised under this Act.

(b) In order to assure consistency in policies and actions under this Act with other related Federal programs and activities, and to assure coordination of the planning, acquisition, and development assistance to States under this Act with other related Federal programs and activities, the President may issue such regulations with respect thereto as he deems desirable, and such assistance may be provided only in accordance with such regulations.

SEC. 105. The beneficiary of assistance under this Act shall keep such records as the Secretary shall prescribe, including records which fully disclose the disposition by the beneficiary of the proceeds of such assistance, the total cost of the project or undertaking in connection with which such assistance is given or used, and the amount and nature of that portion of the cost of the project or undertaking supplied by other sources, and such other records as will facilitate an effective audit.

SEC. 106. The head of any Federal agency having direct or indirect jurisdiction over a proposed Federal or federally assisted undertaking in any State and the head of any Federal department or independent agency having authority to license any undertaking shall, prior to the approval of the expenditure of any Federal funds on the undertaking or prior to the issuance of any license, as the case may be, take into account the effect of the undertaking on any district, site, building, structure, or object that is included in the National Register. The head of any such Federal agency shall afford the Advisory Council on Historic Preservation established under title II of this Act a reasonable opportunity to comment with regard to such undertaking.

SEC. 107. Nothing in this Act shall be construed to be applicable to the White House and its grounds, the Supreme Court building and its grounds, or the United States Capitol and its related buildings and grounds.

SEC. 108. There are authorized to be appropriated not to exceed $2,000,000 to carry out the provisions of this Act for the fiscal year 1967, and not more than $10,000,000 for each of the three succeeding fiscal years. Such appropriations shall be available for the financial assistance authorized by this title and for the administrative expenses of the Secretary in connection therewith, and shall remain available until expended.

TITLE II

SEC. 201. (a) There is established an Advisory Council on Historic Preservation (hereinafter referred to as the "Council") which shall be composed of seventeen members as follows:

(1)   The Secretary of the Interior.
(2)   The Secretary of Housing and Urban Development.
(3)   The Secretary of Commerce.
(4)   The Administrator of the General Services Administration.
(5)   The Secretary of the Treasury.
(6)   The Attorney General.
(7)   The Chairman of the National Trust for Historic Preservation.

(8)   Ten appointed by the President from outside the Federal Government. In making these appointments, the President shall give due consideration to the selection of officers of State and local governments and individuals who are significantly interested and experienced in the matters to be considered by the Council.

(b)   Each member of the Council specified in paragraphs (1) through (6) of subsection (a) may designate another officer of his department or agency to serve on the Council in his stead.

(c)   Each member of the Council appointed under paragraph (8) of subsection (a) shall serve for a term of five years from the expiration of his predecessor's term; except that the members first appointed under that paragraph shall serve for terms of from one to five years, as designated by the President at the time of appointment, in such manner as to insure that the terms of not less than one nor more than two of them will expire in any one year.

(d)   A vacancy in the Council shall not affect its powers, but shall be filled in the same manner as the original appointment (and for the balance of the unexpired term).

(e)   The Chairman of the Council shall be designated by the President.

(f)   Eight members of the Council shall constitute a quorum.

SEC. 202. (a) The Council shall—

(1)   advise the President and the Congress on matters relating to historic preservation; recommend measures to coordinate activities of Federal, State, and local agencies and private institutions and individuals relating to historic preservation; and advise on the dissemination of information pertaining to such activities;

(2)   encourage, in cooperation with the National Trust for Historic Preservation and appropriate private agencies, public interest and participation in historic preservation;

(3)   recommend the conduct of studies in such areas as the adequacy of legislative and administrative statutes and regulations pertaining to historic preservation activities of State and local governments and the effects of tax policies at all levels of government on historic preservation;

(4)   advise as to guidelines for the assistance of State and local governments in drafting legislation relating to historic preservation; and

(5)   encourage, in cooperation with appropriate public and private agencies and institutions, training and education in the field of historic preservation.

(b)   The Council shall submit annually a comprehensive report of its activities and the results of its studies to the President and the Congress and shall from time to time submit such additional and special reports as it deems advisable. Each report shall propose such legislative enactments and other actions as, in the judgment of the Council, are necessary and appropriate to carry out its recommendations.

SEC. 203. The Council is authorized to secure directly from any department, bureau, agency, board, commission, office, independent establishment or instrumentality of the executive branch of the Federal Government information, suggestions, estimates, and statistics for the purpose of this title; and each such department, bureau, agency, board, commission, office, independent establishment or instrumentality is authorized to furnish such information, suggestions, estimates, and statistics to the extent permitted by law and within available funds.

SEC. 204. The members of the Council specified in paragraphs (1) through (7) of section 201 (a) shall serve without additional compensation. The members of the Council appointed under paragraph (8) of section 201 (a) shall receive $100 per diem when engaged in the performance of the duties of the Council. All members of the Council shall receive reimbursement for necessary traveling and subsistence expenses incurred by them in the performance of the duties of the Council.

SEC. 205. (a) The Director of the National Park Service or his designee shall be the Executive Director of the Council. Financial and administrative services (including those related to budgeting, accounting, financial reporting, personnel and procurement) shall be provided the Council by the Department of the Interior, for which payments shall be made in advance, or by reimbursement, from funds of the Council in such amounts as may be agreed upon by the Chairman of the Council and the Secretary of the Interior: *Provided,* That the regulations of the Department of the Interior for the collection of indebtness of personnel resulting from erroneous payments (5 U.S.C. 46e) shall apply to the collection of erroneous payments made to or on behalf of a Council employee, and regulations of said Secretary for the administrative control of funds (31 U.S.C. 665(g)) shall apply to appropriations of the Council; *And provided further,* That the Council shall not be required to prescribe such regulations.

(b) That the Council shall have power to appoint and fix the compensation of such additional personnel as may be necessary to carry out its duties, without regard to the provisions of the civil service laws and the Classification Act of 1949.

(c) The Council may also procure, without regard to the civil service laws and the Classification Act of 1949, temporary and intermittent services to the same extent as is authorized for the executive departments by section 15 of the Administrative Expenses Act of 1946 (5 U.S.C. 55a) but at rates not to exceed $50 per diem for individuals.

(d) The members of the Council specified in paragraphs (1) through (6) of section 201 (a) shall provide the Council, on a reimbursable basis, with such facilities and services under their jurisdiction and control as may be needed by the Council to carry out its duties, to the extent that such facilities and services are requested by the Council and are otherwise available for that purpose. To the extent of available appropriations, the Council may obtain, by purchase, rental, donation, or otherwise, such additional property, facilities, and services as may be needed to carry out its duties.

Approved October 15, 1966.

16 U.S.C. 470 et seq.

AN ACT TO PROVIDE FOR A NATIONAL
WILD AND SCENIC RIVERS SYSTEM,
AND FOR OTHER PURPOSES,

1968 (82 Stat. 906)

Be it enacted by the Senate and House of Representatives of the United States of America in Congress assembled, That (a) this Act may be cited as the "Wild and Scenic Rivers Act."

(b)   It is hereby declared to be the policy of the United States that certain selected rivers of the Nation which, with their immediate environments, possess outstandingly remarkable scenic, recreational, geologic, fish and wildlife, historic, cultural, or other similar values, shall be preserved in free-flowing condition, and that they and their immediate environments shall be protected for the benefit and enjoyment of present and future generations. The Congress declares that the established national policy of dam and other construction at the appropriate sections of the rivers of the United States needs to be complemented by a policy that would preserve other selected rivers or sections thereof in their free-flowing condition to protect the water quality of such rivers and to fulfill other vital national conservation purposes.

(c)   The purpose of this Act is to implement this policy by instituting a national wild and scenic rivers system, by designating the initial components of that system, and by prescribing the methods by which and standards according to which additional components may be added to the system from time to time.

SEC. 2. (a) The national wild and scenic rivers system shall comprise rivers (i) that are authorized for inclusion therein by Act of Congress, or (ii) that are designated as wild, scenic or recreational rivers by or pursuant to an act of the legislature of the State or States through which they flow, that are to be permanently administered as wild, scenic or recreational rivers by an agency or political subdivision of the State or States concerned without expense to the United States, that are found by the Secretary of the Interior, upon application of the Governor of the State or the Governors of the States concerned, or a person or persons thereunto duly appointed by him or them, to meet the criteria established in this Act and such criteria supplementary thereto as he may prescribe, and that are approved by him for inclusion in the system, including, upon application of the Governor of the State concerned, the Allagash Wilderness Waterway, Maine, and that segment of the Wolf River, Wisconsin, which flows through Langlade County.

(b)   A wild, scenic or recreational river area eligible to be included in the system is a free-flowing stream and the related adjacent land area that possesses one or more of the values referred to in section 1, subsection (b) of this Act. Every wild, scenic or recreational river in its free-flowing condition, or upon restoration to this condition, shall be considered eligible for inclusion in the national wild and scenic rivers system and, if included, shall be classified, designated, and administered as one of the following:

(1)   Wild river areas—Those rivers or sections of rivers that are free of impoundments and generally inaccessible except by trail with watersheds or shorelines essentially primitive and waters unpolluted. These represent vestiges of primitive America.

(2)   Scenic river areas—Those rivers or sections of rivers that are free of impoundments, with shorelines or watersheds still largely primitive and shorelines largely undeveloped, but accessible in places by roads.

(3)   Recreational river areas—those rivers or sections of rivers that are readily accessible by road or railroad, that may have some development along their shorelines, and that may have undergone some impoundment or diversion in the past.

SEC. 3. (a) The following rivers and the land adjacent thereto are hereby designated as components of the national wild and scenic rivers system:

(1)   CLEARWATER, MIDDLE FORK, IDAHO.—The Middle Fork from the town of Kooskia upstream to the town of Lowell; the Lochsa River from its junction with the Selway at Lowell forming the Middle Fork, upstream to the Powell Ranger Station; and the Selway River from Lowell upstream to its origin; to be administered by the Secretary of Agriculture.

(2)   ELEVEN POINT, MISSOURI.—The segment of the river extending downstream from Thomasville to State Highway 142; to be administered by the Secretary of Agriculture.

(3)   FEATHER, CALIFORNIA.—The entire Middle Fork; to be administered by the Secretary of Agriculture.

(4) RIO GRANDE, NEW MEXICO.—The segment extending from the Colorado State line downstream to the State Highway 96 crossing, and the lower four miles of the Red River; to be administered by the Secretary of the Interior.

(5)   ROGUE, OREGON.—The segment of the river extending from the mouth of the Applegate River downstream to the Lobster Creek Bridge; to be administered by agencies of the Departments of the Interior or Agriculture as agreed upon by the Secretaries of said Departments or as directed by the President.

(6)   SAINT CROIX, MINNESOTA AND WISCONSIN.—The

segment between the dam near Taylors Falls, Minnesota, and the dam near Gordon, Wisconsin, and its tributary, the Namekagon, from Lake Namekagon downstream to its confluence with the Saint Croix; to be administered by the Secretary of the Interior: *Provided* That except as may be required in connection with items (a) and (b) of this paragraph, no funds available to carry out the provisions of this Act may be expended for the acquisition or development of lands in connection with, or for administration under this Act of, that portion of the Saint Croix River between the dam near Taylors Falls, Minnesota, and the upstream end of Big Island in Wisconsin, until sixty days after the date on which the Secretary has transmitted to the President of the Senate and Speaker of the House of Representatives a proposed cooperative agreement between the Northern States Power Company and the United States (a) whereby the company agrees to convey to the United States, without charge, appropriate interests in certain of its lands between the dam near Taylors Falls, Minnesota, and the upstream end of Big Island in Wisconsin, including the company's right, title, and interest to approximately one hundred acres per mile, and (b) providing for the use and development of other lands and interests in land retained by the company between said points adjacent to the river in a manner which shall complement and not be inconsistent with the purposes for which the lands and interests in land donated by the company are administered under this Act. Said agreement may also include provision for State or local governmental participation as authorized under subsection (e) of section 10 of this Act.

(7)   SALMON, MIDDLE FORK, IDAHO.—From its origin to its confluence with the main Salmon River; to be administered by the Secretary of Agriculture.

(8)   WOLF, WISCONSIN.—From the Langlade-Menominee County line downstream to Keshena Falls; to be administered by the Secretary of the Interior.

(b)   The agency charged with the administration of each component of the national wild and scenic rivers system designated by subsection (a) of this section shall, within one year from the date of this Act, establish detailed boundaries therefor (which shall include an average of not more than three hundred and twenty acres per mile on both sides of the river); determine which of the classes outlined in section 2, subsection (b), of this Act best fit the river or its various segments; and prepare a plan for necessary developments in connection with its administration in accordance with such classification. Said boundaries, classification, and development plans shall be published in the Federal Register and shall not become effective until ninety days after they have been forwarded to the President of the Senate and the Speaker of the House of Representatives.

SEC. 4. (a) The Secretary of the Interior or, where national forest lands are involved, the Secretary of Agriculture or, in appropriate cases, the two Secretaries jointly shall study and from time to time submit to the President and the Congress proposals for the addition to the national wild and scenic rivers system of rivers which are designated herein or hereafter by the Congress as potential additions to such system; which, in his or their judgment, fall within one or more of the classes set out in section 2, subsection (b), of this Act; and which are proposed to be administered, wholly or partially, by an agency of the United States. Every such study and plan shall be coordinated with any water resources planning involving the same river which is being conducted pursuant to the Water Resources Planning Act (79 Stat. 244; 42 U.S.C. 1962 et seq.).

Each proposal shall be accompanied by a report, including maps and illustrations, showing among other things the area included within the proposal; the characteristics which make the area a worthy addition to the system; the current status of land ownership and use in the area; the reasonably foreseeable potential uses of the land and water which would be enhanced, foreclosed, or curtailed if the area were included in the national wild and scenic rivers system; the Federal agency (which in the case of a river which is wholly or substantially within a national forest, shall be the Department of Agriculture) by which it is proposed the area be administered; the extent to which it is proposed that administration, including the costs thereof, be shared by State and local agencies; and the estimated cost to the United States of acquiring necessary lands and of administering the area as a component of the system. Each such report shall be printed as a Senate or House document.

(b)   Before submitting any such report to the President and the Congress, copies of the proposed report shall, unless it was prepared jointly by the Secretary of the Interior and the Secretary of Agriculture, be submitted by the Secretary of the Interior to the Secretary of Agriculture or by the Secretary of Agriculture to the Secretary of the Interior, as the case may be, and to the Secretary of the Army, the Chairman of the Federal Power Commission, the head of any other affected Federal department or agency and, unless the lands proposed to be included in the area are already owned by the United States or have already been authorized for acquisition by Act of Congress, the Governor of the State or States in which they are located or an officer designated by the Governor to receive the same. Any recommendations or comments on the proposal which the said officials furnish the Secretary or Secretaries who prepared the report within ninety days of the date on which the report is submitted to them, together with the Secretary's or Secretaries' comments thereon, shall be included with the transmittal to the President and the Congress. No river or portion of any river shall be added to the national wild and scenic rivers system subsequent to enactment of this Act until the close of the next full session of the State legislature, or legislatures in case more than one State is involved, which begins following the submission of any recommendation to the President with respect to such addition as herein provided.

   (c)   Before approving or disapproving for inclusion in the national wild and scenic rivers system any river designated as a wild, scenic or recreational river by or pursuant to an act of a State legislature, the Secretary of the Interior shall submit the proposal to the Secretary of Agriculture, the Secretary of the Army, the Chairman of the Federal Power Commission, and the head of any other affected department or agency and shall evaluate and give due weight to any recommendations or comments which the said officials furnish him within ninety days of the date on which it is submitted to them. If he approves the proposed inclusion, he shall publish notice thereof in the Federal Register.

   SEC. 5. (a) The following rivers are hereby designated for potential addition to the national wild and scenic rivers system:

   (1)   Allegheny, Pennsylvania: The segment from its mouth to the town of East Brady, Pennsylvania.

   (2)   Bruneau, Idaho: The entire main stem.

   (3)   Buffalo, Tennessee: The entire river.

   (4)   Chattooga, North Carolina, South Carolina, and Georgia: The entire river.

   (5)   Clarion, Pennsylvania: The segment between Ridgway and its confluence with the Allegheny River.

   (6)   Delaware, Pennsylvania and New York: The segment from Hancock, New York, to Matamoras, Pennsylvania.

   (7)   Flathead, Montana: The North Fork from the Canadian border downstream to its confluence with the Middle Fork; the Middle Fork from its headwaters to its confluence with the South Fork; and the South Fork from its origin to Hungry Horse Reservoir.

   (8)   Gasconade, Missouri: The entire river.

   (9)   Illinois, Oregon: The entire river.

   (10) Little Beaver, Ohio: The segment of the North and Middle Forks of the Little Beaver River in Columbiana County from a point in the vicinity of Negly and Elkton, Ohio, downstream to a point in the vicinity of East Liverpool, Ohio.

   (11) Little Miami, Ohio: That segment of the main stem of the river, exclusive of its tributaries, from a point at the Warren-Clermont County line at Loveland, Ohio, upstream to the sources of Little Miami including North Fork.

(12) Maumee, Ohio and Indiana: The main stem from Perrsyburg, Ohio, to Fort Wayne, Indiana, exclusive of its tributaries in Ohio and inclusive of its tributaries in Indiana.

(13) Missouri, Montana: The segments between Fort Benton and Ryan Island.

(14) Moyie, Idaho: The segment from the Canadian border to its confluence with the Kootenai River.

(15) Obed, Tennessee: The entire river and its tributaries, Clear Creek and Daddy Creek.

(16) Penobscot, Maine: Its east and west branches.

(17) Pere Marquette, Michigan: The entire river.

(18) Pine Creek, Pennsylvania: The segment from Ansonia to Waterville.

(19) Priest, Idaho: The entire main stem.

(20) Rio Grande, Texas: The portion of the river between the west boundary of Hudspeth County and the east boundary of Terrell County on the United States side of the river: *Provided*, That before undertaking any study of this potential scenic river, the Secretary of the Interior shall determine, through the channels of appropriate Executive agencies, that Mexico has no objection to its being included among the studies authorized by this Act.

(21) Saint Croix, Minnesota and Wisconsin: The segment between the dam near Taylors Falls and its confluence with the Mississippi River.

(22) Saint Joe, Idaho: The entire main stem.

(23) Salmon, Idaho: The segment from the town of North Fork to its confluence with the Snake River.

(24) Skagit, Washington: The segment from the town of Mount Vernon to and including the mouth of Bacon Creek; the Cascade River between its mouth and the junction of its North and South Forks; the South Fork to the boundary of the Glacier Peak Wilderness Area; the Suiattle River from its mouth to the Glacier Peak Wilderness Area boundary at Milk Creek; the Sauk River from its mouth to its junction with Elliott Creek; the North Fork of the Sauk River from its junction with the South Fork of the Sauk to the Glacier Peak Wilderness Area boundary.

(25) Suwannee, Georgia and Florida: The entire river from its source in the Okefenokee Swamp in Georgia to the gulf and the outlying Ichetucknee Springs, Florida.

(26) Upper Iowa, Iowa: The entire river.

(27) Youghiogheny, Maryland and Pennsylvania: The segment from Oakland, Maryland, to the Youghiogheny Reservoir, and from the Youghiogheny Dam downstream to the town of Connellsville, Pennsylvania.

(b)   The Secretary of the Interior and, where national forest lands are involved, the Secretary of Agriculture shall proceed as expeditiously as possible to study each of the rivers named in subsection (a) of this section in order to determine whether it should be included in the national wild and scenic rivers system. Such studies shall be completed and reports made thereon to the President and the Congress, as provided in section 4 of this Act, within ten years from the date of this Act: *Provided however*, That with respect to the Suwannee River, Georgia and Florida, and the Upper Iowa River, Iowa, such study shall be completed and reports made thereon to the President and the Congress, as provided in section 4 of this Act, within two years from the date of enactment of this Act. In conducting these studies the Secretary of the Interior and the Secretary of Agriculture shall give priority to those rivers with respect to which there is the greatest likelihood of developments which, if undertaken, would render them unsuitable for inclusion in the national wild and scenic rivers system.

(c)   The study of said rivers shall be pursued in as close cooperation with appropriate agencies of the affected State and its political subdivisions as possible, shall be carried on jointly with such agencies if request for such joint study is made by the State, and shall include a determination of the degree to which the State or its political subdivisions might participate in the preservation and administration of the river should it be proposed for inclusion in the national wild and scenic rivers system.

(d)   In all planning for the use and development of water and related land resources, consideration shall be given by all Federal agencies involved to potential national wild, scenic and recreational river areas, and all river basin and project plan reports submitted to the Congress shall consider and discuss any such potentials. The Secretary of the Interior and the Secretary of Agriculture shall make specific studies and investigations to determine which additional wild, scenic and recreational river areas within the United States shall be evaluated in planning reports by all Federal agencies as potential alternative uses of the water and related land resources involved.

SEC. 6. (a) The Secretary of the Interior and the Secretary of Agriculture are each authorized to acquire lands and interests in land within the authorized boundaries of any component of the national wild and scenic rivers system designated in section 3 of this Act, or hereafter designated for inclusion in the system by Act of Congress, which is administered by him, but he shall not acquire fee title an average of more than 100 acres per mile on both sides of the river. Lands owned by a State may be acquired only by donation, and lands owned by an Indian tribe or a political subdivision of a State may not be acquired without the consent of the appropriate governing body thereof as long as the Indian tribe or political subdivision is following a plan for management and protection of the lands which the Secretary finds protects the land and assures its use for purposes consistent with this Act. Money appropriated for Federal purposes from the land and water conservation fund shall, without prejudice to the use of appropriations from other sources, be available to Federal departments and agencies for the acquisition of property for the purposes of this Act.

(b)   If 50 per centum or more of the entire acreage within a federally administered wild, scenic or recreational river area is owned by the United States, by the State or States within which it lies, or by political subdivisions of those States, neither Secretary shall acquire fee title to any lands by condemnation under authority of this Act. Nothing contained in this section, however, shall preclude the use of condemnation when necessary to clear title or to acquire scenic easements or such other easements as are reasonably necessary to give the public access to the river and to permit its members to traverse the length of the area or of selected segments thereof.

(c)   Neither the Secretary of the Interior nor the Secretary of Agriculture may acquire lands by condemnation, for the purpose of including such lands in any national wild, scenic or recreational river area, if such lands are located within any incorporated city, village, or borough which has in force and applicable to such lands a duly adopted, valid zoning ordinance that conforms with the purposes of this Act. In order to carry out the provisions of this subsection the appropriate Secretary shall issue guidelines, specifying standards for local zoning ordinances, which are consistent with the purposes of this Act. The standards specified in such guidelines shall have the object of (A) prohibiting new commercial or industrial uses other than commercial or industrial uses which are consistent with the purposes of this Act, and (B) the protection of the bank lands by means of acreage, frontage, and setback requirements on development.

(d)   The appropriate Secretary is authorized to accept title to non-Federal property within the authorized boundaries of any federally administered component of the national wild and scenic rivers system designated in section 3 of this Act or hereafter for inclusion in the system by Act of Congress and, in exchange therefor, convey to the grantor any federally owned property which is under his jurisdiction within the State in which the component lies and which he classifies as suitable for exchange or other disposal. The values of the properties so exchanged either shall be approximately equal or, it they are not approximately equal, shall be equalized by the payment of cash to the grantor or to the Secretary as the circumstances require.

(e) The head of any Federal department or agency having administrative jurisdiction over any lands or interests in land within the authorized boundaries of any federally administered component of the national wild and scenic rivers system designated in section 3 of this Act or hereafter designated for inclusion in the system by Act of Congress is authorized to transfer to the appropriate secretary jurisdiction over such lands for administration in accordance with the provisions of this Act. Lands acquired by or transferred to the Secretary of Agriculture for the purposes of this Act within or adjacent to a national forest shall upon such acquisition or transfer become national forest lands.

(f) The appropriate Secretary is authorized to accept donations of lands and interests in land, funds, and other property for use in connection with his administration of the national wild and scenic rivers system.

(g) (1) Any owner or owners (hereinafter in this subsection referred to as "owner") of improved property on the date of its acquisition, may retain for themselves and their successors or assigns a right of use and occupancy of the improved property for noncommercial residential purposes for a definite term not to exceed twenty-five years or, in lieu thereof, for a term ending at the death of the owner, or the death of his spouse, or the death of either or both of them. The owner shall elect the term to be reserved. The appropriate Secretary shall pay to the owner the fair market value of the property on the date of such acquisition less the fair market value on such date of the right retained by the owner.

(2) A right of use and occupancy retained pursuant to this subsection shall be subject to termination whenever the appropriate Secretary is given reasonable cause to find that such use and occupancy is being exercised in a manner which conflicts with the purposes of this Act. In the event of such a finding, the Secretary shall tender to the holder of that right an amount equal to the fair market value of the portion of the right which remains unexpired on the date of termination. Such right of use or occupancy shall terminate by operation of law upon tender of the fair market price.

(3) The term "improved property", as used in this Act, means a detached, one-family dwelling (hereinafter referred to as "dwelling"), the construction of which was begun before January 1, 1967, together with so much of the land on which the dwelling is situated, the said land being in the same ownership as the dwelling, as the appropriate Secretary shall designate to be reasonably necessary for the enjoyment of the dwelling for the sole purpose of noncommer-

cial residential use, together with any structures accessory to the dwelling which are situated on the land so designated.

SEC. 7. (a) The Federal Power Commission shall not license the construction of any dam, water conduit, reservoir, powerhouse, transmission line, or other project works under the Federal Power Act (41 Stat. 1063), as amended (16 U.S.C. 791a et seq.), on or directly affecting any river which is designated in section 3 of this Act as a component of the national wild and scenic rivers system or which is hereafter designated for inclusion in that system, and no department or agency of the United States shall assist by loan, grant, license, or otherwise in the construction of any water resources project that would have a direct and adverse effect on the values for which such river was established, as determined by the Secretary charged with its administration. Nothing contained in the foregoing sentence, however, shall preclude licensing of, or assistance to, developments below or above a wild, scenic or recreational river area or on any stream tributary thereto which will not invade the area or unreasonably diminish the scenic, recreational, and fish and wildlife values present in the area on the date of approval of this Act. No department or agency of the United States shall recommend authorization of any water resources project that would have a direct and adverse effect on the values for which such river was established, as determined by the Secretary charged with its administration, or request appropriations to begin construction of any such project, whether heretofore or hereafter authorized, without advising the Secretary of the Interior or the Secretary of Agriculture, as the case may be, in writing of its intention so to do at least sixty days in advance, and without specifically reporting to the Congress in writing at the time it makes its recommendation or request in what respect construction of such project would be in conflict with the purposes of this Act and would affect the component and the values to be protected by it under this Act.

(b)   The Federal Power Commission shall not license the construction of any dam, water conduit, reservoir, powerhouse, transmission line, or other project works under the Federal Power Act, as amended, on or directly affecting any river which is listed in section 5, subsection (a), of this Act, and no department or agency of the United States shall assist by loan, grant, license, or otherwise in the construction of any water resources project that would have a direct and adverse effect on the values for which such river might be designated, as determined by the Secretary responsible for its study or approval—

(i)   during the five-year period following enactment of this Act unless, prior to the expiration of said period, the Secretary of the

Interior and, where national forest lands are involved, the Secretary of Agriculture, on the basis of study, conclude that such river should not be included in the national wild and scenic rivers system and publish notice to that effect in the Federal Register, and

(ii) during such additional period thereafter as, in the case of any river which is recommended to the President and the Congress for inclusion in the national wild and scenic rivers system, is necessary for congressional consideration thereof or, in the case of any river recommended to the Secretary of the Interior for inclusion in the national wild and scenic rivers system under section 2(a)(ii) of this Act, is necessary for the Secretary's consideration thereof, which additional period, however, shall not exceed three years in the first case and one year in the second.

Nothing contained in the foregoing sentence, however, shall preclude licensing of, or assistance to, developments below or above a potential wild, scenic or recreational river area or on any stream tributary thereto which will not invade the area or diminish the scenic, recreational, and fish and wildlife values present in the potential wild, scenic or recreational river areas on the date approval of this Act. No department or agency of the United States shall, during the periods hereinbefore specified, recommend authorization of any water resources project on any such river or request appropriations to begin construction of any such project, whether heretofore or hereafter authorized, without advising the Secretary of the Interior and, where national forest lands are involved, the Secretary of Agriculture in writing of its intention so do to at least sixty days in advance of doing so and without specifically reporting to the Congress in writing at the time it makes its recommendation or request in what respect construction of such project would be in conflict with the purposes of this Act and would affect the component and the values to be protected by it under this Act.

(c)   The Federal Power Commission and all other Federal agencies shall, promptly upon enactment of this Act, inform the Secretary of the Interior and, where national forest lands are involved, the Secretary of Agriculture, of any proceeding, studies, or other activities within their jurisdiction which are now in progress and which affect or may affect any of the rivers specified in section 5, subsection (a), of this Act. They shall likewise inform him of any such proceedings, studies, or other activities which are hereafter commenced or resumed before they are commenced or resumed.

(d)   Nothing in this section with respect to the making of a loan or grant shall apply to grants made under the Land and Water Conservation Fund Act of 1965 (78 Stat. 897; 16 U.S.C. 4601-5 et seq.).

SEC. 8. (a) All public lands within the authorized boundaries of any component of the national wild and scenic rivers system which is designated in section 3 of this Act or which is hereafter designated for inclusion in that system are hereby withdrawn from entry, sale, or other disposition under the public land laws of the United States.

(b)   All public lands, which constitute the bed or bank, or are within one-quarter mile of the bank, of any river which is listed in section 5, subsection (a), of this Act are hereby withdrawn from entry, sale, or other disposition under the public land laws of the United States for periods specified in section 7, subsection (b), of this Act.

SEC. 9. (a) Nothing in this Act shall affect the applicability of the United States mining and mineral leasing laws within components of the national wild and scenic rivers system except that—

(i)   all prospecting, mining operations, and other activities on mining claims which, in the case of a component of the system designated in section 3 of this Act, have not heretofore been perfected or which, in the case of a component hereafter designated pursuant to this Act or any other Act of Congress, are not perfected before its inclusion in the system and all mining operations and other activities under a mineral lease, license, or permit issued or renewed after inclusion of a component in the system shall be subject to such regulations as the Secretary of the Interior or, in the case of national forest lands, the Secretary of Agriculture may prescribe to effectuate the purposes of this Act;

(ii) subject to valid existing rights, the perfection of, or issuance of a patent to, any mining claim affecting lands within the system shall confer or convey a right or title only to the mineral deposits and such rights only to the use of the surface and the surface resources as are reasonably required to carrying on prospecting or mining operations and are consistent with such regulations as may be prescribed by the Secretary of the Interior or, in the case of national forest lands, by the Secretary of Agriculture; and

(iii) subject to valid existing rights, the minerals in Federal lands which are part of the system and constitute the bed or bank or are situated within one-quarter mile of the bank of any river designated a wild river under this Act or any subsequent Act are hereby withdrawn from all forms of appropriation under the mining laws and from operation of the mineral leasing laws including, in both cases, amendments thereto.

Regulations issued pursuant to paragraphs (i) and (ii) of this subsection shall, among other things, provide safeguards against pollution of the river involved and unnecessary impairment of the scenery within the component in question.

(b)   The minerals in any Federal lands which constitute the bed or bank or are situated within one-quarter mile of the bank of any river which is listed in section 5, subsection (a) of this Act are hereby withdrawn from all forms of appropriation under the mining laws during the periods specified in section 7, subsection (b) of this Act. Nothing contained in this subsection shall be construed to forbid prospecting or the issuance or leases, licenses, and permits under the mineral leasing laws subject to such conditions as the Secretary of the Interior and, in the case of national forest lands, the Secretary of Agriculture find appropriate to safeguard the area in the event it is subsequently included in the system.

SEC. 10. (a) Each component of the national wild and scenic rivers system shall be administered in such manner as to protect and enhance the values which caused it to be included in said system without, insofar as is consistent therewith, limiting other uses that do not substantially interfere with public use and enjoyment of these values. In such administration primary emphasis shall be given to protecting its esthetic, scenic, historic, archeologic, and scientific features. Management plans for any such component may establish varying degrees of intensity for its protection and development, based on the special attributes of the area.

(b)   Any portion of a component of the national wild and scenic rivers system that is within the national wilderness preservation system, as established by or pursuant to the Act of September 3, 1964 (78 Stat. 890; 16 U.S.C., ch. 23), shall be subject to the provisions of both the Wilderness Act and this Act with respect to preservation of such river and its immediate environment and in case of conflict between the provisions of these Acts the more restrictive provisions shall apply.

(c)   Any component of the national wild and scenic rivers system that is administered by the Secretary of the Interior through the National Park Service shall become a part of the national park system, and any such component that is administered by the Secretary through the Fish and Wildlife Service shall become part of the national wildlife refuge system. The lands involved shall be subject to the provisions of this Act and the Acts under which the national park system or national wildlife system, as the case may be, is administered, and in case of conflict between the provisions of these Acts, the more restrictive provisions shall apply. The Secretary of the Interior, in his administration of any component of the national wild and scenic rivers system, may utilize such general statutory authorities relating to areas of the national park system and such general statutory authorities otherwise available to him for recreation and preservation purposes and for the conservation and management of natural resources as he deems appropriate to carry out the purposes of this Act.

(d) The Secretary of Agriculture, in his administration of any component of the national wild and scenic rivers system area, may utilize the general statutory authorities relating to the national forests in such manner as he deems appropriate to carry out the purposes of this Act.

(e) The Federal agency charged with the administration of any component of the national wild and scenic rivers system may enter into written cooperative agreements with the Governor of a State, the head of any State agency, or the appropriate official of a political subdivision of a State for State or local governmental participation in the administration of the component. The States and their political subdivisions shall be encouraged to cooperate in the planning and administration of components of the system which include or adjoin State- or county-owned lands.

SEC. 11. (a) The Secretary of the Interior shall encourage and assist the States to consider, in formulating and carrying out their comprehensive statewide outdoor recreation plans and proposals for financing assistance for State and local projects submitted pursuant to the Land and Water Conservation Fund Act of 1965 (78 Stat. 897), needs and opportunities for establishing State and local wild, scenic and recreational river areas. He shall also, in accordance with the authority contained in the Act of May 28, 1963 (77 Stat. 49), provide technical assistance and advice to, and cooperate with, States, political subdivisions, and private interests, including nonprofit organizations, with respect to establishing such wild, scenic and recreational river areas.

(b) The Secretaries of Agriculture and of Health, Education, and Welfare shall likewise, in accordance with the authority vested in them, assist, advise, and cooperate with State and local agencies and private interests with respect to establishing such wild, scenic and recreational river areas.

SEC. 12. (a) The Secretary of the Interior, the Secretary of Agriculture, and heads of other Federal agencies shall review administrative and management policies, regulations, contracts, and plans affecting lands under their respective jurisdictions which include, border upon, or are adjacent to the rivers listed in subsection (a) of section 5 of this Act in order to determine what actions should be taken to protect such rivers during the period they are being considered for potential addition to the national wild and scenic rivers system. Particular attention shall be given to scheduled timber harvesting, road construction, and similar activities which might be contrary to the purposes of this Act.

(b) Nothing in this section shall be construed to abrogate any existing rights, privileges, or contracts affecting Federal lands held by any private party without the consent of said party.

(c) The head of any agency administering a component of the national wild and scenic rivers system shall cooperate with the Secretary of the Interior and with the appropriate State water pollution control agencies for the purpose of eliminating the pollution of the waters of the river.

SEC. 13. (a) Nothing in this Act shall affect the jurisdiction or responsibilities of the States with respect to fish and wildlife. Hunting and fishing shall be permitted on lands and waters administered as parts of the system under applicable State and Federal laws and regulations unless, in the case of hunting, those lands or waters are within a national park or monument. The administering Secretary may, however, designate zones where, and establish periods when, no hunting is permitted for reasons of public safety, administration, or public use and enjoyment and shall issue appropriate regulations after consultation with the wildlife agency of the State or States affected.

(b) The jurisdiction of the States and the United States over waters of any stream included in a national wild, scenic or recreational river area shall be determined by established principles of law. Under the provisions of this Act, any taking by the United States of a water right which is vested under either State or Federal law at the time such river is included in the national wild and scenic rivers system shall entitle the owner thereof to just compensation. Nothing in this Act shall constitute an express or implied claim or denial on the part of the Federal Government as to exemption from State water laws.

(c) Designation of any stream or portion thereof as a national wild, scenic or recreational river area shall not be construed as a reservation of the waters of such streams for purposes other than those specified in this Act, or in quantities greater than necessary to accomplish these purposes.

(d) The jurisdiction of the States over waters of any stream included in a national wild, scenic or recreational river area shall be unaffected by this Act to the extent that such jurisdiction may be exercised without impairing the purposes of this Act or its administration.

(e) Nothing contained in this Act shall be construed to alter, amend, repeal, interpret, modify, or be in conflict with any interstate compact made by any States which contain any portion of the national wild and scenic rivers system.

(f) Nothing in this Act shall affect existing rights of any State, including the right of access, with respect to the beds of navigable streams, tributaries, or rivers (or segments thereof) located in a national wild, scenic or recreational river area.

(g) The Secretary of the Interior or the Secretary of Agriculture, as the case may be, may grant easements and rights-of-way upon, over, under, across, or through any component of the national wild and scenic rivers system in accordance with the laws applicable to the national park system and the national forest system, respectively: *Provided*, That any conditions precedent to granting such easements and rights-of-way shall be related to the policy and purpose of this Act.

SEC. 14. The claim and allowance of the value of an easement as a charitable contribution under section 170 of the title 26, United States Code, or as a gift under section 2522 of said title shall constitute an agreement by the donor on behalf of himself, his heirs, and assigns that, if the terms of the instrument creating the easement are violated, the donee or the United States may acquire the servient estate at its fair market value as of the time the easement was donated minus the value of the easement claimed and allowed as a charitable contribution or gift.

SEC. 15. As used in this Act, the term—

(a) "River" means a flowing body of water or estuary or a section, portion, or tributary thereof, including rivers, streams, creeks, runs, kills, rills, and small lakes.

(b) "Free-flowing", as applied to any river or section of a river, means existing or flowing in natural condition without impoundment, diversion, straightening, rip-rapping, or other modification of the waterway. The existence, however, of low dams, diversion works, and other minor structures at the time any river is proposed for inclusion in the national wild and scenic rivers system shall not automatically bar its consideration for such inclusion: *Provided*, That this shall not be construed to authorize, intend, or encourage future construction of such structures within components of the national wild and scenic rivers system.

(c) "Scenic easement" means the right to control the use of land (including the air space above such land) for the purpose of protecting the scenic view from the river, but such control shall not affect without the owner's consent, any regular use exercised prior to the acquisition of the easement.

SEC. 16. There are hereby authorized to be appropriated such sums as may be necessary, but not more than $17,000,000, for the acquisition of lands and interests in land under the provisions of this Act.

Approved October 2, 1968.

16 U.S.C. 1271–1287.

AN ACT TO ESTABLISH A NATIONAL
TRAILS SYSTEM, AND FOR OTHER PURPOSES,

1968 (82 Stat. 919)

Be it enacted by the Senate and House of Representatives of the United States of America in Congress assembled,

SHORT TITLE

SECTION 1. This Act may be cited as the "National Trails System Act".

STATEMENT OF POLICY

SEC. 2. (a) In order to provide for the ever-increasing outdoor recreation needs of an expanding population and in order to promote public access to, travel within, and enjoyment and appreciation of the open-air, outdoor areas of the Nation, trails should be established (i) primarily, near the urban areas of the Nation, and (ii) secondarily, within established scenic areas more remotely located.

(b)   the purpose of this Act is to provide the means for attaining these objectives by instituting a national system of recreation and scenic trails, by designating the Appalachian Trail and the Pacific Crest Trail as the initial components of that system, and by prescribing the methods by which, and standards according to which, additional components may be added to the system.

NATIONAL TRAILS SYSTEM

SEC. 3. The national system of trails shall be composed of—

(a) National recreation trails, established as provided in section 4 of this Act, which will provide a variety of outdoor recreation uses in or reasonably accessible to urban areas.

(b) National scenic trails, established as provided in section 5 of this Act, which will be extended trails so located as to provide for maximum outdoor recreation potential and for the conservation and enjoyment of the nationally significant scenic, historic, natural, or cultural qualities of the areas through which such trails may pass.

(c) Connecting or side trails, established as provided in section 6 of this Act, which will provide additional points of public access to national recreation or national scenic trails or which will provide connections between such trails.

The Secretary of the Interior and the Secretary of Agriculture, in consultation with appropriate governmental agencies and public and private organizations, shall establish a uniform marker for the national trails system.

NATIONAL RECREATION TRAILS

SEC. 4. (a) The Secretary of the Interior, or the Secretary of Agriculture, where lands administered by him are involved, may establish and designate national recreation trails, with the consent of the Federal agency, State, or political subdivision having jurisdiction over the lands involved, upon finding that—

(i) such trails are reasonably accessible to urban areas, and, or

(ii) such trails meet the criteria established in this Act and such supplementary criteria as he may prescribe.

(b) As provided in this section, trails within park, forest, and other recreation areas administered by the Secretary of the Interior or the Secretary of Agriculture or in other federally administered areas may be established and designated as "National Recreation Trails" by the appropriate Secretary and, when no Federal land acquisition is involved—

(i) trails in or reasonably accessible to urban areas may be designated as "National Recreation Trails" by the Secretary of the Interior with the consent of the States, their political subdivisions, or other appropriate administering agencies, and

(ii) trails within park, forest, and other recreation areas owned or administered by States may be designated as "National Recreation Trails" by the Secretary of the Interior with the consent of the State.

NATIONAL SCENIC TRAILS

SEC. 5. (a) National scenic trails shall be authorized and designated only by Act of Congress. There are hereby established as the initial National Scenic Trails:

(1) The Appalachian Trail, a trail of approximately two thousand miles extending generally along the Appalachian Mountains from Mount Katahdin, Maine, to Springer Mountain, Georgia. Insofar as practicable, the right-of-way for such trail shall comprise the trail depicted on the maps identified as "Nationwide System of Trails, Proposed Appalachian Trail, NST-AT-101-May 1967", which shall be on file and available for public inspection in the office of the

Director of the National Park Service. Where practicable, such rights-of-way shall include lands protected for it under agreements in effect as of the date of enactment of this Act, to which Federal agencies and States were parties. The Appalachian Trail shall be administered primarily as a footpath by the Secretary of the Interior, in consultation with the Secretary of Agriculture.

(2)   The Pacific Crest Trail, a trail of approximately two thousand three hundred fifty miles, extending from the Mexican-California border northward generally along the mountain ranges of the west coast States to the Canadian-Washington border near Lake Ross, following the route as generally depicted on the map, identified as "Nationwide System of Trails, Proposed Pacific Crest Trail, NST-PC-103-May 1967" which shall be on file and available for public inspection in the office of the Chief of the Forest Service. The Pacific Crest Trail shall be administered by the Secretary of Agriculture, in consultation with the Secretary of the Interior.

(3)   The Secretary of the Interior shall establish an advisory council for the Appalachian National Scenic Trail, and the Secretary of Agriculture shall establish an advisory council for the Pacific Crest National Scenic Trail. The appropriate Secretary shall consult with such council from time to time with respect to matters relating to the trail, including the selection of rights-of-way, standards, of the erection and maintenance of markers along the trail, and the administration of the trail. The members of each advisory council, which shall not exceed thirty-five in number, shall serve without compensation or expense to the Federal Government for a term of five years and shall be appointed by the appropriate Secretary as follows:

(i)   A member appointed to represent each Federal department or independent agency administering lands through which the trail route passes and each appointee shall be the person designated by the head of such department or agency;

(ii)   A member appointed to represent each State through which the trail passes and such appointments shall be made from recommendations of the Governors of such States;

(iii)   One or more members appointed to represent private organizations, including landowners and land users, that, in the opinion of the Secretary, have an established and recognized interest in the trail and such appointments shall be made from recommendations of the heads of such organizations: *Provided*, That the Appalachian Trail Conference shall be represented by a sufficient number of persons to represent the various sections of the country through which the Appalachian Trail passes; and

(iv)   The Secretary shall designate one member to be chairman and shall fill vacancies in the same manner as the original appointment.

(b)   The Secretary of the Interior, and the Secretary of Agriculture where lands administered by him are involved, shall make such additional studies as are herein or may hereafter be authorized by the Congress for the purpose of determining the feasibility and desirability of designating other trails as national scenic trails. Such studies shall be made in consultation with the heads of other Federal agencies administering lands through which such additional proposed trails would pass and in cooperation with interested interstate, State, and local governmental agencies, public and private organizations, and landowners and land users concerned. When completed, such studies shall be the basis of appropriate proposals for additional national scenic trails which shall be submitted from time to time to the President and to the Congress. Such proposals shall be accompanied by a report, which shall be printed as a House or Senate document, showing among other things—

(1)   the proposed route of such trail (including maps and illustrations);

(2)   the areas adjacent to such trails, to be utilized for scenic, historic, natural, cultural, or developmental, purposes;

(3)   the characteristics which, in the judgment of the appropriate Secretary, make the proposed trail worthy of designation as a national scenic trail;

(4)   the current status of land ownership and current and potential use along the designated route;

(5)   the estimated cost of acquisition of lands or interest in lands, if any;

(6)   the plans for developing and maintaining the trail and the cost thereof;

(7)   the proposed Federal administering agency (which, in the case of a national scenic trail wholly or substantially within a national forest, shall be the Department of Agriculture);

(8)   the extent to which a State or its political subdivisions and public and private organizations might reasonably be expected to participate in acquiring the necessary lands and in the administration thereof; and

(9)   the relative use of the lands involved, including: the number of anticipated visitor days for the entire length of, as well as for segments of, such trail; the number of months which such trail, or segments thereof, will be open for recreation purposes; the economic and social benefits which might accrue from alternate land uses; and the estimated man-years of civilian employment and expenditures expected for the purposes of maintenance, supervision, and regulation of such trail.

(c)   The following routes shall be studied in accordance with the objectives outlined in subsection (b) of this section:

(1)   Continental Divide Trail, a three-thousand-one-hundred-mile trail extending from near the Mexican border in southwestern New Mexico northward generally along the Continental Divide to the Canadian border in Glacier National Park.

(2)   Potomac Heritage Trail, an eight-hundred-and-twenty-five mile trail extending generally from the mouth of the Potomac River to its sources in Pennsylvania and West Virginia, including the one-hundred-and-seventy-mile Chesapeake and Ohio Canal towpath.

(3)   Old Cattle Trails of the Southwest from the vicinity of San Antonio, Texas, approximately eight hundred miles through Oklahoma via Baxter Springs and Chetopa, Kansas, to Fort Scott, Kansas, including the Chisholm Trail, from the vicinity of San Antonio or Cuero, Texas, approximately eight hundred miles north through Oklahoma to Abilene, Kansas.

(4)   Lewis and Clark Trail, from Wood River, Illinois, to the Pacific Ocean in Oregon, following both the outbound and inbound routes of the Lewis and Clark Expedition.

(5)   Natchez Trace, from Nashville, Tennessee, approximately six hundred miles to Natchez, Mississippi.

(6)   North Country Trail, from the Appalachian Trail in Vermont, approximately three thousand two hundred miles through the States of New York, Pennsylvania, Ohio, Michigan, Wisconsin, and Minnesota, to the Lewis and Clark Trail in North Dakota.

(7)   Kittanning Trail from Shirleysburg in Huntingdon County to Kittanning, Armstrong County, Pennsylvania.

(8)   Oregon Trail, from Independence, Missouri, approximately two thousand miles to near Fort Vancouver, Washington.

(9)   Santa Fe Trail, from Independence, Missouri, approximately eight hundred miles to Santa Fe, New Mexico.

(10)   Long Trail, extending two hundred and fifty-five miles from the Massachusetts border northward through Vermont to the Canadian border.

(11)   Mormon Trail, extending from Nauvoo, Illinois, to Salt Lake City, Utah, through the States of Iowa, Nebraska, and Wyoming.

(12)   Gold Rush Trails in Alaska.

(13)   Mormon Battalion Trail, extending two thousand miles from Mount Pisgah, Iowa, through Kansas, Colorado, New Mexico, and Arizona to Los Angeles, California.

(14)   El Camino Real from St. Augustine to San Mateo, Florida, approximately 20 miles along the southern boundary of the St. Johns River from Fort Caroline National Memorial to the St. Augustine National Park Monument.

SEC. 6. Connecting or side trails within park, forest, and other recreation areas administered by the Secretary of the Interior or Secretary of Agriculture may be established, designated, and marked as components of a national recreation or national scenic trail. When no Federal land acquisition is involved, connecting or side trails may be located across lands administered by interstate, State, or local governmental agencies with their consent: *Provided*, That such trails provide additional points of public access to national recreation or scenic trails.

ADMINISTRATION AND DEVELOPMENT

SEC. 7. (a) Pursuant to section 5(a), the appropriate Secretary shall select the rights-of-way for National Scenic Trails and shall publish notice thereof in the Federal Register, together with appropriate maps and descriptions: *Provided*, That in selecting the rights-of-way full consideration shall be given to minimizing the adverse effects upon the adjacent landowner or user and his operation. Development and management of each segment of the National Trails System shall be designated to harmonize with and complement any established multiple-use plans for that specific area in order to insure continued maximum benefits from the land. The location and width of such rights-of-way across Federal lands under the jurisdiction of another Federal agency shall be by agreement between the head of that agency and the appropriate Secretary. In selecting rights-of-way for trail purposes, the Secretary shall obtain the advice and assistance of the States, local governments, private organizations, and landowners and land users concerned.

(b) After publication of notice in the Federal Register, together with appropriate maps and descriptions, the Secretary charged with the administration of a national scenic trail may relocate segments of a national scenic trail right-of-way, with the concurrence of the head of the Federal agency having jurisdiction over the lands involved, upon a determination that: (i) such a relocation is necessary to preserve the purposes for which the trail was established, or (ii) the relocation is necessary to promote a sound land management program in accordance with established multiple-use principles: *Provided*, That a substantial relocation of the rights-of-way for such trail shall be by Act of Congress.

(c) National scenic trails may contain campsites, shelters and related-public-use facilities. Other uses along the trail, which will not substantially interfere with the nature and purposes of the trail, may be permitted by the Secretary charged with the administration of the trail. Reasonable efforts shall be made to provide sufficient access opportunities to such trails and, to the extent practicable, efforts shall be made to avoid activities incompatible with the purposes for which such trails were established. The use of motorized vehicles by the general public along any national scenic trail shall be prohibited and nothing in this Act shall be construed as authorizing the use of motorized vehicles within the natural and historical areas of the national park system, the national wildlife refuge system, the national wilderness preservation system where they are presently prohibited or on other Federal lands where trails are designated as being closed to such use by the appropriate Secretary: *Provided*, That the Secretary charged with the administration of such trail shall establish regulations which shall authorize the use of motorized vehicles when, in his judgment, such vehicles are necessary to meet emergencies or to enable adjacent landowners or land users to have reasonable access to their lands or timber rights: *Provided, further*, That private lands included in the national recreation or scenic trails by cooperative agreement of a landowner shall not preclude such owner from using motorized vehicles on or across such trails or adjacent lands from time to time in accordance with regulations to be established by the appropriate Secretary. The Secretary of the Interior and the Secretary of Agriculture, in consultation with appropriate governmental agencies and public and private organizations, shall establish a uniform marker, including thereon an appropriate and distinctive symbol for each national recreation and scenic trail. Where the trails cross lands administered by Federal agencies such markers shall be erected at appropriate points along the trails and maintained by the Federal agency administering the trail in accordance with standards established by the appropriate Secretary and where the trails cross non-Federal lands, in accordance with written cooperative agreements, the appropriate Secretary shall provide such uniform markers to cooperating agencies and shall require such agencies to erect and maintain them in accordance with the standards established.

(d)   Within the exterior of areas under their administration that are included in the right-of-way selected for a national recreation or scenic trail, the heads of Federal agencies may use lands for trail purposes and may acquire lands or interests in lands by written cooperative agreement, donation, purchase with donated or appropriated funds or exchange: *Provided*, That not more than twenty-five acres in any one mile may be acquired without the consent of the owner.

(e)   Where lands included in a national scenic trail right-of-way are outside of the exterior boundaries of federally administered areas, the Secretary charged with the administration of such trail shall encourage the States or local governments involved (1) to enter into written cooperative agreements with landowners, private organizations, and individuals to provide the necessary trail right-of-way, or (2) to acquire such lands or interests therein to be utilized as segments of the national scenic trail: *Provided*, That if the State or local governments fail to enter into such written cooperative agreements or to acquire such lands or interests therein with two years after notice of the selection of the right-of-way is published, the appropriate Secretary may (i) enter into such agreements with landowners, States, local governments, private organizations, and individuals for the use of lands for trail purposes, or (ii) acquire private lands or interests therein by donation, purchase with donated or appropriated funds or exchange in accordance with the provisions of subsection (g) of this section. The lands involved in such rights-of-way should be acquired in fee, if other methods of public control are not sufficient to assure their use for the purpose for which they are acquired: *Provided*, That if the Secretary charged with the administration of such trail permanently relocates the right-of-way and disposes of all title or interest in the land, the original owner, or his heirs or assigns, shall be offered, by notice given at the former owner's last known address, the right of first refusal at the fair market price.

(f)	The Secretary of the Interior, in the exercise of his exchange authority, may accept title to any non-Federal property within the right-of-way and in exchange therefor he may convey to the grantor of such property any federally owned property under his jurisdiction which is located in the State wherein such property is located and which he classifies as suitable for exchange or other disposal. The values of the properties so exchanged either shall be approximately equal, or if they are not approximately equal the values shall be equalized by the payment of cash to the grantor or to the Secretary as the circumstances require. The Secretary of Agriculture, in the exercise of his exchange authority, may utilize authorities and procedures available to him in connection with exchanges of national forest lands.

(g)	The appropriate Secretary may utilize condemnation proceedings without the consent of the owner to acquire private lands or interests therein pursuant to this section only in cases where, in his judgment, all reasonable efforts to acquire such lands or interests therein by negotiation have failed, and in such cases he shall acquire only such title as, in his judgment, is reasonably necessary to provide passage across such lands: *Provided*, That condemnation proceedings may not be utilized to acquire fee title or lesser interests to more than twenty-five acres in any one mile and when used such authority shall be limited to the most direct or practicable connecting right-of-way: *Provided further*, That condemnation is prohibited with respect to all acquisition of lands or interest in lands for the purposes of the Pacific Crest Trail. Money appropriated for Federal purposes from the land and water conservation fund shall, without prejudice to appropriations from other sources, be available to Federal departments for the acquisition of lands or interests in lands for the purposes of this Act.

(h)   The Secretary charged with the administration of a national recreation or scenic trail shall provide for the development and maintenance of such trails within federally administered areas and shall cooperate with and encourage the States to operate, develop, and maintain portions of such trails which are located outside the boundaries of federally administered areas. When deemed to be in the public interest, such Secretary may enter written cooperative agreements with the States or their political subdivisions, landowners, private organizations, or individuals to operate, develop, and maintain any portion of a national scenic trail either within or outside a federally administered area.

Whenever the Secretary of the Interior makes any conveyance of land under any of the public land laws, he may reserve a right-of-way for trails to the extent he deems necessary to carry out the purposes of this Act.

(i)   The appropriate Secretary, with the concurrences of the heads of any other Federal agencies administering lands through which a national recreation or scenic trail passes, and after consultation with the States, local governments, and organizations concerned, may issue regulations, which may be revised from time to time, governing the use, protection, management, development, and administration of trails of the national trails system. In order to maintain good conduct on and along the trails located within federally administered areas and to provide for the proper government and protection of such trails, the Secretary of the Interior and the Secretary of Agriculture shall prescribe and publish such uniform regulations as they deem necessary and any person who violates such regulations shall be guilty of a misdemeanor, and may be punished by a fine of not more than $500, or by imprisonment not exceeding six months, or by both such fine and imprisonment.

STATE AND METROPOLITAN AREA TRAILS

SEC. 8. (a) The Secretary of the Interior is directed to encourage States to consider, in their comprehensive statewide outdoor recreation plans and proposals for financial assistance for State and local projects submitted pursuant to the Land and Water Conservation Fund

Act, needs and opportunities for establishing park, forest, and other recreation trails on lands owned, or administered by States, and recreation trails on lands in or near urban areas. He is further directed, in accordance with the authority contained in the Act of May 28, 1963 (77 Stat. 49), to encourage States, political subdivisions, and private interests, including nonprofit organizations, to establish such trails.

(b) The Secretary of Housing and Urban Development is directed, in administering the program of comprehensive urban planning and assistance under section 701 of the Housing Act of 1954, to encourage the planning of recreation trails in connection with the recreation and transportation planning for metropolitan and other urban areas. He is further directed, in administering the urban open-space program under title VII of the Housing Act of 1961, to encourage such recreation trails.

(c) The Secretary of Agriculture is directed, in accordance with authority vested in him, to encourage States and local agencies and private interests to establish such trails.

(d) Such trails may be designated and suitably marked as parts of the nationwide system of trails by the States, their political subdivisions, or other appropriate administering agencies with the approval of the Secretary of the Interior.

RIGHTS-OF-WAY AND OTHER PROPERTIES

SEC. 9. (a) The Secretary of the Interior or the Secretary of Agriculture as the case may be, may grant easements and rights-of-way upon, over, under, across, or along any component of the national trails system in accordance with the laws applicable to the national park system and the national forest system, respectively: *Provided*, That any conditions contained in such easements and rights-of-way shall be related to the policy and purposes of this Act.

(b) The Department of Defense, the Department of Transportation, the Interstate Commerce Commission, the Federal Communications Commission, the Federal Power Commission, and other Federal agencies having jurisdiction or control over or information concerning the use, abandonment, or disposition of roadways, utility rights-of-way, or other properties which may be suitable for the purpose of improving or expanding the national trails system shall cooperate with the Secretary of the Interior and the Secretary of Agriculture in order to assure, to the extent practicable, that any such properties having values suitable for trail purposes may be made available for such use.

SEC. 10. There are hereby authorized to be appropriated for the acquisition of lands or interests in lands not more than $5,000,000 for the Appalachian National Scenic Trail and not more than $500,000 for the Pacific Crest National Scenic Trail.

Approved October 2, 1968.

16 U.S.C. 1241 et seq.

ADMINISTRATIVE POLICIES FOR
RECREATION AREAS, 1968

DISCUSSION

The Recreational Area Category of the National Park System includes a wide variety of areas having diverse resource values. In some of these areas, the natural or historical resource significance (seashores, scenic rivers) is the primary basis for their inclusion in the National Park System. In others, manmade features (reservoirs, parkways) are the central features for recreational use. Each characterization, of course, is at best only an approximation. For example, areas in which manmade features constitute the central recreation resources may also have significant natural and historical values. Moreover, areas known widely for their scenic splendor or unique features may likewise include significant manmade features.

In the management of recreation areas, outdoor recreational pursuits "shall be recognized as the dominant or primary resource objective." Managing an area to emphasize its recreational values, however, does not mean that its natural and historical values are to be ignored. On the contrary, management must provide for the conservation of natural or historical features when they are of such value as to enhance the recreational opportunities of the area.

Consistent with the recognition of outdoor recreation as the dominant resource management objective, other resources within recreation areas shall be managed for such additional uses as are compatible with fulfilling the recreation mission of the area. Such additional uses, in appropriate circumstances, may involve the management of forest lands on a sustained basis, mineral exploration and mineral leasing, grazing, etc.

RESOURCES MANAGEMENT PLAN

A resources management plan will be prepared for each recreation area. It will be guided by the approved Master Plan, administrative policy, agreements with other agencies involving management of the area, and area legislation. The purpose of this plan is to spell out the details of resource management for public use and enjoyment.

RESEARCH

Resource management programs will be based on adequate knowledge, obtained through appropriate investigation and research. Research is recognized as a tool of resource management and Service research activities will be mission oriented for achievement of resource management programs. Research by others will be encouraged in recreation areas for the increase and dissemination of knowledge.

WATER POLLUTION ABATEMENT AND CONTROL

The Service will strive to maintain quality of all waters (1) originating within the boundaries of recreation areas through
    (a)   Provision of adequate sewage treatment and disposal for all public-use facilities, including self-contained boat sewage storage units;
    (b)   control of erosion;
    (c)   regulation and control, as necessary, of fuel-burning water craft;
    (d)   avoidance of contamination by lethal substances, such as certain insecticides;
    (e)   regulation of the intensity of use in certain areas and at certain times when determined as being necessary based on water quality monitoring;
and (2) flowing through or bounding on recreation areas
    (a)   by applying the methods listed under 1 (a) to (e) above; and
    (b)   by entering into cooperative agreements or compacts with other agencies and governing bodies for cooperative measures to avoid water pollution.

AIR POLLUTION

The Service will work with others within the regional air shed to reduce air pollution from sources within the area and elsewhere in the air shed. Fumes and smoke from campfires, refuse burning, and other kinds of combustion will be controlled in public-use areas to the extent necessary to maintain clean air.

SOLID-WASTE DISPOSAL

Wastes generated within a recreation area may be disposed of within or outside the area so long as disposal does not (1) pollute water or air, (2) result in the defacement of public recreation areas, or (3) result in destruction or impairment of important natural or cultural resources.

RESOURCE MANIPULATION

Because of the wide variety of resource uses acceptable in recreation areas, active resource manipulation, such as management of habitat for fish and wildlife, agricultural uses, forest management, and maintenance of meadowlands, is required to achieve desired results. In these cases, however, effective management requires the application of sound ecological principles to permit the achievement and maintenance of the desired conditions.

CONSERVATION OF SIGNIFICANT
RESOURCES—SCENIC

Many recreation areas contain significant scenic or scientific resources. Scenic resources may or may not be due to the existence of completely natural conditions (i.e., a pastoral or mixed agricultural-natural scene may possess great scenic appeal based on the combination of features). Management of scenic areas will be governed accordingly. If significant scenic features are the result of natural conditions, they will be managed to retain these conditions, at the same time making appropriate recreational, interpretive, educational, and research use of them. If other than natural conditions are involved, management may perpetuate these cultural conditions involved.

CONSERVATION OF SIGNIFICANT
RESOURCES—CULTURAL

Where significant cultural resources are present in a recreation area and are worthy of preservation of their historical value, they shall be protected and presented for public understanding, appreciation, and enjoyment to the extent compatible with the primary purpose of the area. In such cases, the management and use of the cultural resources will be patterned after the management and use of similar resources in historical areas.

SOIL AND MOISTURE CONSERVATION

Programs will be conducted for the prevention and correction of erosion and soil or vegetation deterioration. A recreation area may participate in the program of a Grasslands Conservation District or Soil Conservation District when the purposes, plans, programs, and operation of the district are consistent with the purposes of the recreation area and the policies for its management and use.

LANDSCAPE MANAGEMENT

Programs of landscape management may be carried out in recreation areas (except as designated otherwise) for purposes of enhancing aesthetics generally which may include, but not be limited to:

1. Encouragement of certain species of plants.
2. Increasing the ability of certain areas to absorb recreational use through vegetative management.
3. Maintaining a certain state of plant succession.
4. Retention of provision of open areas, meadows, vistas, etc., or the planting of open areas to trees or shrubs.
5. Enhancement of roadside vegetation.
6. Management of landscape for educational or interpretive purpose.
7. Rearrangement as necessary of land contours, particularly in areas formerly denuded, mined, or excavated.

EXOTIC PLANT SPECIES

Exotic species of plants may be controlled, eradicated, or introduced into recreation areas as part of various management programs for purposes of public recreational use and enjoyment except that no species, particularly those new to the country or region, may be introduced unless there are reasonable assurances from the U.S. Department of Agriculture and responsible State agencies that the species will not become a pest or disrupt desirable natural plant and animal communities and associations of particular scenic significance.

FIRE

The presence or absence of natural fire within a given habitat is recognized as one of the ecological factors contributing to the perpetuation of plants and animals native to that habitat.

Fires in vegetation resulting from natural causes are recognized as natural phenomena and may be allowed to run their course when such burning can be contained within predetermined fire management units and such burning will contribute to the accomplishment of approved vegetation and/or wildlife management objectives.

Prescribed burning to achieve approved vegetation and/or wildlife management objectives may be employed as a substitute for natural fire.

FIRE CONTROL

Any fire threatening cultural resources or physical facilities of a recreation area or any fire burning within a recreation area and posing a threat to any resources or physical facilities outside that area will be controlled and extinguished.

The Service will cooperate in programs to control or extinguish any fire originating on lands adjacent to a recreation area and posing a threat to natural or cultural resources or physical facilities of that area.

Any fire in a recreation area other than one employed in the management of vegetation and/or wildlife of that area will be controlled and extinguished.

GRAZING AND AGRICULTURAL USES

Agricultural uses, such as demonstration farms, and grazing and raising of domestic livestock, may be permitted when they contribute to maintaining a desired condition or scene; contribute to visitor use and enjoyment in terms of recreation, interpretation, and education; or, do not impair significant scientific, scenic, and cultural resources that contribute to the recreational opportunities of the area.

TIMBER HARVESTING

Harvesting of timber, in accordance with sound forest management principles, is permitted in recreation areas where compatible with fulfilling the area's recreation mission or is not significantly detrimental to it. In recreation areas, resources management plans will be prepared on the basis of identifying a primary zone and a secondary zone for recreation.

In the primary zone, forest management will consist mostly of removing timber and utilizing the logs commercially in the following circumstances:

1. Salvage of hazardous trees in public-use areas or trees with insect or disease infestation that cannot otherwise be controlled which endanger adjacent healthy plants.

2. Salvage of blowdown or fire-killed timber which might precipitate insect outbreaks or create serious fire hazards.

3. Harvesting of timber for vista clearing and similar cultural treatment along roads, parking areas, lakeshores, and developed sites, keeping in mind the scenic, aesthetic, and ecological considerations.

4. Selective harvesting of timber in development and maintenance

of recreational sites such as roads, trails, campgrounds, picnic areas, boat ramps, winter use areas, and visitor centers, as well as maintenance, residential, and administrative sites.

The removal of timber in the foregoing situations is incidental to the more important job of facilitating management of the area for recreational use as the dominant purpose of the area.

In the secondary zone where less intensive recreational activities, such as public recreational hunting and back-country trail use, are prevalent, forest utilization shall consist of:

1. Removal of trees when desirable to enhance the wildlife resource for public recreational hunting; and

2. Harvesting of timber pursuant to the best forest management practices in other designated areas to maintain a dynamic, healthy forest when harvesting will promote or is compatible with, or does not significantly impair, public recreation and conservation of the scenic, scientific, historic, or other values contributing to public enjoyment.

Moreover, the programs mentioned above for the primary zones may also be applicable in the secondary zone in connection with trail construction, vista clearing, etc.

MINERAL EXPLORATION
AND MINERAL LEASING

Mineral prospecting and the extraction of minerals or the removal of soil, sand, gravel, and rock may be authorized by permit issued pursuant to applicable regulations, where such use will promote or is compatible with, and does not significantly impair public recreation and the conservation of scenic, scientific, historic, or other values contributing to public enjoyment. As a specific example, such activities might be carried out in reservoir areas below the conservation pool prior to flooding. Where combined with adequate controls over depths, contours, etc., surface removal of material may also serve for landscape enhancement, fish and wildlife management, or otherwise further the purpose of the recreation area.

QUALITY OF ENVIRONMENT

To achieve the purpose of a recreation area, planning and management should be related to the total environment in which the area is located. Such planning and management recognize the need for transportation arteries; utility and communication corridors; consumptive resource uses; and residential, commercial, and recreation land uses inside and within the vicinity of the area as parts of a systematic plan assuring viability and good health of the area and the surrounding region.

The Service should be alert to peripheral use and development proposals that impinge on the environment of a recreation area. Moreover, it should cooperate with and encourage joint and regional planning among public agencies, organizations, and individuals having responsibility for maintaining the quality and aesthetics of the environment surrounding recreation areas.

INSECTS AND DISEASES

Control operations of native insects and diseases will be limited to (1) outbreaks threatening to eliminate the host from the ecosystem or posing a direct threat to resources outside the area; (2) preservation of scenic values; (3) preservation of rare or scientifically valuable specimens or communities; (4) maintenance of shade trees in developed areas; and (5) preservation of historic scenes. Where non-native insects or diseases have become established or threaten invasion of a recreation area, an appropriate management plan will be developed to control or eradicate them when feasible.

National Park Service, Washington, D.C.: GPO, 1968, 17–22.

ADMINISTRATIVE POLICIES
FOR HISTORIC AREAS, 1968

DISCUSSION

The preservation of historic structures, objects and sites (grounds or terrain) is fundamental to their continued use and benefit. Hence, preservation is a prerequisite to use. In actual practice, the two objectives usually complement rather than conflict with each other. Occasionally, however, use, such as at a historic building, must be regulated and, indeed, limited in order to preserve the resource.

Management of historical areas also encourages appropriate uses of such natural and recreational resources as may be within a historical area when such uses can be accommodated without detriment to the preservation and use of the historical resources.

Much of the success in preserving and interpreting the historic resources within an area depends upon the quality of the environment surrounding the area. Management, therefore, is desirous of cooperating with adjoining owners and agencies responsible for planning and managing properties within the vicinity of a historical area which may influence the environment of the area.

In its management of historic properties, the National Park Service uses the term *historic* in a broad sense to include prehistoric as well as historic periods, or a combination of the two. Likewise, for management purposes, historic resources are defined as follows:

HISTORIC SITES (GROUNDS OR TERRAIN)

A historic site is a distinguishable piece of ground or area upon which occurred some important historic event, or which is importantly associated with historic events or persons, or which was subjected to sustained activity of man—historic, prehistoric, or both. The topography itself may have been shaped by the activity of man. Examples of historic sites (grounds or terrain) are battlefields, historic campgrounds, historic trails, and historic farms.

HISTORIC STRUCTURES

A historic structure is a work of man, either prehistoric or historic, consciously created to serve some form of human activity. A historic structure is usually, by nature or design, immovable. Besides buildings of various kinds, the term includes engineered works such as dams, canals, bridges, stockades, forts and associated earthworks serving a similar purpose, Indian mounds, gardens, historic roads, mill races and ponds.

HISTORIC OBJECTS

Historic objects are material things of functional, esthetic, cultural, or scientific value that are usually, by nature or design, movable. They are ordinarily regarded as museum specimens. If, however, they are large and not readily portable, they are ordinarily treated as structures (e.g., nautical vessels, statues).

HISTORIC RESOURCES

Historic sites (grounds or terrain), structures, and objects are the prime resources within the historic areas of the National Park System. In addition, such historic resources may exist, in varying degree, in those units of the System classified as natural areas and recreational areas. Regardless of the location of such historic resources in the System, the administrative policies in this section apply to their preservation, management, and use.

All of these resources enrich and illuminate the cultural heritage of our Nation. Accordingly, it is appropriate and desirable that these historic resources be made available for public use to the greatest extent practicable. To achieve this objective, however, it is neither necessary nor practicable that each resource, especially structures, be accorded the same detailed research and expensive effort required for an exact full restoration.

As to a historic structure, it is often better to retain genuine old work of several periods, which may have cultural values in themselves, than to restore the whole to its aspect at a single period.

Moreover, some historic structures, occasionally, are included within the National Park System incidental to the establishment of an area for another purpose, e.g., nature preservation or commemoration of a significant event with which a building may not be directly associated. Often these structures are already in an advanced state of deterioration. Their preservation or restoration, in these circumstances, may not be warranted by their significance and the cost of preservation or restoration. In such cases, appropriate examples should be recorded by the Historic American Buildings Survey whenever possible. On the other hand, when sound structures of intrinsic artistic merit in themselves or that are valuable in illustrating the history of the Nation, a State, or locality are included in similar circumstances, their retention and use is encouraged. Appropriate examples may be restored to one of the degrees included below.

Consistent with the congressional policy enunciated in the Historic Sites Act of 1935 and the National Historic Preservation Act of 1966, the historic structures within the areas of the National Park System are classified according to the following definitions of significance:

First Order of Significance. Those structures which, in terms of uniqueness, antiquity, or historical, architectural, or cultural associations as assessed against the criteria of national significance applied in evaluating potential National Historic Landmarks, are significant in the preservation and interpretation of the history of the Nation.

Second Order of Significance. Those structures significant primarily in the presentation and interpretation of the history of a region or State.

Third Order of Significance. Those structures significant primarily in the presentation and interpretation of the history of a community or locality.

TREATMENT OF PROPERTIES

The historic structures within the areas of the National Park System are accorded a variety of treatments depending upon their significance. The types of treatment which may be accorded these historic structures are described as follows:

Preservation. Application of measures designed to sustain the form and extent of a structure essentially as existing when the National Park Service assumes responsibility. Preservation aims at halting further deterioration and providing structural safety but does not contemplate significant rebuilding. Preservation includes:

(a)   Techniques of arresting or slowing the deterioration of a structure;

(b)   Improvement of structural conditions to make a structure safe, habitable, or otherwise useful;

(c)   Normal maintenance and minor repairs that do not change or adversely affect the fabric or historic appearance of a structure.

Restoration. The process of accurately recovering, by the removal of later work and the replacement of missing original work, the form and details of a structure or part of a structure, together with its setting, as it appeared at some period in time. Restoration includes:

(a)   Full restoration—both exterior and interior.

(b)   Partial restoration—exterior, interior, or any partial combination. Partial restoration is adopted when only parts of a structure—external, internal, or in combination—are important in illustrating cultural values at its level of historic significance, or contribute to the values for which the area was designated.

(c) Adaptive restoration—all or a portion (façade, for example) of the exterior restored, with interior adapted to modern functional use. Adaptive restoration is the treatment for structures that are visually important in the historic scene but do not otherwise qualify for exhibition purposes. In such cases, the façade or so much of the exterior as is necessary, should be authentically restored to achieve the management purpose so that it will be properly understood from the public view. The interior, in these circumstances, is usually converted to a modern, functional use. The restored portion of the exterior should be faithfully preserved in its restored form and detail.

Reconstruction. The process of accurately reproducing by new construction the form and details of a vanished structure, or part of it, as it appeared at some period in time. Reconstruction includes:

(a) Full reconstruction.

(b) Partial reconstruction.

LIST OF CLASSIFIED STRUCTURES

Consistent with the legislation involving a particular area and the primary purpose of the area, all historic structures in areas of the National Park System that may be worthy and practicable of preservation should be retained for public use. All such properties should be recorded on the List of Classified Structures. The List of Classified Structures should reflect the order of significance of the properties recorded, as determined by the appropriate Regional Director, with professional assistance from the Office of Archeology and Historic Preservation.

PRESERVATION

Preservation is the treatment to be considered first. And the important consideration is whether a historic site or structure should be retained in essentially the state in which it came under the control of the National Park Service.

Structures on the List of Classified Structures of either the first, second, or third orders of significance may be preserved on one of the following bases: (1) Preservation is the most desirable treatment; (2) the significance and interpretive value of the structure does not justify the cost of restoration; (3) there are not sufficient data to permit accurate restoration; (4) restoration is indicated but must, for cost or other reasons, be postponed; (5) the structure upon acquisition already possesses the integrity and authenticity required; or (6) the work of a higher treatment has been completed, e.g., once restored, a structure is then preserved.

RESTORATION

When needed to interpret properly the historic values of the area, historic structures may be fully and exactly restored when of the first order of significance or a vital element of a site or complex of structures of the first order of significance. Fully restored structures will usually be maintained for exhibition purposes only. Once restored, they should be faithfully preserved in form and detail.

When needed to interpret properly the historic values of the area, historic structures of the second and third orders of significance are eligible for lesser degrees of restoration, such as adaptive restoration or partial restoration. Moreover, such historic structures should serve living, utilitarian uses, consistent with interpretation of the historic values of the area.

RECONSTRUCTION

Reconstruction should be authorized only when the following conditions are met:

(a) All or almost all traces of a structure have disappeared and its recreation is essential for public understanding and appreciation of the historical associations for which the park was established.

(b) Sufficient historical, archeological, and architectural data exist to permit an accurate reproduction.

(c) The structure can be erected on the original site or in a setting appropriate to the significance of the area, as in a pioneer community or living farm, where exact site of structures may not be identifiable through research.

BUILDING AND FIRE CODES

In the preservation of historic structures, every attempt should be made to comply with local building and fire codes and to cooperate with local officials. However, compliance should not be allowed to destroy or impair the integrity of the structure. Where full compliance is not feasible, occupancy of the structure at any one time should be limited to the capacity of hall, stairways, and exits.

FIRE DETECTION AND SUPPRESSION

Where warranted by the significance or value of a historic structure or its contents, adequate fire warning and suppression systems should be installed. A detection system is preferable to a suppression system, which could do more damage than fire. Where a manned fire station exists near the structure, a detection system providing a signal directly to the local fire authorities should be installed. Also, fire personnel should be advised of any peculiarities or dangers inherent in the structure and the features and contents whose value warrants the greatest care in the event of fire.

Where a detection system of this type is not practicable, a suppression system should be installed. Fog or freon systems are preferable. Sprinkler systems should be used only in structures whose fabric and contents are not likely to be irreparably damaged by water. Foam systems should be used only when the structure can be swiftly vacated.

In planning and installing detection or suppression systems, the integrity of the structure and the requirements of its interpretation will be respected.

ACQUISITION OF HISTORIC STRUCTURES

The purchase or acceptance as gifts of historic structures situated outside historical areas is permitted only when there is available an authentic structure that would otherwise have to be reconstructed for interpretive purposes in the area.

A historic structure that is germane to the interpretive theme of an area and that was formerly located on a site that has been included in an area of the System may be acquired and returned to that site.

MOVING HISTORIC STRUCTURES

Historic structures of the first order of significance bear an important relation to their sites and, therefore, should be preserved *in situ*. If, however, such a structure has been previously moved, it may be returned to its original location if desirable for interpretive purposes.

Historic structures of the second and third orders of significance may be moved when there is no feasible alternative for their preservation, when their importance is other than in direct relation to their location, or when desirable for interpretive purposes.

In moving a historic structure, every effort should be made to reestablish its historic orientation, immediate setting and general relationship to its environment. If it is necessary to move a number of buildings, they may be arranged in an ensemble appropriate to their historic character.

ADDITIONS TO HISTORIC STRUCTURES

Modern additions, such as heating and air-conditioning equipment, are permitted in historic structures of the first order of significance to the extent that they can be concealed within the structure or its setting.

Other modern construction may be added to historic structures of the second or third orders of significance when necessary for their continued use. A modern addition should be readily distinguishable from the older work; however, the new work should be harmonious with the old in scale, proportion, materials, and color. Such additions should be as inconspicuous as possible from the public view and should not intrude upon the important historic scene.

DAMAGED OR DESTROYED HISTORIC STRUCTURES

Historic structures that are damaged or destroyed by fire, storm, earthquake, war, or other accident may be restored or reconstructed in accordance with the restoration and reconstruction policies stated herein.

RUINS

By definition, ruins are classified as historic structures and will be accorded treatment as indicated herein for the several classes of historic structures.

The preservation techniques designed to arrest further deterioration of ruins are encompassed by the term "ruins stabilization."

Ruins on unexcavated sites should be stabilized only to the extent necessary to preserve them for further investigation. Sites should not be excavated until adequate provisions have been made for the stabilization of ruins as they are exposed. In cases where ruins are too fragile for direct contact, or where deterioration would result from contact, visitor use should be strictly limited or prohibited. The deliberate creation of ruins out of whole structures that come under the care of the National Park Service is prohibited.

HISTORIC GARDENS

Historic gardens, by definition, are classified as historic structures and will be accorded treatment as indicated for the several classes of historic structures. When restored, gardens should be provided intensive maintenance to preserve their correct historic character and prevent overgrowth.

HISTORIC OBJECTS

Historic objects related directly to the history of the area may be acquired by gift, loan, exchange, or purchase, in conformance with legal authorizations and existing procedures and preserved in the area for study and interpretive purposes. A reasonable number of specimens not related directly to the history of the area, also, may be included in the collection for purposes of comparative study. The original fabric of historic structures should not be mutilated to secure specimens for museum collections. Where some of the original fabric is removed incidental to structural repair, such portions of the building may be kept in museum collections if they reveal significant facts about the structure. All historic objects for which the Service is responsible should be properly documented and recorded in accordance with prescribed procedures, and receive the curatorial care needed for optimum preservation.

Historic objects that are excess to the management needs of the Service may be disposed of in accordance with applicable laws and procedures.

PROTECTION OF ANTIQUITIES

The Federal Antiquities Act of 1906 (34 Stat. 225; 16 U.S.C. 431) makes it a Federal offense for any person to appropriate, excavate, injure, or destroy any historic or prehistoric ruin or monument, or any object of antiquity, situated on lands owned or controlled by the United States. The act, however, does authorize the Secretary of the Interior to issue permits for examination and excavation of ruins to properly qualified institutions subject to prescribed rules and regulations. The collecting of antiquities in historical areas is, therefore, not permitted, except by institutions under permit from the Secretary or by Service employees in the performance of their duties.

HISTORIC SITES

In the preservation and use of historic sites (grounds or terrain), manmade features introduced after the date or period of the event commemorated that are compatible with the historic scene may be retained, except where they hamper visitor understanding of the event commemorated or are incongruous intrusions on the historic scene. Natural accretions of time, such as forest growth, also may be retained unless it hampers visitor understanding of the event commemorated. To the extent necessary for visitor understanding, elements of the historic scene may be restored, including restoration of man-made features, vegetative growth, and historic land uses.

AGRICULTURAL USES

Agricultural uses, including demonstration farms, are encouraged in historical areas where they conform to those in practice in the historical period of the area.

Agricultural uses, including domestic livestock grazing, that do not conform to those in practice in the historic period of the area are permitted where they contribute to the maintenance of a historic scene, are sanctioned by law, or are incidental to visitor use. Where grazing has been permitted and its continuation is not specifically covered by the aforestated conditions, it should be eliminated through orderly and cooperative procedures with the individuals concerned.

Grazing by Service or concessioner pack-and-saddle stock may be permitted, also, where it contributes to the maintenance of a historic scene; otherwise, it should be limited to those locations where dry feeding is clearly impracticable.

SPECIMEN TREES

Woods, forests, and individual specimen trees contributing to the historical integrity of a historical area should be managed intensively to maintain the historic scene. Cutting of trees as "living history," as at Hopewell Village, should be encouraged. Trees that pose a safety hazard should be removed. Diseased, dying, or dead trees that threaten to disturb the ecology of the area may be removed provided the total ecological effects of removal will be more desirable than other management actions could produce. Every effort, however, should be made to extend the lives of specimen trees dating from the historic period of a historical area.

VISITOR FACILITIES

Visitor facilities should be planned, designed and located so as to cause the least possible disturbance to and intrusion on the historic features and the historic scene. Where such facilities already exist as intrusions, their removal should be accomplished as soon as feasible.

QUALITY OF ENVIRONMENT

To achieve the purpose of a historical area, i.e., preservation and appropriate public use, planning and management should be related to the total environment in which the area is located. Such planning and management recognizes the need for transportation arteries, utility and communication corridors, consumptive resource uses, and residential, commercial, and recreation land uses in the environs of the park as parts of a systematic plan assuring viability and good health of the park and the surrounding region.

The Service should be alert to peripheral use and development proposals that impinge on the environment of a historical area. Moreover, it should cooperate with and encourage joint and regional planning among public agencies, organizations, and individuals having responsibility for maintaining the quality and aesthetics of the environment surrounding historical areas.

HISTORIC DISTRICTS

The Service will cooperate in the programs and purposes of historic districts, particularly in urban areas, to encourage the preservation of an environment compatible in character, texture, and productive use with the historic resources of the area.

PLANNING COMMISSIONS, ZONING BOARDS

The Service seeks to cooperate with municipal planning commissions, zoning boards, and other agencies to the extent compatible with the purposes of a historical area in order to promote a viable, orderly environment of which the area is an integral part.

LIVING HISTORICAL INTERPRETATION

Living historical interpretation, costumed guides, authentic craft demonstrations, firing of historic small arms and cannon, use of agricultural and industrial implements and practices, and the like, are encouraged.

COMPATIBLE USE OF HISTORIC STRUCTURES

Use of historic structures for meetings, concerts, and social gatherings helps to deepen the cultural value of the physical structures and gives visitors a more intimate feeling of continuity between the present and the past. Such uses are to be encouraged when compatible with the primary purpose of the area. All traditional and modern communication techniques, including the use of period costumes, living farms, and other demonstrations, may be employed to enhance visitor interest, enjoyment, and understanding of the Nation's history.

Historic structures may be used for appropriate meetings, concerts, dances, social gatherings, celebrations, and the like, consistent with the historical values of the park. Except when such activities have a direct interpretive or traditional role, they must be scheduled to avoid the hours of maximum visitor use.

Historic structures may also be utilized for commercial and residential purposes, when compatible with the primary purpose of the area.

Reasonable fees may be charged for the use of facilities.

National Park Service, Washington, D.C.: GPO, 1968, 19–28.

ADMINISTRATIVE POLICIES
FOR NATURAL AREAS, 1968

DISCUSSION

The *preservation* of natural areas is a fundamental requirement for their continued use and enjoyment as *unimpaired* natural areas. Park management, therefore, looks first to the care and management of the natural resources of a park. The concept of preservation of a total environment, as compared with the protection of an individual feature or species, is a distinguishing feature of national park management.

In earlier times, the establishment of a park and the protection of its forests and wildlife from careless disturbance were sufficient to insure its preservation as a natural area. The impact of man on the natural scene was negligible since the parks were surrounded by vast undeveloped lands, and there were comparatively few visitors. This condition prevails no more, for the parks are fast becoming *islands* of primitive America, increasingly influenced by resource use practices around their borders, and by the impact of increasing millions of visitors.

Passive protection is not enough. Active management of the natural environment, plus a sensitive application of discipline in park planning, use, and development, are requirements for today.

The resource management task thus embraces:

1. Safeguarding forests, wildlife, and natural features against impairment or destruction.

2. The application of ecological management techniques to neutralize the unnatural influences of man, thus permitting the natural environment to be maintained essentially by nature.

3. Master planning for the appropriate allocation of lands to various purposes in a park, and in the character and location of use areas as needed for developments.

PLANT AND ANIMAL RESOURCES

Natural areas shall be managed so as to conserve, perpetuate, and portray as a composite whole the indigenous aquatic and terrestrial fauna and flora and the scenic landscape.

Management will minimize, give direction to, or control those changes in the native environment and scenic landscape resulting from human influences on natural processes of ecological succession. Missing native life forms may be reestablished, where practicable. Native environment complexes will be restored, protected, and maintained, where practicable, at levels determined through historical and ecological research of plant-animal relationships. Non-native species may not be introduced into natural areas. Where they have become established or threaten invasion of a natural area, an appropriate management plan should be developed to control them, where feasible.

Commercial harvesting of timber is not permitted except where the cutting of timber "is required in order to control the attacks of insects or diseases or otherwise to conserve the scenery or the natural or historic objects" in a natural area, such as in the case of severe "blow-downs."

FIRE

The presence or absence of natural fire within a given habitat is recognized as one of the ecological factors contributing to the perpetuation of plants and animals native to that habitat.

Fires in vegetation resulting from natural causes are recognized as natural phenomena and may be allowed to run their course when such burning can be contained within predetermined fire management units and when such burning will contribute to the accomplishment of approved vegetation and/or wildlife management objectives.

Prescribed burning to achieve approved vegetation and/or wildlife management objectives may be employed as a substitute for natural fire.

FIRE CONTROL

Any fire threatening cultural resources or physical facilities of a natural area or any fire burning within a natural area and posing a threat to any resources or physical facilities outside that area will be controlled and extinguished.

The Service will cooperate in programs to control or extinguish any fire originating on lands adjacent to a natural area posing a threat to natural or cultural resources or physical facilities of that area.

Any fire in a natural area other than one employed in the management of vegetation and/or wildlife of that area will be controlled and extinguished.

GRAZING

Domestic livestock grazing competes with native wildlife and impedes the effort in natural areas to achieve an ecological balance.

Accordingly, grazing of domestic livestock in natural areas is permitted only where it is sanctioned by law, incidental to visitor use, or is desirable to preserve and interpret significant historical resources of the area. Where grazing has been permitted and its continuation is not specifically covered by the aforestated conditions, it should be eliminated through orderly and cooperative procedures with the individuals concerned. Support of Service or concessioner pack-and-saddle stock by the use of forage in a natural area shall be limited to locations where dry feeding is clearly impracticable.

AGRICULTURAL USES

Agricultural uses, including domestic livestock raising, may be permitted in natural areas only where they are desirable to perpetuate and interpret significant historical resources, are permitted by law, or are required pursuant to acquisition agreements or similar documents.

SOLID-WASTE DISPOSAL

Refuse generated from operations within a natural area shall be disposed of by approved methods outside the area, where practicable and feasible. Refuse disposal within the area, where necessary, shall be accomplished by incineration or sanitary landfill, or through modification of these methods, as appropriate.

OFF-ROAD USE OF MOTORIZED EQUIPMENT

Public use of motor vehicles shall be confined to designated park roads or other designated overland routes exclusive of foot trails and bridle trails. Public use of portable power equipment, such as generators and powersaws, may be permitted in specifically designated areas.

The off-road use of motorized equipment for official purposes shall be carefully planned and controlled to meet the requirements of area management with due regard for the protection of human life and park resources.

CULTURAL RESOURCES

Where significant cultural resources are present in a natural area and are worthy of preservation for their historical value, they shall be protected and presented for public understanding, appreciation, and enjoyment to the extent compatible with the primary purpose of the area. In such cases, the management and use of the cultural resources will be patterned after the management and use of similar resources in historical areas.

SOIL AND MOISTURE CONSERVATION

Programs will be conducted for the prevention and correction of erosion and soil or vegetation deterioration resulting from unnatural causes.

A natural area may participate in the program of a Grasslands Conservation District or Soil Conservation District when the purposes, plans, programs, and operation of the District are consistent with the purposes of the natural area and the policies for its management and use.

QUALITY OF ENVIRONMENT

To achieve the purpose of a natural area, i.e., preservation and appropriate public use, planning and management should be related to the total environment in which the area is located. Such planning and management recognize the need for transportation arteries; utility and communication corridors; consumptive resource uses; and residential, commercial, and recreation land uses in the environs of the park as parts of a systematic plan assuring viability and good health of the park and the surrounding region.

The Service should be alert to peripheral use and development proposals that impinge on the environment of a natural area. Moreover, it should cooperate with, and encourage joint and regional planning among public agencies, organizations, and individuals having responsibility for maintaining the quality and esthetics of the environment surrounding natural areas.

LANDSCAPE MANAGEMENT

When consistent with and not materially disruptive of the maintenance of natural ecological associations of the area, landscape management will be practiced to erase, ameliorate, or conceal the scars and visual impact of structures, facilities, and construction activities related thereto which impinge on the natural scene.

WATER POLLUTION ABATEMENT AND CONTROL

The Service will strive to maintain quality of all waters (1) originating within the boundaries of natural areas through

(a)   provision of adequate sewage treatment and disposal for all public-use facilities, including self-contained boat sewage storage units;

(b)   control of erosion;

(c)   regulation and control, as necessary, of fuel-burning water craft;

(d)   avoidance of contamination by lethal substances, such as certain insecticides;

(e)   regulation of the intensity of use in certain areas and at certain times when determined as being necessary based on water quality monitoring;

and (2) flowing through or bounding on natural areas

(a)   by applying the methods listed under 1(a) to (e) above;

(b)   consistent with the purposes of the natural area and the policies for its management and use by entering into cooperative agreements or compacts with other agencies and governing bodies for cooperative measures to avoid water pollution.

AIR POLLUTION

The Service will work with others within the regional air shed to reduce air pollution from sources within the area and elsewhere in the air shed. Fumes and smoke from campfires, refuse burning, and other kinds of combustion will be controlled in public-use areas to the extent necessary to maintain clean air.

MINERAL EXPLORATION, MINERAL LEASING, AND MINING

Except where authorized by law or when carried on pursuant to valid existing rights or as part of an interpretive program, mineral prospecting, mining, and the extraction of minerals or the removal of soil, sand, gravel, and rock will not be permitted.

FOREST INSECT AND DISEASE CONTROL

Native forest insects and diseases existing under natural conditions are natural elements of the ecosystem. Accordingly, populations of native insects and the incidence of native diseases will be allowed to function unimpeded, except when control is required (1) to prevent the loss of the host from the ecosystem; (2) to prevent the complete alteration of an environment which is expected to be preserved; (3) to prevent outbreaks of the insect or disease from spreading to forests or trees outside the area; (4) to preserve rare, scientifically valuable, or specimen trees, or unique forest communities; (5) to maintain a suitable overstory, shade, or ornamental trees of Class I and II lands; and (6) to preserve trees significant to the maintenance of historical integrity of Class VI sites.

Where non-native insects or diseases have become established or threaten to invade a natural area, appropriate measures will be taken to control or eradicate them where feasible.

No insect or disease control activities may be undertaken in wilderness areas without the approval of the Director.

Any controls instituted will be those which will be most direct for the target insect or disease and which will have minimal effect upon other components of the ecosystem.

PHYSICAL RESOURCES

To the extent possible, the physical natural resources in a natural area shall be maintained in a natural state for their inherent educational, scientific, and inspirational values, and as a medium for supporting the diversity and the continuation of life processes.

National Park Service, Washington, D.C.: GPO, 1968, 16–21.

A PLAN FOR USE OF FIRE IN
ECOSYSTEM MANAGEMENT

Middle Fork Kings River
February 29, 1968

General
The long range objective of this plan is park-wide in scope and would restore and maintain as nearly as possible the natural environment of the Park in a condition that prevailed when Europeans first visited the vicinity. In furtherance thereof and as a primary management tool, it is proposed to return fire to its environmental role as an important factor in the restoration and maintenance of the natural environment.

Initially, management efforts will be confined to the Middle Fork Kings, and directed toward the mixed conifer and associated communities. Interest in the mixed conifer ecosystem is based on the two following considerations: It is our most extensive and valuable forest community since it contains all sequoia groves. Secondly, the pristine quality, with or without its sequoia component, is a seral type depending upon periodic fire for its continuance.

The Middle Fork Kings has been selected for initial work for reasons listed below:

1. It follows the recommendation of the Secretary's Advisory Board on Wildlife Management, being situated away from roads and areas of high visitation.
2. The fireproof and natural perimeters of the area provide an ideal place to experiment without resorting to elaborate fire lines and breaks. Here natural breaks would be used almost exclusively.
3. Low and middle elevations have long been in need of fire management to restore biotic communities.

Although this is primarily a management plan, there are elements of research in it because of the need to evaluate results.

Prescribed fire will be the primary tool used to achieve the desired ecological conditions. However, use of the term "prescribed fire" here is not intended to imply the strict control of physical and biotic factors implied in its most absolute interpretation. Nor does it imply that we can predict its effect in exact terms. Rather, it implies where, when, and under what range of conditions fire will be ignited and allowed to burn. Further, while prescribed fire implies objectives to be accomplished, these will be described in fairly broad terms.

MANAGEMENT

To initiate this project, several sites in the Middle Fork Kings, ranging from the canyon bottom to the lower ends of hanging valleys, have been tentatively selected for restoration work. All of these are within what is termed the mixed conifer type. In order to assess properly the results of treatment, control areas for each tentative treatment area have been designated.

For prescribed burning to be economically feasible as a management tool in extensive acreages of wildlands, techniques must be used that minimize the need for line preparation and other control work. As a consequence, the best possible combination of natural barriers have been chosen and full use made of fire weather forecasts. Before ignition, a test fire will be ignited in a representative site where immediate control is assured. Only after a satisfactory response will the prescribed burn be ignited.

In almost all cases, it is proposed to employ backing fires burning downslope using area ignition only for certain unique situations. Fuel breaks by burning out from a scratch line will be used where desirable to keep fire out of certain areas or fuel types. Burning may be conducted during fire season but confined to periods of moderate fire load index.

RESEARCH

Two categories of research seem to be indicated. One is allied to the primary objective and consists of determining the effects of fire on the various elements of the environment. The second category deals with research on prescribed burning techniques.

I. Effect on Resources: To provide comparative data, investigation of the first category should be carried out in the control, as well as the treatment area. We feel that the following data should be collected:

    A. Vegetation
        1. Preparation of a detailed type map prior to burning.
        2. Estimate and, on transects, measure herbaceous cover and litter for each type before and after treatment.
        3. Install representative transects in each major cover type before burning and measure plant species composition, age class, and, in the case of brush, condition class.
        4. Set up photo points and take before and after photographs.
    B. Animal Life
        1. Various counts and indices to measure the population response to prescribed burning should be conducted.

Vertebrate and invertebrate populations in both terrestrial and aquatic environments would be observed.

C. Soils

1. Prior to burning, collect soil samples on transects and elsewhere for analysis to determine type, structure, primary chemical constituents, and reaction. Comparative measurements following treatment would be taken.
2. Determine soil moisture percentages before and after burning.

D. Water

1. Pre-treatment water samples would be taken from various sites in the treatment and control areas and in the Middle Fork Kings River proper. Post-treatment sampling immediately after burning and at various periods afterwards would also be necessary.
2. Measurements of spring flows before and after treatment should be made.

II. Investigation of Prescribed Burning Techniques

The techniques of fire behavior research have been well developed. The required instruments will be obtained through the United States Forest Service Western Fire Laboratory at Riverside.

PROJECT COSTS

The determination of costs is very difficult at this time. Our guiding premise, however, is to accomplish the fire management portion of this project with as little manpower, manipulation, or preparation as possible. The use of natural barriers and environmental factors will be used to the utmost to minimize the costs. Research costs will be kept within reason but adequate funds must be provided if meaningful data is to be collected.

Fire management costs can be absorbed for personal services covering planning and supervision, but funds are needed for a small crew necessary for any line building and actual burning. It is planned that some of the research can be accomplished by the Park Research Biologist and other staff members. Listed below are the estimated costs and funds needed to implement the program.

Approximate cost by fiscal year:

1968 Fiscal Year	
Transportation and Per Diem (Personnel)	$ 1,000
Miscellaneous	500
	$ 1,500
1969 Fiscal Year	
Transportation and Per Diem	2,000
Instruments	1,000

Personal Services	2,300
Miscellaneous	1,000
	$ 6,300

SUMMARY

A modest start for prescribed burning is proposed. Preliminary field examinations have been accomplished and tentative sites chosen. Final selection should depend upon on-the-site inspection by those involved in the plan. A Research Biologist who can head up the investigative phase of the plan has been appointed.

FIRE MANAGEMENT

As a result of Service policy expressed in the "Compilation of Administrative Policies for Natural Areas", dated September 13, 1967, Sequoia and Kings Canyon is initiating the following change in its fire control plan:

In the Middle Forks Kings River drainage, most lightning caused fires at or above 8,000 feet elevation will be allowed to run their course. Control measures will be taken only if unacceptable loss of wilderness values, loss of life, or spread to lands outside the Park seem imminent. Fires occurring below 8,000 feet will be suppressed.

It is therefore proposed to return fire to its ecological role in the area for the purposes of environmental restoration with the thought that the resulting burns will provide data complementary to that derived from the prescribed burning program.

J.S. McLaughlin, Superintendent
Sequoia and Kings Canyon National Parks
February 29, 1968

Historic Files, Office of Natural Resources Management, Sequoia National Park.

AN ACT TO ESTABLISH A NATIONAL POLICY FOR THE ENVIRONMENT, TO PROVIDE FOR THE ESTABLISHMENT OF A COUNCIL ON ENVIRONMENTAL QUALITY, AND FOR OTHER PURPOSES,

1969 (83 Stat. 852)

Be it enacted by the Senate and House of Representatives of the United States of America in Congress assembled, That this Act may be cited as the "National Environmental Policy Act of 1969".

PURPOSE

SEC. 2. The purposes of this Act are: To declare a national policy which will encourage productive and enjoyable harmony between man and his environment; to promote efforts which will prevent or eliminate damage to the environment and biosphere and stimulate the health and welfare of man; to enrich the understanding of the ecological systems and natural resources important to the Nation; and to establish a Council on Environmental Quality.

TITLE I

DECLARATION OF NATIONAL ENVIRONMENTAL POLICY

SEC. 101. (a) The Congress, recognizing the profound impact of man's activity on the interrelations of all components of the natural environment, particularly the profound influences of population growth, high-density urbanization, industrial expansion, resource exploitation, and new and expanding technological advances and recognizing further the critical importance of restoring and maintaining environmental quality to the overall welfare and development of man, declares that it is the continuing policy of the Federal Government, in cooperation with State and local governments, and other concerned public and private organizations, to use all practicable means and measures, including financial and technical assistance, in a manner calculated to foster and promote the general welfare, to create and maintain conditions under which man and nature can exist in productive harmony, and fulfill the social, economic, and other requirements of present and future generations of Americans.

(b)   In order to carry out the policy set forth in this Act, it is the continuing responsibility of the Federal Government to use all practicable means, consistent with other essential considerations of national policy, to improve and coordinate Federal plans, functions, programs, and resources to the end that the Nation may—

(1)   fulfill the responsibilities of each generation as trustee of the environment for succeeding generations;

(2)   assure for all Americans safe, healthful, productive, and esthetically and culturally pleasing surroundings;

(3)   attain the widest range of beneficial uses of the environment without degradation, risk to health or safety, or other undesirable and unintended consequences;

(4)   preserve important historic, cultural, and natural aspects of our national heritage, and maintain, wherever possible, an environment which supports diversity and variety of individual choice;

(5)   achieve a balance between population and resource use which will permit high standards of living and a wide sharing of life's amenities; and

(6)   enhance the quality of renewable resources and approach the maximum attainable recycling of depletable resources.

(c)   The Congress recognizes that each person should enjoy a healthful environment and that each person has a responsibility to contribute to the preservation and enhancement of the environment.

SEC. 102. The Congress authorizes and directs that, to the fullest extent possible: (1) the policies, regulations, and public laws of the United States shall be interpreted and administered in accordance with the policies set forth in this Act, and (2) all agencies of the Federal Government shall—

(A)   utilize a systematic, interdisciplinary approach which will insure the integrated use of the natural and social sciences and the environmental design arts in planning and in decisionmaking which may have an impact on man's environment;

(B)   identify and develop methods and procedures, in consultation with the Council on Environmental Quality established by title II of this Act, which will insure that presently unquantified environmental amenities and values may be given appropriate consideration in decisionmaking along with economic and technical considerations;

(C)   include in every recommendation or report on proposals for legislation and other major Federal actions significantly affecting the quality of the human environment, a detailed statement by the responsible official on—

(i)   the environmental impact of the proposed action,

(ii)   any adverse environmental effects which cannot be avoided should the proposal be implemented,

(iii)   alternatives to the proposed action,

(iv)   the relationship between local short-term uses of man's environment and the maintenance and enhancement of long-term productivity, and,

(v)   any irreversible and irretrievable commitments of resources which would be involved in the proposed action should it be implemented.

Prior to making any detailed statement, the responsible Federal official shall consult with and obtain the comments of any Federal agency which has jurisdiction by law or special expertise with respect to any environmental impact involved. Copies of such statement and the comments and views of the appropriate Federal, State, and local agencies, which are authorized to develop and enforce environmental standards, shall be made available to the President, the Council on Environmental Quality and to the public as provided by section 552 of title 5, United States Code, and shall accompany the proposal through the existing agency review process;

(D)   study, develop, and describe appropriate alternatives to recommended courses of action in any proposal which involves unresolved conflicts concerning alternative uses of available resources;

(E)   recognize the worldwide and long-range character of environmental problems and, where consistent with the foreign policy of the United States, lend appropriate support to initiatives, resolutions, and programs designed to maximize international cooperation in anticipating and preventing a decline in the quality of mankind's world environment;

(F)   make available to States, counties, municipalities, institutions, and individuals, advice and information useful in restoring, maintaining, and enhancing the quality of the environment;

(G)   initiate and utilize ecological information in the planning and development of resource-oriented projects; and

(H)   assist the Council on Environmental Quality established by title II of this Act.

SEC. 103. All agencies of the Federal Government shall review their present statutory authority, administrative regulations, and current policies and procedures for the purpose of determining whether there are any deficiencies or inconsistencies therein which prohibit full compliance with the purpose and provisions of this Act and shall propose to the President not later than July 1, 1971, such measures as may be necessary to bring their authority and policies into conformity with the intent, purposes, and procedures set forth in this Act.

SEC. 104. Nothing in Section 102 or 103 shall in any way affect the specific statutory obligations of any Federal agency (1) to comply with criteria or standards of environmental quality, (2) to coordinate or consult with any other Federal or State agency, or (3) to act, or refrain from acting contingent upon the recommendations or certification of any other Federal or State agency.

SEC. 105. The policies and goals set forth in this Act are supplementary to those set forth in existing authorizations of Federal agencies.

TITLE II

COUNCIL ON ENVIRONMENTAL QUALITY

SEC. 201. The President shall transmit to the Congress annually beginning July 1, 1970, an Environmental Quality Report (hereinafter referred to as the "report") which shall set forth (1) the status and condition of the major natural, manmade, or altered environmental classes of the Nation, including, but not limited to, the air, the aquatic, including marine, estuarine, and fresh water, and the terrestrial environment, including, but not limited to, the forest, dryland, wetland, range, urban, suburban, and rural environment; (2) current and foreseeable trends in the quality, management and utilization of such environments and the effects of those trends on the social, economic, and other requirements of the Nation; (3) the adequacy of available natural resources for fulfilling human and economic requirements of the Nation in the light of expected population pressures; (4) a review of the programs and activities (including regulatory activities) of the Federal Government, the State and local governments, and nongovernmental entities or individuals, with particular reference to their effect on the environment and on the conservation, development and utilization of natural resources; and (5) a program for remedying the deficiencies of existing programs and activities, together with recommendations for legislation.

SEC. 202. There is created in the Executive Office of the President a Council on Environmental Quality (hereinafter referred to as the "Council"). The Council shall be composed of three members who shall be appointed by the President to serve at his pleasure, by and with the advice and consent of the Senate. The President shall designate one of the members of the Council to serve as Chairman. Each member shall be a person who, as a result of his training, experience, and attainments, is exceptionally well qualified to analyze and interpret environmental trends and information of all kinds; to appraise programs and activities of the Federal Government in the light of the policy set forth in title I of this Act; to be conscious of and responsive to the scientific, economic, social, esthetic, and cultural needs and interests of the Nation; and to formulate and recommend national policies to promote the improvement of the quality of the environment.

SEC. 203. The Council may employ such officers and employees as may be necessary to carry out its functions under this Act. In addition, the Council may employ and fix the compensation of such experts and consultants as may be necessary for the carrying out of its functions under this Act, in accordance with section 3109 of title 5, United States Code (but without regard to the last sentence thereof).

SEC. 204. It shall be the duty and function of the Council—

(1) to assist and advise the President in the preparation of the Environmental Quality Report required by section 201;

(2) to gather timely and authoritative information concerning the conditions and trends in the quality of the environment both current and prospective, to analyze and interpret such information for the purpose of determining whether such conditions and trends are interfering, or are likely to interfere, with the achievement of the policy set forth in title I of this Act, and to compile and submit to the President studies relating to such conditions and trends;

(3) to review and appraise the various programs and activities of the Federal Government in the light of the policy set forth in title I of this Act for the purpose of determining the extent to which such programs and activities are contributing to the achievement of such policy, and to make recommendations to the President with respect thereto;

(4) to develop and recommend to the President national policies to foster and promote the improvement of environmental quality to meet the conservation, social, economic, health, and other requirements and goals of the Nation;

(5) to conduct investigations, studies, surveys, research, and analyses relating to ecological systems and environmental quality;

(6) to document and define changes in the natural environment, including the plant and animal systems, and to accumulate necessary data and other information for a continuing analysis of these changes or trends and an interpretation of their underlying causes;

(7) to report at least once each year to the President on the state and condition of the environment; and

(8) to make and furnish such studies, reports thereon, and

recommendations with respect to matters of policy and legislation as the President may request.

SEC. 205. In exercising its powers, functions, and duties under this Act, the Council shall—

(1) consult with the Citizens' Advisory Committee on Environmental Quality established by Executive Order numbered 11472, dated May 29, 1969, and with such representatives of science, industry, agriculture, labor, conservation organizations, State and local governments and other groups, as it deems advisable; and

(2) utilize, to the fullest extent possible, the services, facilities, and information (including statistical information) of public and private agencies and organizations, and individuals, in order that duplication of effort and expense may be avoided, thus assuring that the Council's activities will not unnecessarily overlap or conflict with similar activities authorized by law and performed by established agencies.

SEC. 206. Members of the Council shall serve full time and the Chairman of the Council shall be compensated at the rate provided for Level II of the Executive Schedule Pay Rates (5 U.S.C. 5313). The other members of the Council shall be compensated at the rate provided for Level IV of the Executive Schedule Pay Rates (5 U.S.C. 5315).

SEC. 207. There are authorized to be appropriated to carry out the provisions of this Act not to exceed $300,000 for fiscal year 1970, $700,000 for fiscal year 1971, and $1,000,000 for each fiscal year thereafter.

Approved January 1, 1970.

42 U.S.C. 4321 et seq.

7

Transformation and Expansion

1970–1980

The changes set in motion by the Leopold report and the spate of legislative action that followed in the 1960s continued to expand and mature the national park system in the 1970s. The decade saw vast additions to the park system in part arising from a series of omnibus bills disparagingly called "park barrel legislation" by opponents. Also, a series of major environmental protection laws further defined the web of regulatory controls on federal activities including those in the national park system. Finally, specific legislation and policy evaluations pertaining to the national parks further defined the changing and threatened system.

The profile of park system functions further broadened in 1972 with the addition of the first two specifically urban recreation areas. Gateway National Recreation Area in the New York metropolitan zone brought the National Park Service into the nation's largest city with all the distinctive requirements and challenges of that environment. On the same day Congress authorized Golden Gate National Recreation Area in and around San Francisco. Management and protection of such areas would become radically different from that in the traditional parks of the old system.

Returning to the more classic, natural parks in 1978, Congress grappled with the problems of parks as natural islands encroached upon by adjacent activities. Redwood National Park, approved in 1968, protected the largest remaining stands of coastal redwoods primarily in the lower watershed of Redwood Creek. Logging outside the boundary, however, resulted in siltation deemed harmful to the park trees. In a controversial action, legislators expanded the park boundaries to encompass the remaining watershed and protect the endemic ecosystem. In the process they agreed to compensate those unemployed by the loss of logging jobs.

During the same year President Jimmy Carter addressed the preservation of lands in Alaska, the last great frontier of America and an area under increasing threat of consumptive use. As miners, loggers, and others maneuvered for the opportunities to exploit vast new areas, Carter hearkened back to Theodore Roosevelt in using the Antiquities Act to unilaterally proclaim seventeen huge national monuments (summarized in the Appendix). This was by no means the first time the act had been used as an emergency device for protecting wild lands, but it was certainly its most extensive use. Two years later in the waning days of his presidency, Carter signed the Alaska National Interest Lands Conservation Act of 1980 (see the Appendix). This sweeping legislation more than doubled the size of the national park system, anchoring the earlier monuments and many other lands as new parks, national forests, wildlife refuges, and wild and scenic rivers.

The vast expansion of the park system transpired amidst ever tighter federal environmental controls. In 1971 President Richard Nixon strengthened the process of historic preservation with Executive Order No. 11593 stipulating steps to be taken in expanding and maintaining the National Register of Historic Places and in protecting threatened historic resources. Most of this document was subsequently written into the National Historic Preservation Act Amendments of 1980. And 1972 saw passage of the Federal Water Pollution Control Act and the Coastal Zone Management Act (see the Appendix). The former established provisions to ensure clean water while the latter emphasized coordinated action between states and the federal government aimed at restoring and protecting coastal areas including the Great Lakes shoreline. A year later passage of the Endangered Species Act of 1973 (see the Appendix) required federal agencies to modify their activities to ensure protection of endangered species of plants and animals. This provision would materially affect construction in the parks which often served as islands of endangered species amidst zones of agricultural or urban land use. Finally in 1979 Congress corrected more than seven decades of inadequate protection for archaeological sites and objects with passage of the Archaeological Resources Protection Act. This law superseded the Antiquities Act as the prime legislative protection for federal archaeological resources by more fully defining them and establishing appropriate penalty provisions for their destruction or theft.

Attention to the status and management of the national park system also surfaced from a variety of sources during the decade. Owing to the extensive diversification of responsibilities delivered to the Park Service in the previous decade, some attention was felt necessary to secure equal protection for all units. In 1970, just two years after release of the separate handbooks for administration of natural, historic, and recreation areas, Congress passed the General Authorities Act. In it legislators specified that all units administered by the National Park Service were part of the same "system" and that they were to be managed according to the provisions of the Organic Act of 1916 and other related laws.

During the 1960s conservation organizations had vastly grown in both numbers and influence. Some had challenged the federal government's management of parks and had become integrally involved in efforts to authorize or enlarge parks. In 1972 one group, the Conservation Foundation, released its report, *Preservation of National Park Values*, promoting its blueprint of how parks should be managed to best preserve the resources they were charged to protect. In addressing issues from visitor and concessions policies to carrying capacity and the need for research, the conservation group publicized the agenda of one powerful clique of park users.

Finally, as the decade ended and the administration of Jimmy Carter neared its expected end, the National Park Service answered a congressional request to report on the state of the parks. The report, released in 1980, resulted from an extensive survey of park administrators throughout the system. What it concluded was disheartening. The parks suffered from a wide variety of threats ranging from internal ones like overcrowding, overbuilding, and insufficient personnel to foreboding external ones such as air and water pollution, accelerated development on park boundaries, and destruction of migratory park species on outside lands. The lists of specific, identified threats ran to dozens for some parks. Clearly the next decade would require extensive action to stem the problems and protect the nation's "crown jewels."

AN ACT TO IMPROVE THE ADMINISTRATION OF THE NATIONAL PARK SYSTEM BY THE SECRETARY OF THE INTERIOR, AND TO CLARIFY THE AUTHORITIES APPLICABLE TO THE SYSTEM, AND FOR OTHER PURPOSES,

1970 (84 Stat. 825)

Be it enacted by the Senate and House of Representatives of the United States of America in Congress assembled, That Congress declares that the national park system, which began with establishment of Yellowstone National Park in 1872, has since grown to include superlative natural, historic, and recreation areas in every major region of the United States, its territories and island possessions; that these areas, though distinct in character, are united through their interrelated purposes and resources into one national park system as cumulative expressions of a single national heritage; that, individually and collectively, these areas derive increased national dignity and recognition of their superb environmental quality through their inclusion jointly with each other in one national park system preserved and managed for the benefit and inspiration of all the people of the United States; and that it is the purpose of this Act to include all such areas in the System and to clarify the authorities applicable to the system.

SEC. 2. (a) Section 1 of the Act of August 8, 1953 (67 Stat. 496; 16 U.S.C. 1b), is amended by deleting "and miscellaneous areas administered in connection therewith" and "and miscellaneous areas" wherever they appear.

(b) Section 2 of the Act of August 8, 1953 (67 Stat. 496; 16 U.S.C. 1c), is amended to read as follows:

"Sec. 2. (a) The 'national park system' shall include any area of land and water now or hereafter administered by the Secretary of the Interior through the National Park Service for park, monument, historic, parkway, recreational, or other purposes.

"(b) Each area within the national park system shall be administered in accordance with the provisions of any statute made specifically applicable to that area. In addition, the provisions of this Act, and the various authorities relating to the administration and protection of areas under the administration of the Secretary of the Interior through the National Park Service, including but not limited to the Act of August 25, 1916 (39 Stat. 535), as amended (16 U.S.C. 1, 2-4), the Act of March 4, 1911 (36 Stat. 1253), as amended (16 U.S.C.

5) relating to rights-of-way, the Act of June 5, 1920 (41 Stat. 917), as amended (16 U.S.C. 6), relating to donation of land and money, sections 1, 4, 5, and 6 of the Act of April 9, 1924 (43 Stat. 90), as amended (16 U.S.C. 8 and 8a-8c), relating to roads and trails, the Act of March 4, 1931 (46 Stat. 1570; 16 U.S.C. 8d) relating to approach roads to national monuments, the Act of June 3, 1948 (62 Stat. 334), as amended (16 U.S.C. 8e-8f), relating to conveyance of roads to States, the Act of August 31, 1954 (68 Stat. 1037), as amended (16 U.S.C. 452a), relating to acquisitions of inholdings, section 1 of the Act of July 3, 1926 (44 Stat. 900), as amended (16 U.S.C. 12), relating to aid to visitors in emergencies, the Act of March 3, 1905 (33 Stat. 873; 16 U.S.C. 10), relating to arrests, sections 3, 4, 5, and 6 of the Act of May 26, 1930 (46 Stat. 381), as amended (16 U.S.C. 17b, 17c, 17d, and 17e), relating to services or other accommodations for the public, emergency supplies and services to concessioners, acceptability of travelers checks, care and removal of indigents, the Act of October 9, 1965 (79 Stat. 696; 16 U.S.C. 20-20g), relating to concessions, the Land and Water Conservation Fund Act of 1965, as amended, and the Act of July 15, 1968 (82 Stat. 355), shall, to the extent such provisions are not in conflict with any such specific provision, be applicable to all areas within the national park system and any reference in such Act to national parks, monuments, recreation areas, historic monuments, or parkways shall hereinafter not be construed as limiting such Acts to those areas."

Sec. 3. In order to facilitate the administration of the national park system, the Secretary of the Interior is authorized, under such terms and conditions as he may deem advisable, to carry out the following activities:

(a) provide transportation of employees located at isolated areas of the national park system and to members of their families, where (1) such areas are not adequately served by commercial transportation, and (2) such transportation is incidental to official transportation services;

(b) provide recreation facilities, equipment, and services for use by employees and their families located at isolated areas of the national park system;

(c) appoint and establish such advisory committees in regard to the functions of the National Park Service as he may deem advisable, members of which shall receive no compensation for their services as such but who shall be allowed necessary expenses as authorized by section 5 of the Administrative Expenses Act of 1946 (U.S.C. 5703);

(d) purchase field and special purpose equipment required by employees for the performance of assigned functions which shall be regarded and listed as park equipment;

(e)   enter into contracts which provide for the sale or lease to persons, States, or their political subdivisions, of services, resources, or water available within an area of the national park system, if such person, State, or its political subdivision—

(1)   provides public accommodations or services within the immediate vicinity of an area of the national park system to persons visiting the area; and

(2)   has demonstrated to the Secretary that there are no reasonable alternatives by which to acquire or perform the necessary services, resources, or water;

(f)   acquire, and have installed, air-conditioning units for any Government-owned passenger motor vehicles used by the National Park Service, where assigned duties necessitate long periods in automobiles or in regions of the United States where high temperatures and humidity are common and prolonged;

(g)   sell at fair market value without regard to the requirements of the Federal Property and Administrative Services Act of 1949, as amended, products and services produced in the conduct of living exhibits and interpretive demonstrations in areas of the national park system, to enter into contracts including cooperative arrangements with respect to such living exhibits and interpretive demonstrations and park programs, and to credit the proceeds therefrom to the appropriation bearing the cost of such exhibits and demonstrations.

SEC. 4. The Act of March 17, 1948 (62 Stat. 81), is amended by deleting from section 1 thereof the words "over which the United States has, or hereafter acquires, exclusive or concurrent criminal jurisdiction," and changing section 3 to read as follows:

"SEC. 3. For the purposes of this Act, the environs of the District of Columbia are hereby defined as embracing Arlington, Fairfax, Loudoun, Prince William, and Stafford Counties and the city of Alexandria in Virginia, and Prince Georges, Charles, Anne Arundel, and Montgomery Counties in Maryland."

Approved August 18, 1970.

16 U.S.C. 1a et seq.

EXECUTIVE ORDER NO. 11593
PROTECTION AND ENHANCEMENT
OF THE CULTURAL ENVIRONMENT
1971

By virtue of the authority vested in me as President of the United States and in furtherance of the purposes and policies of the National Environmental Policy Act of 1969 (83 Stat. 852, 42 U.S.C. 4321 et seq.), the National Historic Preservation Act of 1966 (80 Stat. 915, 16 U.S.C. 470 et seq.), the Historic Sites Act of 1935 (49 Stat. 666, 16 U.S.C. 461 et seq.), and the Antiquities Act of 1906 (34 Stat. 225, 16 U.S.C. 431 et seq.), it is ordered as follows:

Section 1. Policy

The Federal Government shall provide leadership in preserving, restoring, and maintaining the historic and cultural environment of the Nation. Agencies of the executive branch of the Government (hereinafter referred to as "Federal agencies") shall (1) administer the cultural properties under their control in a spirit of stewardship and trusteeship for future generations, (2) initiate measures necessary to direct their policies, plans and programs in such a way that federally owned sites, structures, and objects of historical, architectural or archaeological significance are preserved, restored and maintained for the inspiration and benefit of the people, and (3), in consultation with the Advisory Council on Historic Preservation (16 U.S.C. 470i), institute procedures to assure that Federal plans and programs contribute to the preservation and enhancement of non-federally owned sites, structures and objects of historical, architectural or archaeological significance.

Section 2. Responsibilities of Federal Agencies

Consonant with the provisions of the acts cited in the first paragraph of this order, the heads of Federal agencies shall:

(a)   no later than July 1, 1973, with the advice of the Secretary of the Interior, and in cooperation with the liaison officer for historic preservation of the State or territory involved, locate, inventory, and nominate to the Secretary of the Interior all sites, buildings, districts, and objects under their jurisdiction or control that appear to qualify for listing on the National Register of Historic Places.

(b)   exercise caution during the interim period until inventories and evaluations required by subsection (a) are completed to assure that any federally owned property that might qualify for nomination is not inadvertently transferred, sold, demolished or substantially altered. The agency head shall refer any questionable actions to the Secretary of the Interior for an opinion respecting the property's eligibility for inclusion on the National Register of Historic Places. The Secretary shall consult with the liaison officer for historic preservation for the State or territory involved in arriving at his opinion. Where, after a reasonable period in which to review and evaluate the property, the Secretary determines that the property is likely to meet the criteria prescribed for listing on the National Register of Historic Places, the Federal agency head shall reconsider the proposal in light of national environmental and preservation policy. Where, after such reconsideration, the Federal agency head proposes to transfer, sell, demolish or substantially alter the property he shall not act with respect to the property until the Advisory Council on Historic Preservation shall have been provided an opportunity to comment on the proposal.

(c) initiate measures to assure that where as a result of Federal action or assistance a property listed on the National Register of Historic Places is to be substantially altered or demolished, timely steps be taken to make or have made records, including measured drawings, photographs and maps, of the property, and that copy of such records then be deposited in the Library of Congress as part of the Historic American Buildings Survey or Historic American Engineering Record for future use and reference. Agencies may call on the Department of the Interior for advice and technical assistance in the completion of the above records.

(d) initiate measures and procedures to provide for the maintenance, through preservation, rehabilitation, or restoration, of federally owned and registered sites at professional standards prescribed by the Secretary of the Interior.

(e) submit procedures required pursuant to subsection (d) to the Secretary of the Interior and to the Advisory Council on Historic Preservation no later than January 1, 1972, and annually thereafter, for review and comment.

(f) cooperate with purchasers and transferees of a property listed on the National Register of Historic Places in the development of viable plans to use such property in a manner compatible with preservation objectives and which does not result in an unreasonable economic burden to public or private interests.

Section 3. Responsibilities of the Secretary of the Interior
The Secretary of the Interior shall:

(a) encourage State and local historic preservation officials to evaluate and survey federally owned historic properties and, where appropriate, to nominate such properties for listing on the National Register of Historic Places.

(b) develop criteria and procedures to be applied by Federal agencies in the reviews and nominations required by section 2(a). Such criteria and procedures shall be developed in consultation with the affected agencies.

(c) expedite action upon nomination to the National Register of Historic Places concerning federally owned properties proposed for sale, transfer, demolition or substantial alteration.

(d) encourage State and Territorial liaison officers for historic preservation to furnish information upon request to Federal agencies regarding their properties which have been evaluated with respect to historic, architectural or archaeological significance and which as a result of such evaluations have not been found suitable for listing on the National Register of Historic Places.

(e) develop and make available to Federal agencies and State and local governments information concerning professional methods and techniques for preserving, improving, restoring and maintaining historic properties.

(f) advise Federal agencies in the evaluations, identification, preservation, improvement, restoration and maintenance of historic properties.

(g) review and evaluate the plans of transferees of surplus Federal properties transferred for historic monument purposes to assure that the historic character of such properties is preserved in rehabilitation, restoration, improvement, maintenance and repair of such properties.

(h) review and comment upon Federal agency procedures submitted pursuant to section 2(e) of this order.

Dated: May 13, 1971
RICHARD M. NIXON

CFR Title 3, 1971–1975 Compilation, 559–562.

AN ACT TO ESTABLISH THE GATEWAY NATIONAL RECREATION AREA IN THE STATES OF NEW YORK AND NEW JERSEY, AND FOR OTHER PURPOSES,

1972 (86 Stat. 1308)

Be it enacted by the Senate and House of Representatives of the United States of America in Congress assembled, That in order to preserve and protect for the use and enjoyment of present and future generations an area possessing outstanding natural and recreational features, the Gateway National Recreation Area (hereinafter referred to as "recreation area") is hereby established.

(a)   The recreation area shall comprise the following lands, waters, marshes, and submerged lands in the New York Harbor area generally depicted on the map entitled "Boundary Map, Gateway National Recreation Area," numbered 951-40017 sheets 1 through 3 and dated May, 1972:

(1)   Jamaica Bay Unit—including all islands, marshes, hassocks, submerged lands, and waters in Jamaica Bay, Floyd Bennett Field, the lands generally located between highway route 27A and Jamaica Bay, and the area of Jamaica Bay up to the shoreline of John F. Kennedy International Airport;

(2)   Breezy Point Unit—the entire area between the eastern boundary of Jacob Riis Park and the westernmost point of the peninsula;

(3)   Sandy Hook Unit—the entire area between Highway 36 Bridge and the northernmost point of the peninsula;

(4)   Staten Island Unit—including Great Kills Park, Miller Field (except for approximately 26 acres which are to be made available for public school purposes), Fort Wadsworth, and the waterfront lands located between the streets designated as Cedar Grove Avenue, Seaside Boulevard, and Drury Avenue and the bay from Great Kills to Fort Wadsworth;

(5)   Hoffman and Swinburne Islands; and

(6)   All submerged lands, islands, and waters within one-fourth of a mile of the mean low water line of any waterfront area included above.

(b)   The map referred to in this section shall be on file and available for public inspection in the offices of the National Park Service, Department of the Interior, Washington, District of Columbia.

After advising the Committees on Interior and Insular Affairs of the United States House of Representatives and the United States Senate in writing, the Secretary of the Interior (hereinafter referred to as the "Secretary") is authorized to make minor revisions of the boundaries of the recreation area when necessary by publication of a revised drawing or other boundary description in the Federal Register.

SEC. 2. (a) Within the boundaries of the recreation area, the Secretary may acquire lands and waters or interest therein by donation, purchase or exchange, except that lands owned by the States of New York or New Jersey or any political subdivisions thereof may be acquired only by donation.

(b) With the concurrence of the agency having custody thereof, any Federal property within the boundaries of the recreation area may be transferred, without consideration, to the administrative jurisdiction of the Secretary for administration as a part of the recreation area.

(c) Within the Breezy Point Unit, (1) the Secretary shall acquire an adequate interest in the area depicted on the map referred to in section 1 of this Act to assure the public use of and access to the entire beach. The Secretary may enter into an agreement with any property owner or owners to assure the continued maintenance and use of all remaining lands in private ownership as a residential community composed of single-family dwellings. Any such agreement shall be irrevocable, unless terminated by mutual agreement, and shall specify, among other things:

(A) that the Secretary may designate, establish and maintain a buffer zone on Federal lands separating the public use area and the private community;

(B) that all construction commencing within the community, including the conversion of dwellings from seasonal to year-round residences, shall comply with standards to be established by the Secretary;

(C) that additional commercial establishments shall be permitted only with the express prior approval of the Secretary or his designee.

(2) If a valid, enforceable agreement is executed pursuant to paragraph (1) of this subsection, the authority of the Secretary to acquire any interest in the property subject to the agreement, except for the beach property, shall be suspended.

(3) The Secretary is authorized to accept by donation from the city of New York any right, title, or interest which it holds in the parking lot at Rockaway which is part of the Marine Bridge project at Riis Park. Nothing herein shall be deemed to authorize the United States to extinguish any present or future encumbrance or to autho-

rize the State of New York or any political subdivision or agency thereof to further encumber any interest in the property so conveyed.

(d) Within the Jamaica Bay Unit, (1) the Secretary may accept title to lands donated by the city of New York subject to a retained right to continue existing uses for a specifically limited period of time if such uses conform to plans agreed to by the Secretary, and (2) the Secretary may accept title to the area known as Broad Channel Community only if, within five years after the date of enactment of this Act, all improvements have been removed from the area and a clear title to the area is tendered to the United States.

SEC. 3. (a) The Secretary shall administer the recreation area in accordance with the provisions of the Act of August 25, 1916 (39 Stat. 535; 16 U.S.C. 1, 2-4), as amended and supplemented. In the administration of the recreation area the Secretary may utilize such statutory authority available to him for the conservation and management of wildlife and natural resources as he deems appropriate to carry out the purposes of this Act: *Provided,* That the Secretary shall administer and protect the islands and waters within the Jamaica Bay Unit with the primary aim of conserving the natural resources, fish, and wildlife located therein and shall permit no development or use of this area which is incompatible with this purpose.

(b) The Secretary shall designate the principal visitor center constructed within the recreation area as the "William Fitts Ryan Visitor Center" in commemoration of the leadership and contributions which Representative William Fitts Ryan made with respect to the creation and establishment of this public recreation area.

(c) The Secretary is authorized to enter into cooperative agreements with the States of New York and New Jersey, or any political subdivision thereof, for the rendering, on a reimbursable basis, of rescue, firefighting, and law enforcement services and cooperative assistance by nearby law enforcement and fire preventive agencies.

(d) The authority of the Secretary of the Army to undertake or contribute to water resource developments, including shore erosion control, beach protection, and navigation improvements (including the deepening of the shipping channel from the Atlantic Ocean to the New York harbor) on land and/or waters within the recreation area shall be exercised in accordance with plans which are mutually acceptable to the Secretary of the Interior and the Secretary of the Army and which are consistent with both the purposes of this Act and the purpose of existing statutes dealing with water and related land resource development.

(e) The authority of the Secretary of Transportation to maintain and operate existing airway facilities and to install necessary new facilities within the recreation area shall be exercised in accordance with plans which are mutually acceptable to the Secretary of the Interior and the Secretary of Transportation and which are consistent with both the purpose of this Act and the purpose of existing statutes dealing with the establishment, maintenance, and operation of airway facilities: *Provided,* That nothing in this section shall authorize the expansion of airport runways into Jamaica Bay or air facilities at Floyd Bennett Field.

(f)   The Secretary shall permit hunting, fishing, shell-fishing, trapping, and the taking of specimens on the lands and waters under his jurisdiction within the Gateway National Recreation Area in accordance with the applicable laws of the United States and the laws of the States of New York and New Jersey and political subdivisions thereof, except that the Secretary may designate zones where and established periods when these activities may be not permitted for reasons of public safety, administration, fish or wildlife management, or public use and enjoyment.

(g)   In the Sandy Hook and Staten Island Units, the Secretary shall inventory and evaluate all sites and structures having present and potential historical, cultural, or architectural significance and shall provide for appropriate programs for the preservation, restoration, interpretation, and utilization of them.

(h)   Notwithstanding any other provision of law, the Secretary is authorized to accept donations of funds from individuals, foundations, or corporations for the purpose of providing services and facilities which he deems consistent with the purposes of this Act.

SEC. 4. (a) There is hereby established a Gateway National Recreation Area Advisory Commission (hereinafter referred to as the "Commission"). Said Commission shall terminate ten years after the date of the establishment of the recreation area.

(b) The Commission shall be composed of eleven members each appointed for a term of two years by the Secretary as follows:

(1)   two members to be appointed from recommendations made by the Governor of the State of New York;

(2)   two members to be appointed from recommendations made by the Governor of the State of New Jersey;

(3)   two members to be appointed from recommendations made by the mayor of New York City;

(4)   two members to be appointed from recommendations made by the mayor of Newark, New Jersey; and

(5)   three members to be appointed by the Secretary to represent the general public.

(c)   The Secretary shall designate one member to be Chairman. Any vacancy in the Commission shall be filled in the same manner in which the original appointment was made.

(d)　A member of the Commission shall serve without compensation as such. The Secretary is authorized to pay the expenses reasonably incurred by the Commission in carrying out its responsibility under this Act upon vouchers signed by the Chairman.

(e)　The Commission established by this section shall act and advise by affirmative vote of a majority of the members thereof.

(f)　The Secretary or his designee shall, from time to time, consult with the members of the Commission with respect to matters relating to the development of the recreation area.

SEC. 5. There are hereby authorized to be appropriated such sums as may be necessary to carry out the provisions of this Act, but not more than $12,125,000 for the acquisition of lands and interests in lands and not more than $92,813,000 (July, 1971 prices) for development of the recreation area, plus or minus such amounts, if any, as may be justified by reason of ordinary fluctuations in the construction costs as indicated by engineering cost indices applicable to the type of construction involved herein.

Approved October 27, 1972.

16 U.S.C. 460cc et seq.

PRESERVATION OF
NATIONAL PARK VALUES

**Report of a Task Force Assembled by the
Conservation Foundation as Part of Its
National Parks for the Future Project, 1972**

**Joseph W. Penfold, Chairman, Stanley A. Cain,
Richard A. Estep, Brock Evans, Roderick Nash,
Douglas Schwartz, Patricia Young**

INTRODUCTION

A group of explorers a century ago, sitting around their campfire at Madison Junction in northwestern Wyoming, could hardly have foreseen the vigor of the idea that sprang from their conversation about the Yellowstone country. They could not have anticipated that their idea would flower into a new dimension of the American dream and would capture the imagination of men around the world.

The national park concept, conceived at Madison Junction and given birth by the President of the United States 100 years ago, has been nurtured in the very essence of the democratic principle. This concept, first enunciated for Yellowstone, says that some areas of remarkable value are too special, too precious ever to be reduced to private ownership and exploitation, but that those areas should instead be retained for the enjoyment and inspiration of all the people.

Yet, over the decades, the concept has come under assault at Yellowstone and elsewhere because of the democratic principle—if the asset is publicly owned, it should be accessible and useable by all the people. Certainly, those who first held the Yellowstone Park vision could not have anticipated the practical difficulties of park use a hundred years later—difficulties brought on by an exploding population, new forms of transportation, and new wealth and leisure made available through the hard labor of a people dedicated to conquering and subverting the wilderness.

With the end of World War II, a booming economy, greater mobility, and longer vacations combined to power a move to the outdoors such as America had never experienced before. Restraints heretofore imposed by geography, time, distance and cost were, for the most part, swept aside and with them the original simple principle from which the national park ideal was born. The national park visit became a casual thing—of little more significance to many than a visit to any other place that provides a scenic backdrop for everyone's outdoor thing. Appreciated? Of course, in some way—one of a dozen vacation stops, one more decal on the window, one more place for later comparison as to efficiency of trailer hook-up, quality of cafeteria, variety of souvenirs and congestion of highway and campground. The national park was fast becoming a playground, a bland experience little different from what the visitor can and does find at a thousand other areas.

The visitor has almost lost something else of enormous importance, a crucial ingredient of the democratic ideal—the opportunity for choice. He is in danger of losing the opportunity to choose the remarkable experience which the national parks were established to save for him, because it is in danger of disappearing.

But the opportunity has not been wholly lost. There remains a spark of the original concept. More and more people are leaving the gadgetry and comforts of technology and striking out for the wilderness to find solitude and adventure with what they can carry on their backs. There in the backcountry of our natural area parks the wilderness persists, little changed in a century. There man can find and be a partner once again in the elementary processes of an undisturbed ecosystem and recapture the awe, the spiritual exaltation, the acute awareness of the very roots of life from which he sprang. The basic choice remains with us—whether we circle back to the original concept, or permit further spin-off into stultifying mediocrity. The choice is ours, whether the parks shall remain the "crown jewels" of our outdoor heritage to be cherished, protected, preserved and worthy of our rigorous self-imposed restraints, or permitted to degenerate into the commonplace.

It is a difficult choice, but it must be made. And nobody else can make it. The choice is ours alone.

SUMMARY OF RECOMMENDATIONS

1.	Management of national parks must conform to their dominant values. Basically, this means that recreation in and enjoyment of the parks must be in terms of their preservation function, that preservation of remaining wilderness should be given top priority in all policy decisions, and that criteria other than numbers of visitors should be used as measurements of a national park's value to society.

2.	High priority should be given to research directed at finding the physical, ecological, and psychological carrying capacity of every unit of the National Park System. This information should be used to determine user quotas for each unit, but care should be taken that a quota system does not discriminate against certain segments of society. The National Park Service should give concentrated attention to educating the public about the meaning of carrying capacity and a need for quota system.

3.   The park visitor should be encouraged to get away from the sight, sound and smell of mechanized civilization—and to hike and camp in the backcountry.

4.   Hotel-type accommodations, private automobiles, and car-camping should be phased out of national park units.

5.   The National Park Service should seek to establish federally financed scientific and social research units on the nation's campuses. A citizens' organization, Friends of Park Research, should be formed to support an expanded research effort.

6.   Expanded attention to the National Park Service's historical and archaeological programs is required. An accelerated research effort is a priority item, as is use of the Wilderness Act of 1964 to protect archaeological sites.

7.   The National Park Service should work toward the elimination of private ownership of buildings and facilities by concessionaires as a preliminary step to moving all such facilities outside the parks.

8.   National park boundaries should, wherever possible, include entire ecosystems. Neighboring political jurisdictions should be encouraged to conduct their land planning and regulatory activities in ways which support the purposes of a park unit.

Reprint of pages one to three of the report with permission of the World Wildlife Fund.

AN ACT TO AMEND THE ACT OF OCTOBER 2, 1968, AN ACT TO ESTABLISH A REDWOOD NATIONAL PARK IN THE STATE OF CALIFORNIA, AND FOR OTHER PURPOSES,

1978 (92 Stat. 163)

Be it enacted by the Senate and House of Representatives of the United States of America in Congress assembled,

TITLE I

SEC. 101. (a) In order to protect existing irreplaceable Redwood National Park resources from damaging upslope and upstream land uses, to provide a land base sufficient to insure preservation of significant examples of the coastal redwood in accordance with the original intent of Congress, and to establish a more meaningful Redwood National Park for the use and enjoyment of visitors, the Act entitled "An Act to establish a Redwood National Park in the State of California, and for other purposes", approved October 2, 1968 (82 Stat. 931), is amended as follows:

(1) In subsection 2(a) after "September 1968," insert "and the area indicated as 'Proposed Additions' on the map entitled 'Additional Lands, Redwood National Park, California', numbered 167-80005-D and dated March 1978.".

(2) In section 2, subsection (a), delete "fifty-eight thousand" and substitute "one hundred and six thousand" and delete the period at the end of the subsection and add "and publicly owned highways and roads." In section 2, subsection (b), delete "by donation only". At the end of section 2, insert the following new subsection "(c)":

"(c) Within the area outside the boundaries of Redwood National Park indicated as the 'Park Protection Zone' on the map entitled 'Proposed Additions, Redwood National Park, California', numbered 167-80005-D and dated March 1978, the Secretary is authorized to acquire lands and interests in land: *Provided,* That lands may be acquired from a willing seller or upon a finding by the Secretary that failure to acquire all or a portion of such lands could result in physical damage to park resources and following notice to the Committee on Energy and Natural Resources of the United States Senate and the Committee on Interior and Insular Affairs of the House of Representatives. Any lands so acquired shall be managed in a manner which will maximize the protection of the resources of Redwood

National Park, and in accordance with the Act of October 21, 1976 (90 Stat. 2743). Acquisition of a parcel of land under the authority of this subsection shall not as a result of such acquisition diminish the right of owners of adjacent lands to the peaceful use and enjoyment of their land and shall not confer authority upon the Secretary to acquire additional lands except as provided in this subsection.".

(3) In subsection 3(a), delete the period at the end of the second sentence and add the following: "which donation of lands or interest in lands may be accepted in the discretion of the Secretary subject to such preexisting reverters and other conditions as may appear in the title to these lands held by the State of California, and such other reverters and conditions as may be consistent with the use and management of the donated lands as a portion of Redwood National Park. Notwithstanding any other provision of law, the Secretary may expend appropriated funds for the management of and for the construction, design, and maintenance of permanent improvements on such lands and interests in land as are donated by the State of California in a manner not inconsistent with such reverters and other conditions.".

(4) In subsection 3(b)(1), after "NPS-RED-7114-B," insert "and effective on the date of enactment of this phrase, there is hereby vested in the United States all right, title, and interest in, and the right to immediate possession of, all real property within the area indicated as 'Proposed Additions' on the map entitled 'Additional Lands, Redwood National Park, California', numbered 167-80005-D and dated March 1978, and all right, title, and interest in, and the right to immediate possession of the down tree personal property (trees severed from the ground by man) severed prior to January 1, 1975, or subsequent to January 31, 1978, within the area indicated as 'Proposed Additions' on the map entitled 'Additional Lands, Redwood National Park, California', numbered 167-80005-D and dated March 1978,".

At the end of subsection 3(b)(1), insert the following new paragraphs: "Down tree personal property severed subsequent to December 31, 1974, and prior to February 1, 1978 may be removed in accordance with applicable State and Federal law, or other applicable licenses, permits, and existing agreements, unless the Secretary determines that the removal of such down timber would damage second growth resources or result in excessive sedimentation in Redwood Creek: *Provided, however,* That down timber lying in stream beds may not be removed without permission of the Secretary: *Provided,* That such removal shall also be subject to such reasonable conditions as may be required by the Secretary to insure the continued availability of raw materials to Redwoods United, Incorporated, a nonprofit corporation located in Manila, California.

"The Secretary shall permit, at existing levels and extent of access and use, continued access and use of each acquired segment of the B line, L line, M line, and K and K roads by each current affected woods employer or its successor in title and interest: *Provided,* That such use is limited to forest and land management and protection purposes, including timber harvesting and road maintenance. The Secretary shall permit, at existing levels and extent of access and use, continued access and use of acquired portions of the Bald Hills road by each current affected woods employer or its successor in title and interest: *Provided further,* That nothing in this sentence shall diminish the authority of the Secretary to otherwise regulate the use of the Bald Hills road.".

(5) In subsection 3(b)(2), delete the last sentence and add the following sentences at the end of the paragraph: "Any action against the United States with regard to the provisions of this Act and for the recovery of just compensation for the lands and interests therein taken by the United States, and for the down tree personal property taken, shall be brought in the United States district court for the district where the land is located without regard to the amount claimed. The United States may initiate proceedings at any time seeking a determination of just compensation in the district court in the manner provided by sections 1358 and 1403 of title 28, United States Code, and may deposit in the registry of the court the estimated just compensation, or a part thereof, in accordance with the procedure generally described by section 258a of title 40, United States Code. Interest shall not be allowed on such amounts as shall have been paid into the court. In the event that the Secretary determines that the fee simple title to any property (real or personal) taken under this section is not necessary for the purposes of this Act, he may, with particular attention to minimizing the payment of severance damages and to allow for the orderly removal of down timber, revest title to such property subject to such reservations, terms, and conditions, if any, as he deems appropriate to carry out the purposes of this Act, and may compensate the former owner for no more than the fair market value of the rights so reserved, except that the Secretary may not revest title to any property for which just compensation has been paid; or, the Secretary may sell at fair market value without regard to the requirements of the Federal Property and Administrative Services Act of 1949, as amended, such down timber as in his judgment may be removed without damage to the park, the proceeds from such sales being credited to the Treasury of the United States. If the State of California designates a right-of-way for a bypass highway around the eastern boundary of Prairie Creek Redwood State Park prior to October 1, 1984, the Secretary is authorized and directed to

acquire such lands or interests in lands as may be necessary for such a highway and, subject to such conditions as the Secretary may determine are necessary to assure the adequate protection of Redwood National Park, shall thereupon donate the designated right-of-way to the State of California for a new bypass highway from a point south of Prairie Creek Redwood State Park through the drainage of May Creek and Boyes Creek to extend along the eastern boundary of Prairie Creek Redwood State Park within Humboldt County. Such acreage as may be necessary in the judgment of the Secretary for this conveyance, and for a buffer thereof, shall be deemed to be a publicly owned highway for purposes of section 101(a)(2) of this amendment effective on the date of enactment of this section.".

(6) In subsection 3(e), delete "sixty days" in the last sentence and add the following sentences at the end of the subsection: "Effective on the date of enactment of this sentence, there are made available from the amounts provided in section 10 herein or as may be hereafter provided such sums as may be necessary for the acquisition of interests in land. Effective on October 1, 1978, there are authorized to be appropriated such sums as may be necessary for implementation of contracts and cooperative agreements pursuant to this subsection: *Provided,* That it is the express intent of Congress that the Secretary shall to the greatest degree possible insure that such contracts and cooperative agreements provide for the maximum retention of senior employees by such owners and for their utilization in rehabilitation and other efforts. The Secretary, in consultation with the Secretary of Agriculture, is further authorized, pursuant to contract or cooperative agreement with agencies of the Federal Executive, the State of California, any political or governmental subdivision thereof, any corporation, not-for-profit corporation, private entity or person, to initiate, provide funds, equipment, and personnel for the development and implementation of a program for the rehabilitation of areas within and upstream from the park contributing significant sedimentation because of past logging disturbances and road conditions, and, to the extent feasible, to reduce risk of damage to streamside areas adjacent to Redwood Creek and for other reasons: *Provided further,* That authority to make payments under this subsection shall be effective only to such extent or in such amounts as are provided in advance in appropriation Acts. Such contracts or cooperative agreements shall be subject to such other conditions as the Secretary may determine necessary to assure the adequate protection of Redwood National Park generally, and to provide employment opportunities to those individuals affected by this taking and to contribute to the economic revival of Del Norte and Humboldt Counties in northern California. The Secretary shall un-

dertake and publish studies on erosion and sedimentation originating within the hydrographic basin of Redwood Creek with particular effort to identify sources and causes, including differentiation between natural and man-aggravated conditions, and shall adapt his general management plan to benefit from the results of such studies. The Secretary, or the Secretary of Agriculture, where appropriate, shall also manage any additional Federal lands under his jurisdiction that are within the hydrographic basin of Redwood Creek in a manner which will minimize sedimentation which could affect the park, and in coordination with plans for sediment management within the basin. To effectuate the provisions of this subsection, and to further develop scientific and professional information and data concerning the Redwood Forest ecosystem, and the various factors that may affect it, the Secretary may authorize access to the area subject to this subsection by designated representatives of the United States.".

(b)   The first section of the Act of August 18, 1970 (84 Stat. 825), is amended by adding the following: "Congress further reaffirms, declares, and directs that the promotion and regulation of the various areas of the National Park system, as defined in section 2 of this Act, shall be consistent with and founded in the purpose established by the first section of the Act of August 25, 1916, to the common benefit of all the people of the United States. The authorization of activities shall be construed and the protection, management, and administration of these areas shall be conducted in light of the high public value and integrity of the National Park System and shall not be exercised in derogation of the values and purposes for which these various areas have been established, except as may have been or shall be directed and specifically provided by Congress.".

(c)   Notwithstanding any provision of the Act of October 2, 1968, supra, the vesting in the United States of all right, title, and interest in, and the right to immediate possession of, all real property and all down tree personal property within the area indicated as "Proposed Additions" on the map entitled, "Additional Lands, Redwood National Park, California," numbered 167-80005-D and dated March 1978, as established by subsection (a)(4) of the first section of this Act, shall be effective on the date of enactment of this section. The provisions of subsection 3(b)(3) of the Act of October 2, 1968, supra, shall also relate to the effective date of this section. From the appropriations authorized for fiscal year 1978 and succeeding fiscal years such sums as may be necessary may be expended for the acquisition of lands and interests in lands, and down tree personal property, authorized to be acquired, or acquired, pursuant to the provisions of this Act.

SEC. 102. (a) The Secretary, in consultation with the Secretaries of Agriculture, Commerce, and Labor, shall conduct an analysis of appropriate Federal actions that may be necessary or desirable to mitigate any adverse economic impacts to public and private segments of the local economy, other than the owners of properties taken by this Act, as a result of the addition of property to Redwood National Park under the first section of this Act. The Secretaries shall also consider the benefits of making grants or entering into contracts or cooperative agreements with the State of California or Del Norte and Humboldt Counties as provided by subsection (b) for the purpose of development and implementation of a program of forest resource improvement and utilization, including, but not limited to, reforestation, erosion control, and other forest land conservation measures, fisheries and fish and wildlife habitat improvements, and wood energy facilities. Not later than January 1, 1979, the Secretary shall submit to the Speaker of the House of Representatives and the President of the Senate a report of his analysis, including his recommendations with respect to actions that should be taken to mitigate any significant short-term and long-term adverse effects on the local economy caused by such addition.

(b) The Secretary of Commerce and the Secretary of Labor, in consultation with the Secretary, and pursuant to his study, shall apply such existing programs as are necessary and appropriate to further mitigate identified employment and other adverse economic impacts on public and private segments of the local economy, other than with regard to the payment of just compensation to the owners of properties taken by this Act and by the Act of October 2, 1968, supra. In addition to the land rehabilitation and employment provisions of this Act, which should have a substantial positive economic effect on the local economy, the Secretaries of Commerce and Labor are further authorized and directed to implement existing authorities to establish employment programs, pursuant to such grants, contracts and cooperative agreements with agencies of the Federal Executive, the State of California, any political or governmental subdivision thereof, any corporation, not-for-profit corporation, private entity or person, for the development and implementation of such programs, as, in the discretion of the Secretaries of Commerce and Labor, may be necessary to provide employment opportunities to those individuals affected by this taking and to contribute to the economic revival of Del Norte and Humboldt Counties, in northern California. Effective on October 1, 1978, there are authorized such sums as may be necessary to carry out the employment and economic mitigation provisions of this Act: *Provided,* That the authority to make payments under this section shall be effective only to such extent or in such amounts as are provided in advance in appropriation Acts.

(c)   The Secretary of Agriculture within one year after the date of enactment of this Act, shall prepare and transmit to Congress a study of timber harvest scheduling alternatives for the Six Rivers National Forest. Such alternatives shall exclude the timber inventories now standing on units of the Wilderness Preservation System and shall be consistent with laws applicable to management of the national forests. In developing the alternatives, the Secretary shall take into consideration economic, silvicultural, environmental, and social factors.

[There follows an extensive section on fair hiring practices.]

Approved March 27, 1978.

16 U.S.C. 79b-c, 79k, 1a-1.

AN ACT TO PROTECT ARCHAEOLOGICAL RESOURCES ON PUBLIC LANDS AND INDIAN LANDS, AND FOR OTHER PURPOSES,

1979 (93 Stat. 721)

Be it enacted by the Senate and House of Representatives of the United States of America in Congress assembled,

SHORT TITLE

SECTION 1. This Act may be cited as the "Archaeological Resources Protection Act of 1979".

FINDINGS AND PURPOSE

SEC. 2. (a) The Congress finds that—
(1) archaeological resources on public lands and Indian lands are an accessible and irreplaceable part of the Nation's heritage;
(2) these resources are increasingly endangered because of their commercial attractiveness;
(3) existing Federal laws do not provide adequate protection to prevent the loss and destruction of these archaeological resources and sites resulting from uncontrolled excavations and pillage; and
(4) there is a wealth of archaeological information which has been legally obtained by private individuals for noncommercial purposes and which could voluntarily be made available to professional archaeologists and institutions.
(b) The purpose of this Act is to secure, for the present and future benefit of the American people, the protection of archaeological resources and sites which are on public lands and Indian lands, and to foster increased cooperation and exchange of information between governmental authorities, the professional archaeological community, and private individuals having collections of archaeological resources and data which were obtained before the date of the enactment of this Act.

DEFINITIONS

SEC. 3. As used in this Act—
(1) The term "archaeological resource" means any material remains of past human life or activities which are of archaeologi-

cal interest, as determined under uniform regulations promulgated pursuant to this Act. Such regulations containing such determination shall include, but not be limited to: pottery, basketry, bottles, weapons, weapon projectiles, tools, structures or portions of structures, pit houses, rock paintings, rock carvings, intaglios, graves, human skeletal materials, or any portion or piece of any of the foregoing items. Nonfossilized and fossilized paleontological specimens, or any portion or piece thereof, shall not be considered archaeological resources, under the regulations under this paragraph, unless found in an archaeological context. No item shall be treated as an archaeological resource under regulations under this paragraph unless such item is at least 100 years of age.

(2) The term "Federal land manager" means, with respect to any public lands, the Secretary of the department, or the head of any other agency or instrumentality of the United States, having primary management authority over such lands. In the case of any public lands or Indian lands with respect to which no department, agency, or instrumentality has primary management authority, such term means the Secretary of the Interior. If the Secretary of the Interior consents, the responsibilities (in whole or in part) under this Act of the Secretary of any department (other than the Department of the Interior) or the head of any other agency or instrumentality may be delegated to the Secretary of the Interior with respect to any land managed by such other Secretary or agency head, and in any such case, the term "Federal land manager" means the Secretary of the Interior.

(3) The term "public lands" means—

(A) lands which are owned and administered by the United States as part of—

(i) the national park system,

(ii) the national wildlife refuge system, or

(iii) the national forest system; and

(B) all other lands the fee title to which is held by the United States, other than lands on the Outer Continental Shelf and lands which are under the jurisdiction of the Smithsonian Institution.

(4) The term "Indian lands" means lands of Indian tribes, or Indian individuals, which are either held in trust by the United States or subject to a restriction against alienation imposed by the United States, except for any subsurface interests in lands not owned or controlled by an Indian tribe or an Indian individual.

(5) The term "Indian tribe" means any Indian tribe, band, nation, or other organized group or community, including any Alaska Native village or regional or village corporation as defined in,

or established pursuant to, the Alaska Native Claims Settlement Act (85 Stat. 688).

(6) The term "person" means an individual, corporation, partnership, trust, institution, association, or any other private entity or any officer, employee, agent, department, or instrumentality of the United States, of any Indian tribe, or of any State or political subdivision thereof.

(7) The term "State" means any of the fifty States, the District of Columbia, Puerto Rico, Guam, and the Virgin Islands.

EXCAVATION AND REMOVAL

SEC. 4. (a) Any person may apply to the Federal land manager for a permit to excavate or remove any archaeological resource located on public lands or Indian lands and to carry out activities associated with such excavation or removal. The application shall be required, under uniform regulations under this Act, to contain such information as the Federal land manager deems necessary, including information concerning the time, scope, and location and specific purpose of the proposed work.

(b) A permit may be issued pursuant to an application under subsection (a) if the federal land manager determines, pursuant to uniform regulations under this Act, that—

(1) the applicant is qualified to carry out the permitted activity,

(2) the activity is undertaken for the purpose of furthering archaeological knowledge in the public interest,

(3) the archaeological resources which are excavated or removed from public lands will remain the property of the United States, and such resources and copies of associated archaeological records and data will be preserved by a suitable university, museum, or other scientific or educational institution, and

(4) the activity pursuant to such permit is not inconsistent with any management plan applicable to the public lands concerned.

(c) If a permit issued under this section may result in harm to, or destruction of, any religious or cultural site, as determined by the Federal land manager, before issuing such permit, the Federal land manager shall notify any Indian tribe which may consider the site as having religious or cultural importance. Such notice shall not be deemed a disclosure to the public for purposes of section 9.

(d) Any permit under this section shall contain such terms and conditions, pursuant to uniform regulations promulgated under this Act, as the Federal land manager concerned deems necessary to carry out the purposes of this Act.

(e)   Each permit under this section shall identify the individual who shall be responsible for carrying out the terms and conditions of the permit and for otherwise complying with this Act and other law applicable to the permitted activity.

(f)   Any permit issued under this section may be suspended by the Federal land manager upon his determination that the permittee has violated any provision of subsection (a), (b), or (c) of section 6. Any such permit may be revoked by such Federal land manager upon assessment of a civil penalty under section 7 against the permittee or upon the permittee's conviction under section 6.

(g)(1) No permit shall be required under this section or under the Act of June 8, 1906 (16 U.S.C. 431), for the excavation or removal by any Indian tribe or member thereof of any archaeological resource located on Indian lands of such Indian tribe, except that in the absence of tribal law regulating the excavation or removal of archaeological resources on Indian lands, an individual tribal member shall be required to obtain a permit under this section.

(2)   In the case of any permits for the excavation or removal of any archaeological resource located on Indian lands, the permit may be granted only after obtaining the consent of the Indian or Indian tribe owning or having jurisdiction over such lands. The permit shall include such terms and conditions as may be requested by such Indian or Indian tribe.

(h)(1) No permit or other permission shall be required under the Act of June 8, 1906 (16 U.S.C. 431-433), for any activity for which a permit is issued under this section.

(2)   Any permit issued under the Act of June 8, 1906, shall remain in effect according to its terms and conditions following the enactment of this Act. No permit under this Act shall be required to carry out any activity under a permit issued under the Act of June 8, 1906, before the date of the enactment of this Act which remains in effect as provided in this paragraph, and nothing in this Act shall modify or affect any such permit.

(i)   Issuance of a permit in accordance with this section and applicable regulations shall not require compliance with section 106 of the Act of October 15, 1966 (80 Stat. 917, 16 U.S.C. 470f).

(j)   Upon the written request of the Governor of any State, the Federal land manager shall issue a permit, subject to the provisions of subsections (b)(3), (b)(4), (c), (e), (f), (g), (h), and (i) of this section for the purpose of conducting archaeological research, excavation, removal, and curation, on behalf of the State or its educational institutions, to such Governor or to such designee as the Governor deems qualified to carry out the intent of this Act.

CUSTODY OF RESOURCES

SEC. 5. The Secretary of the Interior may promulgate regulations providing for—

(1) the exchange, where appropriate, between suitable universities, museums, or other scientific or educational institutions, of archaeological resources removed from public lands and Indian lands pursuant to this Act, and

(2) the ultimate disposition of such resources and other resources removed pursuant to the Act of June 27, 1960 (16 U.S.C. 469-469c) or the Act of June 8, 1906 (16 U.S.C. 431-433).

Any exchange or ultimate disposition under such regulation of archaeological resources excavated or removed from Indian lands shall be subject to the consent of the Indian or Indian tribe which owns or has jurisdiction over such lands. Following promulgation of regulations under this section, notwithstanding any other provision of law, such regulations shall govern the disposition of archaeological resources removed from public lands and Indian lands pursuant to this Act.

PROHIBITED ACTS AND CRIMINAL PENALTIES

SEC. 6. (a) No person may excavate, remove, damage, or otherwise alter or deface any archaeological resource located on public lands or Indian lands unless such activity is pursuant to a permit issued under section 4, a permit referred to in section 4(h)(2), or the exemption contained in section 4(g)(1).

(b) No person may sell, purchase, exchange, transport, receive, or offer to sell, purchase, or exchange any archaeological resource if such resource was excavated or removed from public lands or Indian lands in violation of—

(1) the prohibition contained in subsection (a), or

(2) any provision, rule, regulation, ordinance, or permit in effect under any other provision of Federal law.

(c) No person may sell, purchase, exchange, transport, receive, or offer to sell, purchase, or exchange, in interstate or foreign commerce, any archaeological resource excavated, removed, sold, purchased, exchanged, transported, or received in violation of any provision, rule, regulation, ordinance, or permit in effect under State or local law.

(d) Any person who knowingly violates, or counsels, procures, solicits, or employs any other person to violate, any prohibition contained in subsection (a), (b), or (c) of this section shall, upon conviction, be fined not more than $10,000 or imprisoned not more than one year, or both: *Provided, however,* That if the commercial or archaeological value of the archaeological resources involved and the cost of restoration and repair of such resources exceeds the sum of $5,000, such person shall be fined not more than $20,000 or imprisoned not more than two years, or both. In the case of a second or subsequent such violation upon conviction such person shall be fined not more than $100,000, or imprisoned not more than five years, or both.

(e)　The prohibitions contained in this section shall take effect on the date of the enactment of this Act.

(f)　Nothing in subsection (b)(1) of this section shall be deemed applicable to any person with respect to an archaeological resource which was in the lawful possession of such person prior to the date of the enactment of this Act.

(g)　Nothing in subsection (d) of this section shall be deemed applicable to any person with respect to the removal of arrowheads located on the surface of the ground.

CIVIL PENALTIES

SEC. 7. (a)(1) Any person who violates any prohibition contained in an applicable regulation or permit issued under this Act may be assessed a civil penalty by the Federal land manager concerned. No penalty may be assessed under this subsection unless such person is given notice and opportunity for a hearing with respect to such violation. Each violation shall be a separate offense. Any such civil penalty may be remitted or mitigated by the Federal land manager concerned.

(2)　The amount of such penalty shall be determined under regulations promulgated pursuant to this Act, taking into account, in addition to other factors—

(A)　the archaeological or commercial value of the archaeological resource involved, and

(B)　the cost of restoration and repair of the resources and the archaeological site involved.

Such regulations shall provide that, in the case of a second or subsequent violation by any person, the amount of such civil penalty may be double the amount which would have been assessed if such violation were the first violation by such person. The amount of any penalty assessed under this subsection for any violation shall not exceed an amount equal to double the cost of restoration and repair of resources and archaeological sites damaged and double the fair market value of resources destroyed or not recovered.

(3)　No penalty shall be assessed under this section for the removal of arrowheads located on the surface on the ground.

(b)(1) Any person aggrieved by an order assessing a civil penalty under subsection (a) may file a petition for judicial review of such order with the United States District Court for the District of Columbia or for any other district in which such a person resides or transacts business. Such a petition may only be filed within the 30-day period beginning on the date the order making such assessment was issued. The court shall hear such action on the record made before the Federal land manager and shall sustain his action if it is supported by substantial evidence on the record considered as a whole.

(2)   If any person fails to pay an assessment of a civil penalty—

(A) after the order making the assessment has become a final order and such person has not filed a petition for judicial review of the order in accordance with paragraph (1), or

(B) after a court in an action brought under paragraph (1) has entered a final judgment upholding the assessment of a civil penalty, the Federal land managers may request the Attorney General to institute a civil action in a district court of the United States for any district in which such person is found, resides, or transacts business to collect the penalty and such court shall have jurisdiction to hear and decide any such action. In such action, the validity and amount of such penalty shall not be subject to review.

(c)   Hearings held during proceedings for the assessment of civil penalties authorized by subsection (a) shall be conducted in accordance with section 554 of title 5 of the United States Code. The Federal land manager may issue subpenas for the attendance and testimony of witnesses and the production of relevant papers, books, and documents, and administer oaths. Witnesses summoned shall be paid the same fees and mileage that are paid to witnesses in the courts of the United States. In case of contumacy or refusal to obey a subpena served upon any person pursuant to this paragraph, the district court of the United States for any district in which such person is found or resides or transacts business, upon application by the United States and after notice to such person, shall have jurisdiction to issue an order requiring such person to appear and give testimony before the Federal land manger or to appear and produce documents before the Federal land manager, or both, and any failure to obey such order of the court may be punished by such court as a contempt thereof.

REWARDS; FORFEITURE

SEC. 8. (a) Upon the certification of the Federal land manager concerned, the Secretary of the Treasury is directed to pay from penalties and fines collected under sections 6 and 7 an amount equal to one-half of such penalty or fine, but not to exceed $500, to any person who furnishes information which leads to the finding of a civil violation, or the conviction of criminal violation, with respect to which such penalty or fine was paid. If several persons provided such information, such amount shall be divided among such persons. No officer or employee of the United States or of any State or local government who furnishes information or renders service in the performance of his official duties shall be eligible for payment under this subsection.

(b)   All archaeological resources with respect to which a violation of subsection (a), (b), or (c) of section 6 occurred and which are in the possession of any person, and all vehicles and equipment of any person which were used in connection with such violation, may be (in the discretion of the court or administrative law judge, as the case may be) subject to forfeiture to the United States upon—

(1)   such person's conviction of such violation under section 6,

(2)   assessment of a civil penalty against such person under section 7 with respect to such violation, or

(3)   a determination by any court that such archaeological resources, vehicles, or equipment were involved in such violation.

(c)   In cases in which a violation of the prohibition contained in subsection (a), (b), or (c) of section 6 involve archaeological resources excavated or removed from Indian lands, the Federal land manager or the court, as the case may be, shall provide for the payment to the Indian or Indian tribe involved of all penalties collected pursuant to section 7 and for the transfer to such Indian or Indian tribe of all items forfeited under this section.

CONFIDENTIALITY

SEC. 9. (a) Information concerning the nature and location of any archaeological resource for which the excavation or removal requires a permit or other permission under this Act or under any other provision of Federal law may not be made available to the public under subchapter II of chapter 5 of title 5 of the United States Code or under any other provision of law unless the Federal land manager concerned determines that such disclosure would—

(1)   further the purposes of this Act or the Act of June 27, 1960 (16 U.S.C. 469-469c), and

(2)   not create a risk of harm to such resources or to the site at which such resources are located.

(b)   Notwithstanding the provisions of subsection (a), upon the written request of the Governor of any State, which request shall state—

(1) the specific site or area for which information is sought,

(2) the purpose for which such information is sought,

(3) a commitment by the Governor to adequately protect the confidentiality of such information to protect the resource from commercial exploitation,

the Federal land manager concerned shall provide to the Governor information concerning the nature and location of archaeological resources within the State of the requesting Governor.

REGULATIONS; INTERGOVERNMENTAL COORDINATION

SEC. 10. (a) The Secretaries of the Interior, Agriculture and Defense and the Chairman of the Board of the Tennessee Valley Authority, after consultation with other Federal land managers, Indian tribes, representatives of concerned State agencies, and after public notice and hearing, shall promulgate such uniform rules and regulations as may be appropriate to carry out the purposes of this Act. Such rules and regulations may be promulgated only after consideration of the provisions of the American Indian Religious Freedom Act (92 Stat. 469; 42 U.S.C. 1996). Each uniform rule or regulation promulgated under this Act shall be submitted on the same calendar day to the Committee on Energy and Natural Resources of the United States Senate and to the Committee on Interior and Insular Affairs of the United States House of Representatives, and no such uniform rule or regulation may take effect before the expiration of a period of ninety calendar days following the date of its submission to such Committees.

(b) Each Federal land manager shall promulgate such rules and regulations, consistent with the uniform rules and regulations under subsection (a), as may be appropriate for the carrying out of his functions and authorities under this Act.

COOPERATION WITH PRIVATE INDIVIDUALS

SEC. 11. The Secretary of the Interior shall take such action as may be necessary, consistent with the purposes of this Act, to foster and improve the communication, cooperation, and exchange of information between—

(1) private individuals having collections of archaeological resources and data which were obtained before the date of the enactment of this Act, and

(2) Federal authorities responsible for the protection of archaeological resources on the public lands and Indian lands and

professional archaeologists and associations of professional archaeologists.

In carrying out this section, the Secretary shall, to the extent practicable and consistent with the provision of this Act, make efforts to expand the archaeological data base for the archaeological resources of the United States through increased cooperation between private individuals referred to in paragraph (1) and professional archaeologists and archaeological organizations.

SAVINGS PROVISIONS

SEC. 12. (a) Nothing in this Act shall be construed to repeal, modify, or impose additional restrictions on the activities permitted under existing laws and authorities relating to mining, mineral leasing, reclamation, and other multiple uses of the public lands.

(b)   Nothing in the Act applies to, or requires a permit for, the collection for private purposes of any rock, coin, bullet, or mineral which is not an archaeological resource, as determined under uniform regulations under section 3(1).

(c)   Nothing in this Act shall be construed to affect any land other than public land or Indian land or to affect the lawful recovery, collection, or sale of archaeological resources from land other than public land or Indian land.

REPORT

SEC. 13. As part of the annual report required to be submitted to the specified committees of the Congress pursuant to section 5(c) of the Act of June 27, 1960 (74 Stat. 220; 16 U.S.C. 469-469a), the Secretary of the Interior shall comprehensively report as a separate component on the activities carried out under the provisions of this Act, and he shall make such recommendations as he deems appropriate as to changes or improvements needed in the provisions of this Act. Such report shall include a brief summary of the actions undertaken by the Secretary under section 11 of this Act, relating to cooperation with private individuals.

Approved October 31, 1979.

16 U.S.C. 470aa–470mm.

STATE OF THE PARKS
MAY 1980
A REPORT TO THE CONGRESS

Office of Science and Technology, National Park Service

EXECUTIVE SUMMARY

The National Park Service recently has completed its first Servicewide survey designed to identify and characterize threats that endanger the natural and the cultural resources of the parks. This 1980 State of the Parks report presents the results of that survey. It focuses on three aspects of the threats to the parks problem: First, the Report identifies <u>specific threats</u> which endanger the resources of individual parks; second it identifies <u>sources of threats</u>, whether located internal or external to park boundaries; and third, the Report identifies <u>park resources which are endangered</u> by the threats.

The findings presented in this Report are based upon information submitted by park superintendents, park natural and cultural resource managers, park scientists, and park planners. It should be emphasized that the recognition and the perception of threats may be affected by the professional training and experience of these observers. A park superintendent, for example, may be most sensitive to operational problems associated with overcrowding or vandalism; a park scientist or resource management specialist may be particularly aware of threats to water quality or wildlife habitat; a park planner is more likely to perceive disruptions associated with zoning or land use practices. Further, the respondent to the threats survey may show a greater interest in problem areas with prospective solutions and may fail to emphasize certain issues simply because they appear to be beyond his or her span of influence. Clearly, there are subjective judgments involved in any threats reporting and evaluation process. However, collectively, the results of this systematic park-by-park survey provide important new information concerning a broad spectrum of problems and issues with which the National Park Service must deal.

THREATS SURVEY RESULTS

The term "threats" as used in this Report refers to those pollutants, visitor activities, exotic species, industrial development projects, etc., which have the potential to cause significant damage to park resources or to seriously degrade important park values or park experiences. Without qualification, it can be stated that the cultural and the natural resources of the parks are endangered both from without and from within by a broad range of such threats. In this regard, the most significant findings developed in this study may be summarized as follows:

• Park units representing all size and use categories, and all types of ecosystem, reported a wide range of threats affecting their resources. These threats, which emanate from both internal and external sources, are causing severe degradation of park resources.

• The 63 National Park natural areas greater than 30,000 acres in size reported an average number of threats nearly double that of the Servicewide norm. Included in this category are units such as Yellowstone, Yosemite, Great Smoky Mountains, Everglades and Glacier. Most of these great parks were at one time pristine areas surrounded and protected by vast wilderness regions. Today, with their surrounding buffer zones gradually disappearing, many of these parks are experiencing significant and widespread adverse effects associated with external encroachment.

• The 12 Biosphere Reserve Parks, which are unique natural areas that range in size from 15,000 acres to more than two million acres and which are dedicated to long-term ecosystem monitoring under the UNESCO Man and the Biosphere program, surprisingly reported an average number of threats nearly three times the Servicewide norm. This is particularly disturbing because the Biosphere Reserve Parks are considered to be model ecological control areas. The large number of threats reported by these parks may reflect the greater emphasis directed to monitoring in these areas.

If this is, in fact, the reason for the increased occurrence of reported threats in the Biosphere Reserve parks, it suggests that significant numbers of threats may have been overlooked in other parks which, to date, have received much less monitoring and research attention.

• More than 50 percent of the reported threats were attributed to sources or activities located external to the parks. The most frequently identified external threats included: Industrial and commercial development projects on adjacent lands; air pollutant emissions, often associated with facilities located considerable distances from the affected parks; urban encroachment; and roads and railroads.

• Threats associated with activities or with sources located within park boundaries continue to cause significant degradation of park resources, park values, and visitor park experiences. The most frequently reported internal threats are associated with: Heavy visitor use; park utility access corridors; vehicle noise; soil erosion; and exotic plant and animal species.

• Scenic resources were reported to be significantly threatened in more than 60 percent of the parks. Air quality resources were reported endangered in more than 45 percent of the parks. Mammal, plant, and fresh water resources all were reported threatened in more than 40 percent of the units. Other endangered resources are listed in ranked order of occurrence in the report.

• A surprising 75 percent of the reported threats to park resources have been classified by onsite park observers as inadequately documented by either private or government research. Threats associated with air pollution, water pollution, and visitor related activities most frequently were cited as needing additional monitoring, scientific measurements or research documentation.

CONCLUSIONS

The units of the National Park Service represent all of the major categories of ecosystems within which we live. These parks, individually and collectively, constitute the best of our natural and cultural resources. The lessons that we learn and the progress that we make in our attempts to better manage and protect these resources are of benefit to us all.

The results of this study indicate that no parks of the System are immune to external and internal threats, and that these threats are causing significant and demonstrable damage. There is no question but that these threats will continue to degrade and destroy irreplaceable park resources until such time as mitigation measures are implemented. In many cases, this degradation or loss of resources is irreversible. It represents a sacrifice by a public that, for the most part, is unaware that such a price is being paid.

The diversity and complexity of the problems identified in this report serve to emphasize the need for an expanded program to protect and preserve the resources of the parks. To develop and implement such a program requires that the National Park Service know what natural and cultural resources exist in each park, establish the condition of these resources, and determine how and to what extent these resources are threatened. Essentially we must do the following:

• We must prepare a comprehensive inventory of the important natural and cultural resources of each park and develop a plan at the park level for managing these resources.

• We must establish accurate baseline data on park resources and conduct comprehensive monitoring programs designed to detect and measure changes both in these resources and in the ecosystem environments within which they exist.

• We must pay additional attention to those threats which are associated with sources and activities located external to the parks. These threats today pose unique problems because of the Service's limited ability to deal directly and effectively with such outside influences.

• We must improve our capability to better quantify and document the impacts of various threats, particularly those which are believed to most seriously affect important park resources and park values.

To accomplish these objectives will require that the Service significantly expand its research and resource management capabilities. At the present time, the natural science research program of the National Park Service is base funded at a level of only nine million dollars and is staffed by fewer than 100 scientists; this is an average of less than one researcher for each three units of the System and represents only 1.1 percent of the total Park Service staff. Similarly, there are fewer than 200 personnel in the Service who have resource management training; many of these personnel are senior level Park Rangers and Park Superintendents who are able to devote only a small portion of their time to resource management problems.

The staff and the funding resources currently available within the research and the resource management areas clearly are inadequate to respond to the needs of the Service. Internal actions are being taken to strengthen existing research and resource management programs. However, it must be emphasized that a continuing and expanded commitment is required to address the wide range of issues defined by the threats survey, and that the support of the Congress will be needed to deal with these important problems.

A final comment. The enabling legislation establishing the National Park Service and its individual park units clearly mandates, as the primary objective, the protection, preservation and conservation of park resources, in perpetuity for the use and enjoyment of future generations. The National Park Service recognizes that changes in priorities and reallocation of resources will be required to meet this mandate and is committed to such changes.

Washington, D.C.: GPO, 1980, vii–x.

8

A System Threatened

1981–1992

The period from 1963 to 1980 had seen major changes affecting the national park system. Indeed, one might have expected that vast expansion of the network of parks, monuments, etc., and increasingly biocentric management would continue indefinitely. However, 1981 saw the arrival of economic recession and altered priorities emanating from Washington. Over the next twelve years the national park system would suffer expanded visitation and heightened external threats coupled with serious underfunding and erosion of employee morale. Many aspects of NPS management would be challenged, and politicization of the directorship and erosion of its power promoted. The signal for change came early as a new Interior secretary, James Watt, tendered the traditional secretary's letter on national park system management. He spelled out administration policies that would curtail the system's growth and return to provision of visitor services and pleasures as a primary management goal.

Heightened public interest in touring the parks and in developing commercial and residential complexes on their borders exacerbated the problems noted in 1980. The General Accounting Office issued a report in 1987 that spelled out what park personnel knew — that the threats were not being addressed and that many problems had in fact worsened. It found instead that the parks were not being run consistent with the laws and executive orders established to protect them.

A year later saw the worst fire in national park system history. Approximately half of Yellowstone's acres burned, although the percentage of forest destroyed was a much lower figure. Daily from July through September the public watched in fascination and alarm as the nation's oldest park burned. In the end loud questions from politicians and visitors called into question two decades of fire management. An interagency report released in 1989, however, found the policy basically sound. Ecosystem management survived its toughest test.

In 1990, decades of archaeological business as usual suddenly ended with passage of the Native American Graves Protection and Repatriation Act. Hitherto, excavation of Indian graves and removal of human and ceremonial remains were forbidden. This act imposed further difficulties for museums, including those operated by the National Park Service, for it stipulated that Indian remains be returned to the direct or at least cultural descendants for reburial.

Finally 1992 saw the release of two more reports on the status of the national park system. One, *Science and the National Parks*, echoed the earlier Robbins report (1963) in finding the National Park Service wanting in its pursuit and use of scientific information. It called for far more attention to science as a basis for policy and management. The other report was the end result of a highly publicized and expensive conference on the status and needs of the national parks for the twenty-first century. Called the Vail report for its meeting site, the document reiterated the concerns of the *State of the Parks Report* (1980) and the GAO report (1987). It reaffirmed most of their recommendations and the basic policies developed from the early 1960s through 1980. Although it is unlikely that either report will lead to significant policy changes, they indicate the persistence of longtime problems and of the basic formulas for their solution. It remains to be seen whether the recommendations of these two committees and those that preceded them will finally be adopted in full.

SECRETARY WATT'S LETTER ON
NATIONAL PARK MANAGEMENT

July 6, 1981

To: Director, National Park Service
From: Secretary of the Interior
Subject: Management of the National Park System

Since the National Park Service was established, certain of my predecessors, beginning with Secretary Franklin K. Lane, have outlined the principles which are to guide the Service in the management of the National Park System. I believe that now is a particularly appropriate time, as the Nation enters the 1980's under new leadership and in high expectation of constructive change, to renew that practice.

In his letter of May 13, 1918, to Stephen T. Mather, first Director of the National Park Service, Secretary Lane wrote,

> For the information of the public an outline of the administrative policy to which the new Service will adhere may now be announced. This policy is based on three broad principles: First, that the national parks must be maintained in absolutely unimpaired form for the use of future generations as well as those of our own time; second, that they are set apart for the use, observation, health, and pleasure of the people; and third, that the national interest must dictate all decisions affecting public or private enterprise in the parks.

The principles outlined for the first Director of the Service more than 60 years ago apply with equal force to the management of the National Park System today, and I reaffirm them wholeheartedly. Moreover, the Congress, as the maker of public land policy, has consistently supported these broad principles through the enactment of various laws, capped by the language of the Act of August 18, 1970, as amended:

> Congress further reaffirms, declares, and directs that the promotion and regulation of the various areas of the National Park System . . . shall be consistent with and founded in the purpose established by the first section of the Act of August 25, 1916, to the common benefit of all the people of the United States. The authorization of activities shall be construed and the protection, management, and administration of these areas shall be conducted in light of the high public value and integrity of the National Park System and shall not be exercised in derogation of the values and purposes for which these various

areas have been established, except as may have been or shall be directly and specifically provided by Congress.

I commend to every employee of the National Park Service a re-reading of the first section of the Act of August 25, 1916, which set forth the fundamental purpose of parks: ". . . to conserve the scenery and the natural and historic objects and the wildlife therein and to provide for the enjoyment of the same in such manner and by such means as will leave them unimpaired for the enjoyment of future generations." No agency in Government has a clearer mandate.

I am concerned that in recent years the funding and manpower resources of the Service have not always been directed to fulfilling its basic mission as set forth above. The challenge to the Service in the 1980's, and my charge to you as Director, is to assure that the funding and staffing available to you in a time of reduced Federal expenditures are applied to achieve the most benefit for visitors to the most heavily visited parks, being always mindful of the preservation of park resources. I am particularly concerned that you give priority to protecting and maintaining the "crown jewels," the internationally known, unique natural parks in the system.

Two recent developments have provided the Service with the opportunity to reemphasize its basic responsibilities. First, the "State of the Parks" analysis begun in 1980 should continue, and actions vigorously pursued to monitor, prevent, or mitigate significant threats. Second, the deteriorating condition of physical facilities has been addressed by the additional funds for maintenance provided by Congress in the FY 1981 budget. The Service must put in place the management systems necessary to follow up on these Congressional initiatives in order that the public gets the highest return on its park dollar. It is timely to develop a long-range plan of funding and personnel requirements, so that the National Park System may be developed and operated to meet acceptable standards. You should begin immediately to explore the components of such a long-range plan.

As Director, you should emphasize management of the National Park System. At this time, the attention of all managers within the System needs to be specifically directed to achieving efficiency, serving visitors, and protecting park values, rather than expanding the System. Special emphasis should be placed on bringing the old-line parks up to standard.

The National Park Service is in a unique position to serve people. As such, it is one of the few Federal agencies whose personnel have the opportunity of meeting their true bosses, the thousands of visitors, each of whom has spent considerable time and money just to be in a park. The Service has a tradition of providing a high standard of service to the visitor, and this standard must be maintained. At a time when most families' budgets are being increasingly squeezed, the opportunity of a park visit becomes even more special. Courtesy, patience, and helpfulness on the part of park employees will assure visitors that yours is an organization that lives up to its name, as a Service. As Director, you should ensure that you recruit men and women who like people for those front-line positions where service to the visitor is immediate and direct.

For those visitors who take the opportunity to stay in lodges, rent bicycles or horses, and purchase food and equipment while in the parks, the concessioner is essential to the park experience. Concession operators are partners with the National Park Service in serving people. A successful partnership is based on mutual support and mutual responsibility. Park managers must appreciate the need for the operator to make a reasonable profit, and concession operators need to be aware of the statutory mandate of the Service to protect park resources for people. Both the Service and the concessioners are, above all, responsible for assuring the park visitor an opportunity to use safe, high quality facilities at a reasonable cost. Where, because of deterioration over long periods, this standard is not being met, I believe you should explore with the concessioners alternative ways to find the needed capital and upgrade those facilities that are necessary for the visitor.

In summary, let us get back to the basics. Let us protect the land and its resources. Let us serve the visitor. I am confident that the Service can and will, through professional, innovative, and accountable management, fulfill its statutory mandate, and I pledge my support to that end.

James G. Watt

Personal files of National Park Service Bureau Historian Barry Mackintosh.

LIMITED PROGRESS MADE IN DOCUMENTING
AND MITIGATING THREATS TO THE PARKS

Report by the United States
General Accounting Office, 1987

EXECUTIVE SUMMARY

PURPOSE

In 1980 the National Park Service reported more than 4,000 threats to the natural and cultural resources of the national park system, from both within and outside park borders. The following year, in response to a congressional request, the Park Service developed a strategy to prevent and mitigate the problems identified in its report. The Chairman, Subcommittee on National Parks and Recreation, House Committee on Interior and Insular Affairs, asked GAO to determine, among other things, what progress the Park Service has made in identifying, monitoring, and mitigating threats and how its resource management needs are reflected in the parks' resource management plans and Park Service budgets.

BACKGROUND

In its 1980 <u>State of the Parks</u> report, the Park Service listed about 4,300 threats to the aesthetic qualities, cultural resources, air and water quality, plants, and wildlife of the nation's parks. According to the report, more than half the threats came from sources outside park boundaries and only about 25 percent were adequately documented. The Park Service claimed that it did not have enough staff and funds to adequately identify, monitor, and correct these problems or to give additional attention to external threats.

Following its report, the Park Service developed a servicewide strategy to improve its resource management capabilities. The strategy, to which the Park Service says it is still committed, called for each park to have a resource management plan for both its natural and cultural resources by the end of 1981. These plans were to (1) include an inventory of park resources and a detailed program for monitoring and managing the resources, (2) specify necessary staff and funding, and (3) assign priorities to projects so that resources provided could be allocated toward the most serious problems. The plans also were to be updated annually and used in formulating annual Park Service budgets.

To support the development and use of these plans, the Park Service announced a series of 11 initiatives to improve resource management information and staff capabilities.

RESULTS IN BRIEF

The Park Service's strategy for better managing park resources has yet to be fully implemented. Some parks do not have an approved resource management plan even though they were required to be completed by the end of 1981, others have not updated their plans, and the plans that have been prepared are not being used in formulating the Park Service's annual budgets. Further, many of the 11 initiatives intended to support the development and use of the plans were not followed through.

The Park Service has not kept track of its progress in documenting and mitigating the threats it identified in 1980. The 12 parks GAO visited* have corrected some of the resource problems, but most problems remain and many of those are still not well-understood or documented. Although the parks have proposed projects to address these problems, most were not funded.

PRINCIPAL FINDINGS

Resource Management Plans and Initiatives

Although all units of the national park system were required to prepare resource management plans by the end of 1981 and update them annually, only half met the original deadlines. As of August 1986, 35 units were still without approved cultural plans and 31 without approved natural plans. GAO visited 12 parks in 3 different regions and found that 2 parks had no approved plans and 4 had not updated their plans since they were first approved in 1982 and 1983, respectively. Further, the Park Service had just started developing a process that could be used to analyze park-unit resource management plan data for making regional and servicewide budget and funds allocation decisions.

The Park Service's 11 initiatives were aimed at improving resource information, training staff in resource management, and increasing scientific research. The training initiatives were undertaken and are continuing. Of the remaining initiatives, one was never undertaken and the others were initiated but not carried through. Standards and guidelines for resource inventories and monitoring procedures, for example, were drafted but were not used. Also, plans to expand research programs were dropped for higher priority projects. On the other hand, although not part of its original set of initiatives, the Park Service has put into effect a national air quality monitoring program and established a national inventory of threats to parks from mining and mineral activities.

Documenting and Mitigating Threats

Neither the Park Service nor the individual park units kept track of their progress in addressing the threats identified in the <u>State of</u>

the Parks report. The Park Service's budget for resource management increased considerably between 1980 and 1984, from $44 million to $93 million. Within the 12 national parks GAO visited, additional funds were used to resolve some significant problems, such as the removal of plants and animals harmful to park resources and the repair of deteriorating historic structures. Nevertheless, officials of these 12 parks judged that 255, or 80 percent, of the total 318 threats reported in 1980, were still unresolved as of December 1985. Of these, 111, or about 43 percent of those remaining, were still undocumented—that is, the parks did not know the extent to which these perceived threats were problems, or the dimensions of those that were known problems.

Although the parks have proposed projects to address known and potential resource problems, many projects have not been funded. In the 10 parks GAO visited that had approved resource management plans, nearly 100 projects, intended to deal with deteriorating resources and threats to health and safety and provide more information about potential threats, were proposed to be funded in fiscal year 1986. However, none were funded. For example, at Death Valley National Monument funds were not approved to install protective nets over abandoned mine shafts. At Florissant Fossil Beds National Monument, no funds were provided to prevent further deterioration of petrified tree stumps. Likewise, no funds were provided to study the condition of rare, endangered, or threatened plant species in Hawaii Volcanoes National Park. Two of the 10 parks received about 25 percent of the funds and staff they requested in 1986, one received about 75 percent of its request, and another had only one of 7 projects funded.

RECOMMENDATIONS

To provide the information needed for the Park Service to develop a comprehensive, systemwide approach to protect and manage park resources and provide the basis to make more informed funding decisions, GAO recommends that the Secretary of the Interior direct the Director, National Park Service, to

• enforce the agency's requirement that resource management plans be prepared and updated in accordance with established Park Service guidance and criteria at each park and

• improve procedures on the use of the information provided in the resource management plans to (1) identify and prioritize cultural and natural resource management needs on a regional and service-wide basis and (2) prepare annual budget requests.

To ensure that resource management plans are based on adequate information, GAO is also making recommendations relating to the gathering and monitoring of data on the parks' natural and cultural resources. [These are to]

• develop standards for determining the minimum baseline information needed to properly plan for the management and protection of park resources,

• assess the adequacy of each park's information base in relation to the standards so developed,

• take action to improve park information bases that are found not up to the standards, and

• develop and implement long-term programs to monitor resource condition changes over time.

AGENCY COMMENTS

In its comments on a draft of our report, the Department of the Interior believes that the report fairly addresses the questions the Subcommittee on National Parks and Recreation raised about the Park Service actions since the 1980 State of the Parks report, and it agreed with the thrust of the report's recommendations. The Department did state, however, that it believes the report neglected to emphasize in its recommendations that in taking actions to improve park information bases, the Park Service must not only make a one-time effort to collect baseline information, but must also establish long-term programs to monitor appropriate parameters for changes over time. GAO agrees with Interior and has added a recommendation citing the need for long-term resource monitoring programs.

* Cape Hatteras, Custer Battlefield (now Little Bighorn), Death Valley, Florissant, Glacier, Grant-Kohrs, Great Smoky Mountains, Hawaii Volcanoes, John Muir, Redwood, Stones River and Wright Brothers.

GAO/RCED-87-36, GPO, 1987, 2–5, 37.

INTERAGENCY FINAL REPORT ON FIRE MANAGEMENT POLICY MAY 5, 1989

Departments of the Interior and Agriculture

SUMMARY

The Fire Management Policy Review Team was established on September 28, 1988 to review national policies and their application for fire management in national parks and wilderness and to recommend actions to address the problems experienced during the 1988 fire season. The Team draft report was submitted to the Secretaries of the Interior and Agriculture on December 15, 1988. A 60 day public review and comment period, incorporating a series of public hearings, began with publication of that report in the Federal Register on December 20, 1988. Having reviewed and considered the public comments, this final report is submitted in culmination of the Team's charter.

The Fire Management Policy Review Team finds that:

• The objectives of prescribed natural fire programs in national parks and wildernesses are sound, but the policies need to be refined, strengthened, and reaffirmed. These policies permit fires to burn under predetermined conditions.

• Many current fire management plans do not meet current policies; the prescriptions in them are inadequate; and decision-making needs to be tightened.

• There are risks inherent in managing wildland fires. These risks can be reduced by careful planning and preparation. Use of planned burning and other efforts to reduce hazard fuels near high value structures and to create fuel breaks along boundaries help to reduce risks from both prescribed natural fires and wildfires.

• The ecological effects of prescribed natural fire support resource objectives in parks and wilderness, but in some cases the social and economic effects may be unacceptable. Prescribed natural fires may affect permitted uses of parks and wilderness, such as recreation, and impact outside areas through such phenomena as smoke and stream sedimentation.

• Dissemination of information before and during prescribed natural fires needs to be improved. There needs to be greater public participation in the development of fire management plans.

• Internal management processes, such as training more person-nel, developing uniform terminology, and utilizing similar budget structures, would significantly improve fire management.

• Claims were heard that some managers support "naturalness" above all else, allowing fires to burn outside of prescription require-ments without appropriate suppression actions.

The Team recommends that:

• Prescribed natural fire policies in the agencies be reaffirmed and strengthened.

• Fire management plans be reviewed to assure that current policy requirements are met and expanded to include interagency planning, stronger prescriptions, and additional decision criteria.

• Line officers certify daily that adequate resources are available to ensure that prescribed fires will remain within prescription, given reasonably foreseeable weather conditions and fire behavior.

• Agencies develop contingency plans to constrain the use of pre-scribed fire in the event or anticipation of unfavorable weather or fire conditions, or when necessary to balance competing demands for scarce fire suppression resources.

• Agencies consider opportunities to use management ignited pre-scribed fires to complement prescribed natural fire programs and to reduce hazard fuels.

• Agencies utilize the National Environmental Policy Act require-ments in fire management planning to increase opportunities for public involvement and coordination with state and local government.

• Agencies provide more and better training to assure an ade-quate supply of knowledgeable personnel for fire management pro-grams.

• Agencies review funding methods for prescribed fire programs and fire suppression to improve interagency program effectiveness.

• Additional research and analysis relating to weather, fire be-havior, fire history, fire information integration, post-fire effects, and other topics be carried out so that future fire management pro-grams can be carried out more effectively and with less risk.

• Allegations of misuse of policy be promptly investigated and acted upon as may be appropriate.

Provided by the Superintendent, Yellowstone National Park.

NATIVE AMERICAN GRAVES PROTECTION
AND REPATRIATION ACT,

1990 (104 Stat. 3048)

AN ACT

To provide for the protection of Native American graves, and for other purposes.

Be it enacted by the Senate and House of Representatives of the United States of America in Congress assembled,

SECTION 1. SHORT TITLE.

This Act may be cited as the "Native American Graves Protection and Repatriation Act".

SEC. 2. DEFINITIONS.

For purposes of this Act, the term—

(1) "burial site" means any natural or prepared physical location, whether originally below, on, or above the surface of the earth, into which as a part of the death rite or ceremony of a culture, individual human remains are deposited.

(2) "cultural affiliation" means that there is a relationship of shared group identity which can be reasonably traced historically or prehistorically between a present day Indian tribe or Native Hawaiian organization and an identifiable earlier group.

(3) "cultural items" means human remains and—

(A) "associated funerary objects" which shall mean objects that, as a part of the death rite or ceremony of a culture, are reasonably believed to have been placed with individual human remains either at the time of death or later, and both the human remains and associated funerary objects are presently in the possession or control of a Federal agency or museum, except that other items exclusively made for burial purposes or to contain human remains shall be considered as associated funerary objects.

(B) "unassociated funerary objects" which shall mean objects that, as a part of the death rite or ceremony of a culture, are reasonably believed to have been placed with individual human remains either at the time of death or later, where the remains are not in the possession or control of the Federal agency or museum and the objects can be identified by a preponderance of the evidence as related to specific individuals or families or to known human remains or, by a preponderance of the evidence, as having been removed from a specific burial site of an individual culturally affiliated with a particular Indian tribe,

(C) "sacred objects" which shall mean specific ceremonial objects which are needed by traditional Native American religious leaders for the practice of traditional Native American religions by their present day adherents, and

(D) "cultural patrimony" which shall mean an object having ongoing historical, traditional, or cultural importance central to the Native American group or culture itself, rather than property owned by an individual Native American, and which, therefore, cannot be alienated, appropriated, or conveyed by any individual regardless of whether or not the individual is a member of the Indian tribe or Native Hawaiian organization and such object shall have been considered inalienable by such Native American group at the time the object was separated from such group.

(4) "Federal agency" means any department, agency, or instrumentality of the United States. Such term does not include the Smithsonian Institution.

(5) "Federal lands" means any land other than tribal lands which are controlled or owned by the United States, including lands selected by but not yet conveyed to Alaska Native Corporations and groups organized pursuant to the Alaska Native Claims Settlement Act of 1971.

(6) "Hui Malama I Na Kupuna O Hawai'i Nei" means the nonprofit, Native Hawaiian organization incorporated under the laws of the State of Hawaii by that name on April 17, 1989, for the purpose of providing guidance and expertise in decisions dealing with Native Hawaiian cultural issues, particularly burial issues.

(7) "Indian tribe" means any tribe, band, nation, or other organized group or community of Indians, including any Alaska Native village (as defined in, or established pursuant to, the Alaska Native Claims Settlement Act), which is recognized as eligible for the special programs and services provided by the United States to Indians because of their status as Indians.

(8) "museum" means any institution or State or local government agency (including any institution of higher learning) that receives Federal funds and has possession of, or control over, Native American cultural items. Such term does not include the Smithsonian Institution or any other Federal agency.

(9) "Native American" means of, or relating to, a tribe, people, or culture that is indigenous to the United States.

(10) "Native Hawaiian" means any individual who is a descen-

dant of the aboriginal people who, prior to 1887, occupied and exercised sovereignty in the area that now constitutes the State of Hawaii.

(11) "Native Hawaiian organization" means any organization which—

(A) serves and represents the interests of Native Hawaiians,

(B) has as a primary and stated purpose the provision of services to Native Hawaiians, and

(C) has expertise in Native Hawaiian Affairs, and shall include the Office of Hawaiian Affairs and Hui Malama I Na Kupuna O Hawai'i Nei.

(12) "Office of Hawaiian Affairs" means the Office of Hawaiian Affairs established by the constitution of the State of Hawaii.

(13) "right of possession" means possession obtained with the voluntary consent of an individual or group that had authority of alienation. The original acquisition of a Native American unassociated funerary object, sacred object or object of cultural patrimony from an Indian tribe or Native Hawaiian organization with the voluntary consent of an individual or group with authority to alienate such object is deemed to give right of possession of that object, unless the phrase so defined would, as applied in section 7(c), result in a Fifth Amendment taking by the United States as determined by the United States Claims Court pursuant to 28 U.S.C. 1491 in which event the "right of possession" shall be as provided under otherwise applicable property law. The original acquisition of Native American human remains and associated funerary objects which were excavated, exhumed, or otherwise obtained with full knowledge and consent of the next of kin or the official governing body of the appropriate culturally affiliated Indian tribe or Native Hawaiian organization is deemed to give right of possession to those remains.

(14) "Secretary" means the Secretary of the Interior.

(15) "tribal land" means—

(A) all lands within the exterior boundaries of any Indian reservation;

(B) all dependent Indian communities;

(C) any lands administered for the benefit of Native Hawaiians pursuant to the Hawaiian Homes Commission Act, 1920, and section 4 of Public Law 86-3.

SEC. 3. OWNERSHIP.

(a) NATIVE AMERICAN HUMAN REMAINS AND OBJECTS.—The ownership or control of Native American cultural items which are excavated or discovered on Federal or tribal lands after the date of enactment of this Act shall be (with priority given in the order listed)—

 (1) in the case of Native American human remains and associated funerary objects, in the lineal descendants of the Native American; or

 (2) in any case in which such lineal descendants cannot be ascertained, and in the case of unassociated funerary objects, sacred objects, and objects of cultural patrimony—

 (A) in the Indian tribe or Native Hawaiian organization on whose tribal land such objects or remains were discovered;

 (B) in the Indian tribe or Native Hawaiian organization which has the closest cultural affiliation with such remains or objects and which, upon notice, states a claim for such remains or objects; or

 (C) if the cultural affiliation of the objects cannot be reasonably ascertained and if the objects were discovered on Federal land that is recognized by a final judgment of the Indian Claims Commission or the United States Court of Claims as the aboriginal land of some Indian tribe—

  (1) in the Indian tribe that is recognized as aboriginally occupying the area in which the objects were discovered, if upon notice, such tribe states a claim for such remains or objects, or

  (2) if it can be shown by a preponderance of the evidence that a different tribe has a stronger cultural relationship with the remains or objects that the tribe or organization specified in paragraph (1), in the Indian tribe that has the strongest demonstrated relationship, if upon notice, such tribe states a claim for such remains or objects.

 (b) UNCLAIMED NATIVE AMERICAN HUMAN REMAINS AND OBJECTS.—Native American cultural items not claimed under subsection (a) shall be disposed of in accordance with regulations promulgated by the Secretary in consultation with the review committee established under section 8, Native American groups, representatives of museums and the scientific community.

 (c) INTENTIONAL EXCAVATION AND REMOVAL OF NATIVE AMERICAN HUMAN REMAINS AND OBJECTS.—The intentional removal from or excavation of Native American cultural items from Federal or tribal lands for purposes of discovery, study, or removal of such items is permitted only if—

 (1) such items are excavated or removed pursuant to a permit issued under section 4 of the Archaeological Resources Protection Act of 1979 (93 Stat. 721; 16 U.S.C. 470aa et seq.) which shall be consistent with this Act;

(2) such items are excavated or removed after consultation with or, in the case of tribal lands, consent of the appropriate (if any) Indian tribe or Native Hawaiian organization;

(3) the ownership and right of control of the disposition of such items shall be as provided in subsections (a) and (b); and

(4) proof of consultation or consent under paragraph (2) is shown.

(d) INADVERTENT DISCOVERY OF NATIVE AMERICAN REMAINS AND OBJECTS.—(1) Any person who knows, or has reason to know, that such person has discovered Native American cultural items on Federal or tribal lands after the date of enactment of this Act shall notify, in writing, the Secretary of the Department, or head of any other agency or instrumentality of the United States, having primary management authority with respect to Federal lands and the appropriate Indian tribe or Native Hawaiian organization with respect to tribal lands, if known or readily ascertainable, and, in the case of lands that have been selected by an Alaska Native Corporation or group organized pursuant to the Alaska Native Claims Settlement Act of 1971, the appropriate corporation or group. If the discovery occurred in connection with an activity, including (but not limited to) construction, mining, logging, and agriculture, the person shall cease the activity in the area of the discovery, make a reasonable effort to protect the items discovered before resuming such activity, and provide notice under this subsection. Following the notification under this subsection, and upon certification by the Secretary of the department or the head of any agency or instrumentality of the United States or the appropriate Indian tribe or Native Hawaiian organization that notification has been received, the activity may resume after 30 days of such certification.

(2) The disposition of and control over any cultural items excavated or removed under this subsection shall be determined as provided for in this section.

(3) If the Secretary of the Interior consents, the responsibilities (in whole or in part) under paragraphs (1) and (2) of the Secretary of any department (other than the Department of the Interior) or the head of any other agency or instrumentality may be delegated to the Secretary with respect to any land managed by such other Secretary or agency head.

(e) RELINQUISHMENT.—Nothing in this section shall prevent the governing body of an Indian tribe or Native Hawaiian organization from expressly relinquishing control over any Native American human remains, or title to or control over any funerary object, or sacred object.

SEC. 4. ILLEGAL TRAFFICKING.

(a) ILLEGAL TRAFFICKING.—Chapter 53 of title 18, United States Code, is amended by adding at the end thereof the following new section:

"§ 1170. Illegal Trafficking in Native American Human Remains and Cultural Items

"(a) Whoever knowingly sells, purchases, uses for profit, or transports for sale or profit, the human remains of a Native American without the right of possession to those remains as provided in the Native American Graves Protection and Repatriation Act shall be fined in accordance with this title, or imprisoned not more than 12 months, or both, and in the case of a second or subsequent violation, be fined in accordance with this title, or imprisoned not more than 5 years, or both.

"(b) Whoever knowingly sells, purchases, uses for profit, or transports for sale or profit any Native American cultural items obtained in violation of the Native American Graves Protection and Repatriation Act shall be fined in accordance with this title, imprisoned not more than one year, or both, and in the case of a second or subsequent violation, be fined in accordance with this title, imprisoned not more than 5 years, or both.".

(b) TABLE OF CONTENTS. —The table of contents for chapter 53 of title 18, United States Code, is amended by adding at the end thereof the following new item:

"1170. Illegal Trafficking in Native American Human Remains and Cultural Items.".

SEC. 5. INVENTORY FOR HUMAN REMAINS AND ASSOCIATED FUNERARY OBJECTS.

(a) IN GENERAL.—Each Federal agency and each museum which has possession or control over holdings or collections of Native American human remains and associated funerary objects shall compile inventory of such items and, to the extent possible based on information possessed by such museum or Federal agency, identify the geographical and cultural affiliation of such item.

(b) REQUIREMENTS.—(1) The inventories and identifications required under subsection (a) shall be—

(A) completed in consultation with tribal government and Native Hawaiian organization officials and traditional religious leaders;

(B) completed by not later than the date that is 5 years after the date of enactment of this Act, and

(C) made available both during the time they are being conducted and afterward to a review committee established under section 8.

(2) Upon request by an Indian tribe or Native Hawaiian organization which receives or should have received notice, a museum or Federal agency shall supply additional available documentation to supplement the information required by subsection (a) of this section. The term "documentation" means a summary of existing museum or Federal agency records, including inventories or catalogues, relevant studies, or other pertinent data for the limited purpose of determining the geographical origin, cultural affiliation, and basic facts surrounding acquisition and accession of Native American human remains and associated funerary objects subject to this section. Such term does not mean, and this Act shall not be construed to be an authorization for, the initiation of new scientific studies of such remains and associated funerary objects or other means of acquiring or preserving additional scientific information from such remains and objects.

(c) EXTENSION OF TIME FOR INVENTORY.—Any museum which has made a good faith effort to carry out an inventory and identification under this section, but which has been unable to complete the process, may appeal to the Secretary for an extension of the time requirements set forth in subsection (b)(1)(B). The Secretary may extend such time requirements for any such museum upon a finding of good faith effort. An indication of good faith shall include the development of a plan to carry out the inventory and identification process.

(d) NOTIFICATION.—(1) If the cultural affiliation of any particular Native American human remains or associated funerary objects is determined pursuant to this section, the Federal agency or museum concerned shall, not later than 6 months after the completion of the inventory, notify the affected Indian tribes or Native Hawaiian organizations.

(2) The notice required by paragraph (1) shall include information—

(A) which identifies each Native American human remains or associated funerary objects and the circumstances surrounding its acquisition;

(B) which lists the human remains or associated funerary objects that are clearly identifiable as to tribal origin; and

(C) which lists the Native American human remains and associated funerary objects that are not clearly identifiable as being culturally affiliated with that Indian tribe or Native Hawaiian organization, but which, given the totality of circumstances surrounding acquisition of the remains or objects, are determined by a reasonable belief to be remains or objects culturally affiliated with the Indian tribe or Native Hawaiian organization.

(3) A copy of each notice provided under paragraph (1) shall be sent to the Secretary who shall publish each notice in the Federal Register.

(e) INVENTORY.—For the purposes of this section, the term "inventory" means a simple itemized list that summarizes the information called for by this section.

SEC. 6.   SUMMARY FOR UNASSOCIATED FUNERARY OBJECTS, SACRED OBJECTS, AND CULTURAL PATRIMONY.

(a) IN GENERAL.—Each Federal agency or museum which has possession or control over holdings or collections of Native American unassociated funerary objects, sacred objects, or objects of cultural patrimony shall provide a written summary of such objects based upon available information held by such agency or museum. The summary shall describe the scope of the collection, kinds of objects included, reference to geographical location, means and period of acquisition and cultural affiliation, where readily ascertainable.

(b) REQUIREMENTS.—(1) The summary required under subsection (a) shall be—

(A) in lieu of an object-by-object inventory;

(B) followed by consultation with tribal government and Native Hawaiian organization officials and traditional religious leaders; and

(C) completed by not later than the date that is 3 years after the date of enactment of this Act.

(2) Upon request, Indian Tribes and Native Hawaiian organizations shall have access to records, catalogues, relevant studies or other pertinent data for the limited purposes of determining the geographic origin, cultural affiliation, and basic facts surrounding acquisition and accession of Native American objects subject to this section. Such information shall be provided in a reasonable manner to be agreed upon by all parties.

SEC. 7.   REPATRIATION.

(a) REPATRIATION OF NATIVE AMERICAN HUMAN REMAINS AND OBJECTS POSSESSED OR CONTROLLED BY FEDERAL AGENCIES AND MUSEUMS.—(1) If, pursuant to section 5, the cultural affiliation of Native American human remains and associated funerary objects with a particular Indian tribe or Native Hawaiian organization is established, then the Federal agency or museum, upon the request of a known lineal descendant of the Native American or of the tribe or organization and pursuant to subsections (b) and (e) of this section, shall expeditiously return such remains and associated funerary objects.

(2) If, pursuant to section 6, the cultural affiliation with a particular Indian tribe or Native Hawaiian organization is shown with respect to unassociated funerary objects, sacred objects or objects of cultural patrimony, then the Federal agency or museum, upon the request of the Indian tribe or Native Hawaiian organization and pursuant to subsections (b), (c) and (e) of this section, shall expeditiously return such objects.

(3) The return of cultural items covered by this Act shall be in consultation with the requesting lineal descendant or tribe or organization to determine the place and manner of delivery of such items.

(4) Where cultural affiliation of Native American human remains and funerary objects has not been established in an inventory prepared pursuant to section 5, or the summary pursuant to section 6, or where Native American human remains and funerary objects are not included upon any such inventory, then, upon request and pursuant to subsections (b) and (e) and, in the case of unassociated funerary objects, subsection (c), such Native American human remains and funerary objects shall be expeditiously returned where the requesting Indian tribe or Native Hawaiian organization can show cultural affiliation by a preponderance of the evidence based upon geographical, kinship, biological, archaeological, anthropological, linguistic, folkloric, oral traditional, historical, or other relevant information or expert opinion.

(5) Upon request and pursuant to subsections (b), (c) and (e), sacred objects and objects of cultural patrimony shall be expeditiously returned where—

(A) the requesting party is the direct lineal descendant of an individual who owned the sacred object;

(B) the requesting Indian tribe or Native Hawaiian organization can show that the object was owned or controlled by the tribe or organization; or

(C) the requesting Indian tribe or Native Hawaiian organization can show that the sacred object was owned or controlled by a member thereof, provided that in the case where a sacred object was owned by a member thereof, there are no identifiable lineal descendants of said member or the lineal descendants, upon notice, have failed to make a claim for the object under this Act.

(b) SCIENTIFIC STUDY.—If the lineal descendant, Indian tribe, or Native Hawaiian organization requests the return of culturally affiliated Native American cultural items, the Federal agency or museum shall expeditiously return such items unless such items are indispensable for completion of a specific scientific study, the outcome of which would be of major benefit to the United States. Such items shall be returned by no later than 90 days after the date on which the scientific study is completed.

(c) STANDARD OF REPATRIATION.—If a known lineal descendant or an Indian tribe or Native Hawaiian organization requests the return of Native American unassociated funerary objects, sacred objects or objects of cultural patrimony pursuant to this Act and presents evidence which, if standing alone before the introduction of evidence to the contrary, would support a finding that the Federal agency or museum did not have the right of possession, then such agency or museum shall return such objects unless it can overcome such inference and prove that it has a right of possession to the objects.

(d) SHARING OF INFORMATION BY FEDERAL AGENCIES AND MUSEUMS.—Any Federal agency or museum shall share what information it does possess regarding the object in question with the known lineal descendant, Indian tribe, or Native Hawaiian organization to assist in making a claim under this section.

(e) COMPETING CLAIMS.—Where there are multiple requests for repatriation of any cultural item and, after complying with the requirements of this Act, the Federal agency or museum cannot clearly determine which requesting party is the most appropriate claimant, the agency or museum may retain such item until the requesting parties agree upon its disposition or the dispute is otherwise resolved pursuant to the provisions of this Act or by a court of competent jurisdiction.

(f) MUSEUM OBLIGATION.—Any museum which repatriates any item in good faith pursuant to this Act shall not be liable for claims by an aggrieved party or for claims of breach of fiduciary duty, public trust, or violations of state law that are inconsistent with the provisions of this Act.

SEC. 8. REVIEW COMMITTEE.

(a) ESTABLISHMENT.—Within 120 days after the date of enactment of this Act, the Secretary shall establish a committee to monitor and review the implementation of the inventory and identification process and repatriation activities required under sections 5, 6 and 7.

(b) MEMBERSHIP.—(1) The Committee established under subsection (a) shall be composed of 7 members,

(A) 3 of whom shall be appointed by the Secretary from

nominations submitted by Indian tribes, Native Hawaiian organizations, and traditional Native American religious leaders with at least 2 of such persons being traditional Indian religious leaders;

(B) 3 of whom shall be appointed by the Secretary from nominations submitted by national museum organizations and scientific organizations; and

(C) 1 who shall be appointed by the Secretary from a list of persons developed and consented to by all of the members appointed pursuant to subparagraphs (A) and (B).

(2) The Secretary may not appoint Federal officers or employees to the committee.

(3) In the event vacancies shall occur, such vacancies shall be filled by the Secretary in the same manner as the original appointment within 90 days of the occurrence of such vacancy.

(4) Members of the committee established under subsection (a) shall serve without pay, but shall be reimbursed at a rate equal to the daily rate for GS-18 of the General Schedule for each day (including travel time) for which the member is actually engaged in committee business. Each member shall receive travel expenses, including per diem in lieu of subsistence, in accordance with sections 5702 and 5703 of title 5, United States Code.

(c) RESPONSIBILITIES.—The committee established under subsection (a) shall be responsible for—

(1) designating one of the members of the committee as chairman;

(2) monitoring the inventory and identification process conducted under sections 5 and 6 to ensure a fair, objective consideration and assessment of all available relevant information and evidence;

(3) upon the request of any affected party, reviewing and making findings related to

(A) the identity or cultural affiliation of cultural items, or

(B) the return of such items;

(4) facilitating the resolution of any disputes among Indian tribes, Native Hawaiian organizations, or lineal descendants and Federal agencies or museums relating to the return of such items including convening the parties to the dispute if deemed desirable;

(5) compiling an inventory of culturally unidentifiable human remains that are in the possession or control of each Federal agency and museum and recommending specific actions for developing a process for disposition of such remains;

(6) consulting with Indian tribes and Native Hawaiian organiza-

tions and museums on matters within the scope of the work of the committee affecting such tribes or organizations;

(7) consulting with the Secretary in the development of regulations to carry out this Act;

(8) performing such other related functions as the Secretary may assign to the committee; and

(9) making recommendations, if appropriate, regarding future care of cultural items which are to be repatriated.

(d) Any records and findings made by the review committee pursuant to this Act relating to the identity or cultural affiliation of any cultural items and the return of such items may be admissible in any action brought under section 15 of this Act.

(e) RECOMMENDATIONS AND REPORT.—The committee shall make the recommendations under paragraph (c)(5) in consultation with Indian tribes and Native Hawaiian organizations and appropriate scientific and museum groups.

(f) ACCESS.—The Secretary shall ensure that the committee established under subsection (a) and the members of the committee have reasonable access to Native American cultural items under review and to associated scientific and historical documents.

(g) DUTIES OF SECRETARY.—The Secretary shall—

(1) establish such rules and regulations for the committee as may be necessary, and

(2) provide reasonable administrative and staff support necessary for the deliberations of the committee.

(h) ANNUAL REPORT.—The committee established under subsection (a) shall submit an annual report to the Congress on the progress made, and any barriers encountered, in implementing this section during the previous year.

(i) TERMINATION.—The committee established under subsection (a) shall terminate at the end of the 120-day period beginning on the day the Secretary certifies, in a report submitted to Congress, that the work of the committee has been completed.

SEC. 9. PENALTY.

(a) PENALTY.—Any museum that fails to comply with the requirements of this Act may be assessed a civil penalty by the Secretary of the Interior pursuant to procedures established by the Secretary through regulation. A penalty assessed under this subsection shall be determined on the record after opportunity for an agency hearing. Each violation under this subsection shall be a separate offense.

(b) AMOUNT OF PENALTY.—The amount of a penalty assessed under subsection (a) shall be determined under regulations promulgated pursuant to this Act, taking into account, in addition to other factors—

(1) the archaeological, historical, or commercial value of the item involved;

(2) the damages suffered, both economic and noneconomic, by an aggrieved party, and

(3) the number of violations that have occurred.

(c) ACTIONS TO RECOVER PENALTIES.—If any museum fails to pay an assessment of a civil penalty pursuant to a final order of the Secretary that has been issued under subsection (a) and not appealed or after a final judgment has been rendered on appeal of such order, the Attorney General may institute a civil action in an appropriate district court of the United States to collect the penalty. In such action, the validity and amount of such penalty shall not be subject to review.

(d) SUBPOENAS.—In hearings held pursuant to subsection (a), subpoenas may be issued for the attendance and testimony of witnesses and the production of relevant papers, books, and documents. Witnesses so summoned shall be paid the same fees and mileage that are paid to witnesses in the courts of the United States.

SEC. 10. GRANTS.

(a) INDIAN TRIBES AND NATIVE HAWAIIAN ORGANIZATIONS. —The Secretary is authorized to make grants to Indian tribes and Native Hawaiian organizations for the purpose of assisting such tribes and organizations in the repatriation of Native American cultural items.

(b) MUSEUMS.—The Secretary is authorized to make grants to museums for the purpose of assisting the museums in conducting the inventories and identification required under sections 5 and 6.

SEC. 11. SAVINGS PROVISIONS.

Nothing in this Act shall be construed to—

(1) limit the authority of any Federal agency or museum to

(A) return or repatriate Native American cultural items to Indian tribes, Native Hawaiian organizations, or individuals, and

(B) enter into any other agreement with the consent of the culturally affiliated tribe or organization as to the disposition of, or control over, items covered by this Act;

(2) delay actions on repatriation requests that are pending on the date of enactment of this Act;

(3) deny or otherwise affect access to any court;

(4) limit any procedural or substantive right which may otherwise be secured to individuals or Indian tribes or Native Hawaiian organizations; or

(5) limit the application of any State or Federal law pertaining to theft or stolen property.

SEC. 12. SPECIAL RELATIONSHIP BETWEEN FEDERAL GOVERNMENT AND INDIAN TRIBES.

This Act reflects the unique relationship between the Federal Government and Indian tribes and Native Hawaiian organizations and should not be construed to establish a precedent with respect to any other individual, organization or foreign government.

SEC. 13. REGULATIONS.

The Secretary shall promulgate regulations to carry out this Act within 12 months of enactment.

SEC. 14. AUTHORIZATION OF APPROPRIATIONS.

There is authorized to be appropriated such sums as may be necessary to carry out this Act.

SEC. 15. ENFORCEMENT.

The United States district courts shall have jurisdiction over any action brought by any person alleging a violation of this Act and shall have the authority to issue such orders as may be necessary to enforce the provisions of this Act.

Approved November 16, 1990.

25 U.S.C. 3001–3013.

NATIONAL PARKS FOR THE 21ST CENTURY: THE VAIL AGENDA

Report and Recommendations to the Director of the National Park Service: 1992

THE NATIONAL PARK SERVICE AFTER 75 YEARS: A STATEMENT OF CONDITION

The National Park Service has great strengths—and it has major problems. Without question, its greatest strength is its employees. For the vast majority of its employees, to work for the Park Service is to engage in an ever-renewing project of preserving and protecting some of the nation's and the world's most meaningful and enriching—and, often, most fragile and threatened—natural and cultural resources. Throughout the organization, the individuals who work for the Park Service are precisely those who are drawn to this challenge and who hold forcefully to personal stakes in the units and programs for which they are responsible. They are drawn despite a pay scale that is commonly one or two steps below that of comparably responsible and experienced employees in other sister federal agencies, and despite the common frustrations associated with bureaucracies and politics.

When individuals with this much dedication encounter roadblocks to performance, the result is a weakening of morale and effectiveness. Perceptions exist among many employees and observers—and not without bases in reality—that good job performance is impeded by lowered educational requirements and eroding professionalism; that initiative is thwarted by inadequately trained managers and politicized decision making; that the Park Service lacks the information and resource management/research capability it needs to be able to pursue and defend its mission and resources in Washington, D.C. and in the communities that surround the park units; that the mission and the budget of the Service is being diluted by increasing and tangential responsibilities; that there is a mismatch between the demand that the park units be protected and the tools available when the threats to park resources and values are increasingly coming from outside unit boundaries; and that communication within the Service repeatedly breaks down between field personnel and regional and headquarters management. The result of these perceptions is that the National Park Service faces significant morale and performance problems. These threaten the agency's capacity to manage and protect park resources in the short run, and can impede the agency's future ability to attract and retain employees with the education, skills and dedication of the current workforce. Many of the recommendations of the Working Groups aim accurately at overturning the realities that underlie these perceptions.

Beyond the energy and dedication of its employees, the second great strength of the Park Service is the quality of the heritage and recreational resources under its management. These resources are the foundation of the broad base of public support for the Service, and they are the source of the natural inclination to look to the Park Service to manage new resources that might warrant protection. Notwithstanding their quality, the resources of the Park System now encompass a markedly diffuse range of public values. Citizen support for and interest in individual units varies greatly, as do the contributions each unit makes to the national heritage. Requisite personnel skills, organizational structures, and management demands also vary greatly.

The 359 units of the park system are arrayed in more than 20 separate classifications which aptly describe the system's dispersion, including: national battlefield, national battlefield site, national battlefield park, national historical park, national historic site, national lakeshore, national monument, national memorial, national military park, national park, national preserve, national river, wild and scenic riverway, national recreation area, national seashore, national scenic trail, international historic site, national heritage corridor and national parkway. In addition, the National Park Service is responsible for numerous and valuable external programs of support and assistance which have impact beyond the boundaries of the National Park System and even beyond the United States.

Some specific park units or programmatic responsibilities might, arguably, be better placed with other private, state, local, tribal, or federal agencies. Nevertheless, the broad range of resources and functions now managed by the National Park Service represents a permanent reality. Effective management of such a diffuse system requires the abandonment of any hope for a single, simple management philosophy. This is particularly difficult for an agency with its origins—and its identification in the public's mind—in the management and protection of the nation's most spectacular natural areas, the "crown jewels".

The Symposium process elicited numerous proposals that do not and should not apply to all units of the system: "The parks should be managed as environmental classrooms"; "The parks should be managed for recreation"; "The parks should be managed to teach American history." The challenge for the Park Service is to enunciate objectives which match the breadth of its responsibilities and alleviate intra-agency conflicts which result from the desire for a single, narrowly-focussed management strategy. The National Park Service manages a portfolio of assets; it must learn and implement the strategies of a portfolio manager. This means recognizing that all of the units and programs of the agency contribute to public value, but that the ways that these contributions are made and the forms that they take are varied.

The units and programs of the National Park System, taken together, have an important story to tell—a story that is, at once, interesting, instructive, and inspiring. The National Park System has the potential to bring together the landscapes, places, people and events that contribute in unique ways to the shared national experience and values of an otherwise highly diverse people. Unfortunately, there is widespread concern that the story is going untold; that, without resources, training, research, appropriate facilities and leadership, the Park Service is in danger of becoming merely a provider of "drive through" tourism or, perhaps, merely a traffic cop stationed at scenic, interesting or old places.

There are multiple sources for this concern. Managing and protecting the System's natural, cultural and recreational sites and programs are tasks for professionals—rangers, interpreters, scientists, planners, managers. The same can be said of the tasks of understanding and communicating history, or biology, or cultural significance, or archeology, or geology. Meeting these responsibilities requires education, research and experience in specialized and technical fields. But professionals are expensive, and low grade structures have impaired the ability of the Park Service to attract and retain qualified personnel. They have also gradually forced the weakening of many educational standards for employment. Training budgets, meanwhile, have tended to be focussed on mandated law enforcement and administrative compliance responsibilities. The problems of maintaining a professional workforce are only exacerbated by perceptions that management itself faces the need to enhance its professional competency, or is subject to political interference that dilutes any bolstering sense of mission.

Additionally, as the National Park System has expanded, units and programs have been added that arguably have lacked sufficient national significance to warrant National Park Service designation. Yet, such additions to the system have had sufficient constituent appeal and/ or economic development benefits in selected regions to secure their inclusion in the Park Service portfolio.

At the same time as new responsibilities have been added (and have attracted at least initial funding), the core operational budget of the Park Service has remained flat in real terms since 1983. Meanwhile, recreational visits to park units have risen sharply (25%)

over the same period, reaching almost 260 million in 1990. Clearly, the capability of the Park Service to pursue its most central purposes of resource protection and public enjoyment is being stretched thinner and thinner. These disturbing problems are not the sole responsibility of Congress. The Park Service, partly through its own inaction and partly due to constraints emanating from the Executive Branch during the 1970s and 1980s, has lost the credibility and capability it must possess in order to play a proactive role in charting its own course, in defining and defending its core mission.

The National Park System should be a source of national pride, community, and consensus. It should represent the land, the cultures and the experiences that have defined and sustained the people of the nation in the past, and upon which we must continue to depend in the future. But, today, the ability of the National Park Service to achieve the most fundamental aspects of its mission has been compromised. There is a wide and discouraging gap between the Service's potential and its current state, and the Service has arrived at a crossroads in its history.

The basic facts and dimensions of the issues, problems, opportunities and solutions have been articulated and defined throughout the 75th Anniversary Symposium process. An opportunity for change has been created—nothing more and nothing less. Choices must now be made and action must now be taken by those who are responsible for the future of the National Park System—the Director and employees of the National Park Service, the Administration, Congress, and the concerned and committed public. If we fail to seize this opportunity for change, our common heritage will surely suffer.

STRATEGIC OBJECTIVES AND
RECOMMENDATIONS: THE VAIL AGENDA

Reform and rejuvenation of the National Park Service must begin with leadership that is capable of enunciating and implementing clear and compelling goals for parks policy and Park Service management. But what goals? Both within the Park Service and outside, there exists considerable disagreement over both objectives and means of implementation: Should we promote ecological protection? Recreation? International outreach? Involvement in local out-of-park land use policy? Ease of visitor access? More in-park facilities? No in-park facilities? Aggressive marketing of historic and cultural sites? Technical assistance to private and public partners who might need our resources?

These are the kinds of questions on which reasonable people can easily disagree, particularly in a society as economically, demographically, ethnically, and culturally diverse as ours. The Steering Committee has approached the task of resolving these challenging questions by first addressing the overriding purposes of the National Park Service.

Why would a nation want a system of national parks? If we can answer this question, it will help define the purpose of the National Park Service as it looks beyond its 75th anniversary into the next century. Clearly, the units that make up the National Park System of the United States are beautiful, or interesting, or fun, or restful, or invigorating, or otherwise enjoyable to those who visit them — but such wants can be, and are, satisfied through numerous other public and private sources. What rightfully distinguishes the National Park Service from other providers of aesthetic, cultural, recreational, environmental, and historical experiences and makes it the appropriate focus of a unique status and management philosophy?

The answer lies in the link between the units of the National Park System and those traits of environment, wilderness, landscape, history, and culture that bind Americans together as a distinct people. The units of the National Park System should constitute the sights, the scenery, the environments, the people, the places, the events, the conflicts that have contributed elements of shared national experience, values, and identity to build a national character out of the diversity from which we come.

We may disagree among ourselves as to the worth of the consequences of Columbus' landing in the Western Hemisphere, but we can not seriously deny that his landing shaped life and even landscape in the United States. We can debate the larger issues surrounding Anglo-American expansion that Custer's battlefield at Crow Agency symbolizes, but we cannot deny that it symbolizes a defining time in American history. We can argue whether it is ecologically sound to fight fires in Yellowstone National Park, but we can not fail to see that the very disagreement is an expression of values that we place in what we call "natural" environments. It is the ability of unique places, landscapes, environments, events, and people to become part of the national character that constitutes "national significance" and warrants protection within the National Park System.

The resources protected by the National Park System harbor lessons that the nation wishes and needs to teach itself and replenish in itself, again and again, visitor after visitor. Thus, just as it is the responsibility of the system to protect and nurture resources of significance to the nation, so must it also convey the meanings of those resources/their contributions to the nation/to the public in a continuing process of building the national community.

It is the nature of park resources that their meanings can and should be conveyed in a multitude of ways. For some units, this may occur through acts of restive or active recreation, experiencing the link between park resources and elements of the national identity in ways that words and pictures can not adequately impart. A hike in Glacier National Park arguably conveys our heritage of western wildness better than the necessary lecture on the need for bear bells or any other preparatory introductions. For some park resources, on-site interpretative oral, visual, and/or written communication may be appropriate and necessary. How else to convey to the public the intricate ecology of Everglades National Park or the link between Fort McHenry and all that has followed its period of brief excitement, or Ellis Island and its indelible print on our diverse people? Across the units of the Park System, the methods may vary, but the responsibility to tell each unit's story is inseparable from the reasons we protect that story.

The ability of our national historic sites, cultural symbols, and natural environments to contribute to the public's sense of a shared national identity is at the core of the purpose of the National Park Service. The vision of the Park Service that necessarily follows is one in which the agency's purpose is to preserve, protect, and convey the meaning of those natural, cultural and historical resources that contribute significantly to the nation's values, character, and experience. To fully meet the challenge of this vision in the coming decades, the National Park Service will need uncommon clarity in its policies and compelling leadership in its management. The Steering Committee believes that the Service should be guided in these directions by key strategic objectives that can direct the agency's planning for the future. We have identified six such objectives as paramount.

STRATEGIC OBJECTIVE 1

Resource Stewardship and Protection: The primary responsibility of the National Park Service must be protection of park resources from internal and external impairment.

The mission of preserving and protecting the national treasures that belong in the National Park System can only be met if the Park Service can confront the threats to park resources and has the means of dealing with those threats. The evolving economics and demographics of America are driving economic, social and ecological changes in the regions outside unit boundaries. These changes often can impair park resources. Many formerly remote natural area parks, for example, are seeing increasing suburbanization around their boundaries—often spurred by state and local governments anxious to capitalize on tourism-led regional growth. Similarly, many cultural and historic sites in and near urban areas are fighting to maintain the quality of their park units as their neighborhoods struggle with severe economic and social problems.

Thus, although there is ambivalence and uncertainty among park personnel, the mandate of resource protection means that the prevention of external and transboundary impairment of park resources and their attendant values should be a central objective of Park System policy. Giving force to such a goal will require policies which recognize that:

RECOMMENDATION—The National Park Service should provide technical and planning assistance to public and private parties able to mitigate external and transboundary threats to park unit resources, and to those able to influence the quality of visitor enjoyment and enlightenment through their provision of gateway services.

RECOMMENDATION—The National Park Service should utilize available resources, expertise and cooperative relationships to ensure compliance with applicable law when external activities otherwise endanger park resources.

RECOMMENDATION—Each park unit should be managed to protect unimpaired the special resources and values that constitute its contribution to the national identity and experience. Such values may include a unit's unique historic significance, cultural lessons, wilderness traits, recreational opportunities, and/or ecological systems.

RECOMMENDATION—Natural resources in the park system should be managed under ecological principles that prevent their impairment. Cultural diversity and social and historical contexts should be recognized as significant values in the protection and stewardship of historical and cultural resources.

RECOMMENDATION—The National Park Service should seek active public and private partners engaged in resource protection, research, education, and visitor enjoyment that are consistent with the objectives of protecting park values and conveying their meaning to the public.

RECOMMENDATION—The National Park Service should reinforce its role as a world leader in park affairs through agreements and actions which facilitate the exchange of information, development of environmental and cultural resource preservation strategies, and protection of critical world resources.

RECOMMENDATION—Programs, such as an American Heritage Area program, should be established to preserve and protect natural, cultural and historical resources that are worthy of national recognition, but that do not meet the requirements necessary for full inclusion in the National Park System. Such programs should make use of public and private sector partnerships, technical assistance, and Park Service support.

RECOMMENDATION—The National Park Service should fully implement, and be provided requisite funding for, existing legislative mandates under Public Law 88-29, requiring the Department of Interior to produce at five-year intervals a nationwide recreation plan; the Land and Water Conservation Fund Act; the Urban Parks and Recreation Resources Act; the Historic Preservation Fund Act, and related statutes.

STRATEGIC OBJECTIVE 2
Access and Enjoyment: Each park unit should be managed to provide the nation's diverse public with access to and recreational and educational enjoyment of the lessons contained in that unit, while maintaining unimpaired those unique attributes that are its contribution to the National Park System.

RECOMMENDATION—The National Park Service should minimize the development of facilities within the park boundaries to the extent consistent with the mission of conveying each individual park unit's significance to the public.

RECOMMENDATION—Where wilderness values are present, impairment of those values should not be compromised.

RECOMMENDATION—The repair and maintenance of existing park facilities should be undertaken and designed to fulfill the purpose of conveying park values to the public, while protecting the special qualities of each park unit.

RECOMMENDATION—Facilities that are purely for the convenience of visitors should be provided by the private sector in gateway communities.

STRATEGIC OBJECTIVE 3
Education and Interpretation: It should be the responsibility of the National Park Service to interpret and convey each park unit's and the park system's contributions to the nation's values, character, and experience.

RECOMMENDATION—Each visitor to a park unit should have access to a basic interpretation of the unit's unique features and significance. The Park Service should invest in innovative expansions of its ability to provide interpretation that enhances visitor enjoyment and enlightenment.

RECOMMENDATION—The National Park Service should launch a specific program of educational outreach, directed at schools and community groups and designed to maximize the public's access to the unique ecological, historical, cultural, and geologic lessons contained in the park system.

RECOMMENDATION—The National Park Service should embark upon a systematic, park-by-park, usable inventory of information on park resources and visitor needs.

RECOMMENDATION—Comprehensive information on park unit resources and public needs, acquired by resource professionals and solicited from citizens, should be incorporated directly into the management of park units and other agency programs which serve the public.

STRATEGIC OBJECTIVE 4

Proactive Leadership: The National Park Service must be a leader in local, national and international park affairs, actively pursuing the mission of the National Park System and assisting others in managing their park resources and values.

RECOMMENDATION—The National Park Service should establish a headquarters Office of Legislative and Policy Analysis, and reestablish within this office a corresponding legislative program.

RECOMMENDATION—The National Park Service should establish an Office of Strategic Planning, charged with documenting impediments to the mission of the National Park Service, generating feasible solutions and funding requirements, and communicating these to the Director and the Office of Legislative and Policy Analysis.

RECOMMENDATION—The National Park Service should reestablish an areas study program, covering both natural and heritage resources and charged with initiating and responding to proposals for park system additions. This program could be based within the Office of Strategic Planning.

RECOMMENDATION—The National Park Service should clarify existing legislative and regulatory authorities for addressing external and transboundary resource threats, ensure their use, and seek additional legislative authority where needed.

RECOMMENDATION—The National Park Service should initiate an intensive training program for managers to explain authorities, mechanisms and strategies for addressing external and transboundary issues, and to help managers view the natural, cultural, and historical contexts of their units.

RECOMMENDATION—The management and resources of the National Park Service should be focussed to maximize educational, recreational, and cultural value in the park units and other agency programs which serve the public.

RECOMMENDATION—The National Park Service should assess its capabilities for decentralized management. Effective decentralized organization will require: functions of support and service to the parks, liaison with non-Service parties, systems of accountability and control, training in management principles, and broader grants of authority to superintendents and staff in line operations.

STRATEGIC OBJECTIVE 5

Science and Research: The National Park Service must engage in a sustained and intergrated program of natural, cultural and social science resource management and research aimed at acquiring and using the information needed to manage and protect park resources.

RECOMMENDATION—Secure legislation and funding that support a research mandate for the Park Service.

RECOMMENDATION—Accelerate the training of Park Service managers in information management and the role, use and production of research information.

RECOMMENDATION—Base resource protection, access and interpretation decisions and programs on full consideration of the best available scientific research; where quality information is lacking, initiate it through Park Service resource management professionals.

STRATEGIC OBJECTIVE 6

Professionalism: The National Park Service must create and maintain a highly professional organization and workforce.

RECOMMENDATION—The National Park Service should establish and/or raise educational requirements as appropriate for professional track positions, including those that require strong bases of technical, scientific, interpretative, administrative, and/or managerial knowledge.

RECOMMENDATION—The National Park Service should strengthen recruitment, hiring and retention of a culturally diverse professional workforce.

RECOMMENDATION—The National Park Service should implement a comprehensive program of broad-based, mission-driven employee training.

RECOMMENDATION—All National Park Service employees should receive basic orientation training that covers the agency's objectives, purpose, history, and organization.

RECOMMENDATION—National Park Service training should focus on development of present and future management and leadership capabilities, as well as appropriate professional and technical skills.

RECOMMENDATION—Working with the Office of Personnel Management and the Office of Management and Budget, the National Park Service should undertake a comprehensive review of its existing compensation structure. This review should be conducted under needs criteria derived from the mission of the Park Service and in light of professional compensation structures in related resource agencies.

RECOMMENDATION—The National Park Service should create a Human Resources Management Board with responsibility for senior management assignment, training and development, and for developing the agency's plans for training, career advancement procedures, and educational requirements.

MEETING THE FUNDING NEEDS
OF THE NATIONAL PARK SYSTEM

This report describes a vision of the National Park Service, embodied in six Strategic Objectives, as an assertive, fully capable agency with the ability to manage the National Park System and ensure its protection for future generations. If these Strategic Objectives are the pillars of National Park Service policy and management, adequate funding must be the base of the pillars. There is a cost to ensuring the ongoing protection of America's heritage.

The Steering Committee believes that adequate funding for the National Park Service should continue to be a Federal responsibility, and that Congress is the appropriate source of funding for the operation and management of the System. Public/private partnerships are a valuable tool for maintaining a margin of excellence in Park System programs and for funding special projects of the Service or the park units. Reliance on private funding sources for core functions, however, risks dependency and dilution of the National Park Service's ability to pursue its central purposes. In addition, outside funding can be particularly unstable and insufficient to address core problems.

The National Park Service contributes to the common good. It protects continuing public access to and enjoyment of the resources which symbolize and contribute to our national character and heritage. As such, the National Park System is an important part of America's infrastructure. Like our system of highways, which stretches from shore to shore connecting people and communities, and like the country's bridges, which span vast canyons and waterways, the National Park System ties together the separate elements of environment, history and culture which help to make one nation of the American people. This infrastructure that is the Park System has been steadfastly supported by the American people for many decades. It should not now be allowed to deteriorate.

RECOMMENDATION—The units and programs of the National Park System should be viewed as critical components of the nation's infrastructure. Congressional funding of the National Park Service must be fully adequate to meet the responsibilities of maintaining and enhancing this infrastructure.

RECOMMENDATION—Funding under programs such as the Land and Water Conservation Fund Act should be provided to the National Park Service to the full extent authorized.

RECOMMENDATION—The policy of returning fifty percent of visitor fees to park units should be reaffirmed and implemented.

PLANNING FOR IMPLEMENTATION

RECOMMENDATION—Within twenty-four months, the National Park Service should issue a comprehensive report on the "State of the National Park Service," assessing the progress and prospects for meeting the strategic objectives of the agency.

CONCLUSION

A long and complex process to repair and strengthen the foundations of one of the nation's most prized institutions has begun. The individuals charged with implementing the "Vail Agenda" are accepting an immense responsibility and assuming considerable professional risk. Some of their efforts will succeed, while others will not yield satisfactory results. None of these efforts, however, will be failures. The only failure will be inaction. It is incumbent upon the Director of the National Park Service, the Secretary of the Interior, the Administration, and Congress to provide the support and leadership that is needed. The commitments to a sound future for the National Park Service are strong; expectations are high. The opportunity for progress should not be missed.

National Park Service Document Number D-726, 9–16, objectives and recommendations from pages 17–42. Report also available from the National Park Foundation.

SCIENCE AND THE NATIONAL PARKS

Committee on Improving the Science and Technology Programs of the National Park Service National Academy of Sciences, 1992

RECOMMENDATIONS

In conducting this study of science in the national parks, the National Research Council's Committee on Improving the Science and Technology Programs of the National Park Service originally set out to evaluate the scope and organization of current NPS natural and social science by performing a peer review of NPS research activities. However, the committee soon determined that the crucial problems in the NPS research program are not at the level of individual projects. Instead, they are more fundamental, rooted in the culture of the NPS and in the structure and support it gives to research. Thus, the committee concluded that the real need was for an assessment more broadly focused on the research program and its place within the agency.

The call for change made in this report is not new. But given the lack of response to so many previous calls for change, how can the present report succeed in inspiring action? The members of the committee believe that increased funding or incremental changes alone will not suffice, and they call instead for a fundamental metamorphosis. It is time to move toward a new structure—indeed, toward a new culture—that stresses science in the national park system and guarantees long-term financial, intellectual, and administrative support. There are three key elements:

• There must be an explicit legislative mandate for a research mission of the National Park Service.

• Separate funding and reporting autonomy should be assigned to the science program.

• There must be efforts to enhance the credibility and quality control of the science program. This will require a chief scientist of appropriate stature to provide leadership, cooperation with external researchers, and the formation of an external science advisory board to provide continuing independent oversight.

AN EXPLICIT LEGISLATIVE MANDATE

• To eliminate once and for all any ambiguity in the scientific responsibilities of the Park Service, legislation should be enacted to establish the explicit authority, mission, and objectives of a national park science program.

• The National Park Service should establish a strong, coherent research program, including elements to characterize and gain understanding of park resources and to aid in the development of effective management practices. To provide a scientific basis for protecting and managing the resources entrusted to it, the Park Service should establish, and expand where it already exists, a basic resource information system, and it should establish inventories and monitoring in designated park units. This information should be obtained and stored in ways that are comparable between park units, thereby facilitating access, exchange, integration, and analysis throughout the park system and with other interested research institutions. The NPS should support and develop intensive long-term, ecosystem-level research projects patterned after (and possibly integrated with) the National Science Foundation's Long-Term Ecological Research program and related activities of other federal agencies. The ways resources are used and appreciated by people should be documented. In addition, National Park Service researchers should have more input into the development of resource management plans. Effective interaction between research results and resource management plans cannot take place without both a strong science program and a strong resource management program.

• The National Park Service should also establish and encourage a strong "parks for science" program that addresses major scientific research questions, particularly within those parks that encompass large undisturbed natural areas and wilderness. This effort should include NPS scientists and other scientists in independent and cooperative activities. The goal is to facilitate the use of parks for appropriate scientific inquiry on major natural and related social science questions.

SEPARATE FUNDING AND AUTONOMY

• The National Park Service should revise its organizational structure to elevate and give substantial organizational and budgetary autonomy to the science program, which should include both the planning of research and the resources required to conduct a comprehensive program of natural and social science research. The program should be led by a person with a commitment to its objectives and a thorough understanding of the scientific process and research procedures.

• The National Park Service science program should receive its funds through an explicit, separate (line item) budget. A strategic increase in funding is needed, especially to create and support the needed long-term inventories and the monitoring of park resources.

BUILDING CREDIBILITY AND QUALITY

• To provide leadership and direction, the NPS should elevate and reinvigorate the position of chief scientist, who must be a person of high stature in the scientific community and have as his or her sole responsibilities advocacy for and administration of the science program. The chief scientist would work from the Washington office and report to the Director of the NPS, provide technical direction to the science and resource management staff at the regions and in the parks, and foster interactions with other research agencies and non-government organizations. In addition, the chief scientist should establish a credible program of peer review for NPS science, reaching from the development of research plans through publication of results.

• To help the NPS expand the science program and increase its effectiveness, the Park Service, in cooperation with other agencies, should establish a competitive grants program to encourage more external scientists to conduct research in the national parks. The program should include scientific peer review that involves both NPS scientists and external scientists.

• The National Park Service should enlist the services of a high-level science advisory board to provide long-term guidance in planning, evaluating, and setting policy for the science program. This independent advisory board should report to the director, and its reports should be available to the public.

REALIZING THE VISION

To build a science program that fulfills its potential—that meets the needs of resource managers, helps the public understand and enjoy park resources, and contributes to understanding our changing world—the Park Service must give the science program immediate and aggressive attention. Pressures on these national treasures are increasing rapidly. It is shortsighted to fail to organize and support a science program to protect the parks for future generations. And it is a waste of a unique resource if the parks are not used, with proper safeguards, to help address the scientific challenges faced throughout the biosphere. The current Park Service leadership has expressed its recognition of the need for a reinvigorated science program, as well as the importance of the parks in a broader scientific context. It is time to translate that recognition into action.

The conduct of research is fundamentally different from that of most other NPS functions. It operates on a schedule not determined by the calendar of Congress, but on the calendar of the natural or cultural phenomena being studied. Products from research come with answers frequently surrounded with small or great uncertainty. The design of an experiment and the interpretation of the results often depend on the scientific process as it is conducted in another discipline or in a different part of the world. If the NPS is to meet the scientific and resource management challenges of the twenty-first century, a fundamental metamorphosis must occur within its core. This committee's vision for the NPS science program is ambitious but obtainable. The national parks are, after all, simply too valuable to neglect.

Washington, D.C., National Academy Press, 1992, 9–13. Reprinted with permission of the National Academy of Sciences.

Summaries of Lengthy Documents

ALASKA NATIONAL MONUMENT PROCLAMATIONS

December 1, 1978
(Federal Register, Vol. 43, No. 234,
Tuesday, December 5, 1978)

After several decades of sparse and careful usage of the Antiquities Act reflecting the unwillingness of Congress to allow the president to unilaterally shape conservation policy, Jimmy Carter took abrupt and sweeping action to preserve seventeen enormous and endangered areas of Alaska. Each individual monument proclamation repeated the following key statements after defining and justifying a particular area:

"Now, therefore, I, Jimmy Carter, President of the United States of America, by the authority vested in me by section 2 of the Act of June 8, 1906 (34 Stat. 225, 16 U.S.C. 431), do proclaim that there are hereby set apart and reserved as the [see list below] all lands, including submerged lands, and waters owned or controlled by the United States within the boundaries of the area described on the document entitled [Monument title], attached to and forming a part of this Proclamation.

"All lands, including submerged lands, and all waters within the boundaries of this Monument are hereby appropriated and withdrawn from entry, location, selection, sale or other disposition under the public land laws, other than exchange. There is also reserved all water necessary to the proper care and management of those objects protected by this Monument and for the proper administration of the Monument in accordance with applicable laws.

"The establishment of this Monument is subject to valid existing rights, including, but not limited to, valid selections under the Alaska Native Claims Settlement Act, as amended (43 U.S.C. 1601 *et seq.*), and under or confirmed in the Alaska Statehood Act (48 U.S.C. Note preceding Section 21).

"Nothing in this Proclamation shall be deemed to revoke any existing withdrawal, reservation or appropriation, including any withdrawal under section 17 (d)(1) of the Alaska Native Claims Settlement Act (43 U.S.C. 1616 (d)(1); however, the National Monument shall be the dominant reservation. Nothing in this Proclamation is intended to modify or revoke the terms of the Memorandum of Understanding dated September 1, 1972, entered into between the State of Alaska and the United States as part of the negotiated settlement of *Alaska v. Morton,* Civil No. A-48-72 (D. Alaska, Complaint filed April 10, 1972)."

Aside from vastly increasing park system acreage, these proclamations greatly expanded the use of the Antiquities Act to block resource exploitation and at least temporarily preserve areas until the slow legislative process could act.

The proclamations and areas are as follows:

Proclamation 4611 Admiralty Island National Monument
Proclamation 4612 Aniakchak National Monument
Proclamation 4613 Becharof National Monument
Proclamation 4614 Bering Land Bridge National Monument
Proclamation 4615 Cape Krusenstern National Monument
Proclamation 4616 Denali National Monument
Proclamation 4617 Gates of the Arctic National Monument
Proclamation 4618 Enlarging the Glacier Bay National Monument
Proclamation 4619 Enlarging the Katmai National Monument
Proclamation 4620 Kenai Fjords National Monument
Proclamation 4621 Kobuk Valley National Monument
Proclamation 4622 Lake Clark National Monument
Proclamation 4623 Misty Fiords National Monument
Proclamation 4624 Noatak National Monument
Proclamation 4625 Wrangell-St. Elias National Monument
Proclamation 4626 Yukon-Charley National Monument
Proclamation 4627 Yukon Flats National Monument

CFR Title 3, 1978 Compilation, 69–104.

ALASKA NATIONAL INTEREST
LANDS CONSERVATION ACT

December 2, 1980 (PL 96-487, 94 Stat. 2371)

In the waning days of the Carter Democratic administration, Congress acted to further protect and expand preserved areas in Alaska, many rescued from exploitation two years earlier by presidential proclamation. This complex and lengthy act defines preserved parks, forests, wilderness areas, wildlife refuges, wild and scenic rivers, and Native American corporation lands and the degrees of preservation and usage for each. It prescribes timber, fish, and wildlife protection and use by Native Americans and other citizens.

New areas for the national park system included Aniakchak National Preserve, Cape Krusenstern National Monument, Gates of the Arctic National Park and Preserve, Kenai Fjords National Park, Kobuk Valley National Park, Lake Clark National Park and Preserve, Noatak National Preserve, Wrangell-St. Elias National Park and Preserve, and Yukon-Charley Rivers National Preserve. The act also added new lands to Glacier Bay National Park and Preserve, Katmai National Monument and Preserve, and Denali National Park and Preserve (renamed from Mount McKinley National Park).

New wild and scenic rivers under Park Service administration included Alagnak, Alatna, Aniakchak, Charley, Chilikadrotna, John, Kobuk, Mulchatna, Noatak, North Fork of the Koyukuk, Salmon, Tinayguk, and Tlikakila rivers. Other wild and scenic rivers are designated or expanded in wildlife refuges and in other areas.

The vast majority of acreage in the Denali, Gates of the Arctic, Glacier Bay, Katmai, Kobuk Valley, Lake Clark, Noatak, and Wrangell-St. Elias units is designated wilderness.

CLEAN AIR ACT OF 1967

(PL Chapter 360, 69 Stat. 322)

The main purpose of this act is to protect and enhance the nation's air quality to promote the public health and welfare. The act establishes specific programs that provide special protection for air resources and air quality-related values (AQRVs) associated with NPS units. For example, sections 160–169 of the act establish a program to prevent significant deterioration (PSD) of air quality in clean air regions of the country. The purposes of the PSD program include: to protect resources that might be sensitive to pollutant concentrations lower than the established national standards and "to preserve, protect and enhance the air quality in national parks, national monuments, national seashores, and national or regional natural, recreational, scenic or historic value." In section 169A of the act, Congress also established a national goal of remedying any existing and preventing any future man-made visibility impairment in mandatory class I areas.

Summary taken from NPS–77 Guideline (1993) provided by the Division of Natural Resoures, National Park Service, Washington, D.C.

COASTAL ZONE MANAGEMENT ACT OF 1972

(PL 92-583, 86 Stat. 1280)

This act states a national policy to "preserve, protect, develop, and where possible, to restore or enhance the resources of the nation's coastal zones" (including Great Lakes areas), and to encourage and assist the states in implementing management programs to achieve wise use of land and water resources of coastal zones. Federal agencies are required to comply, as much as possible, with applicable, approved state coastal zone management programs. Indian tribes doing work in federal lands are treated as federal agencies for purposes of this act. (NPS–77 Guideline Summary.)

ENDANGERED SPECIES ACT OF 1973

(PL 93-205, 87 Stat. 884)

This act requires federal agencies to ensure that their activities (authorized, funded, or carried-out) will not jeopardize the existence of any endangered or threatened species of plant or animal (including fish) or result in the destruction or deterioration of critical habitat of such species. It also provides for studies to determine endangered or threatened species and stipulates that it is unlawful for a person to possess, export, or import such species. (Expanded from the NPS–77 Guideline Summary.)

FEDERAL WATER POLLUTION CONTROL ACT
(CLEAN WATER ACT OF 1972)

(PL 92-500, 86 Stat. 816)

This act firmly establishes federal regulation of the nation's waters, and contains provisions designed to "restore and maintain the chemical, physical, and biological integrity of the Nation's waters." The act requires that the states set and enforce water quality standards to meet EPA minimum guidelines. It establishes effluent limitations for point sources of pollution, requires a permit for point source discharge of pollutants (through the National Pollution Discharge Elimination System), requires a permit for discharge of dredged or fill material, and authorizes a "National Wetlands Inventory." Recent changes brought about by the 1987 Federal Water Quality Act place greater emphasis on toxicological-based criteria and in-site biological monitoring. (NPS–77 Guideline Summary.)

Index

About the Editor

Lary Dilsaver is professor of historical geography at the University of South Alabama and consultant to the History Division of the National Park Service. Born and raised in California, he received a Ph.D. from Louisiana State University in 1982. He has co-authored *Challenge of the Big Trees: A Resource History of Sequoia and Kings Canyon National Parks* and co-edited *The Mountainous West* and *The American Environment* (the latter published by Rowman and Littlefield). He has also written some two dozen articles and chapters on land use and management, conservation, and recreation. He and his wife, Robin, live in Spanish Fort, Alabama, adjacent to one of the last battlefields of the Civil War.